MANUFACTURING INTELLIGENCE

MANUFACTURING INTELLIGENCE

Paul Kenneth Wright

THE COURANT INSTITUTE OF MATHEMATICAL SCIENCES
NEW YORK UNIVERSITY

THE ROBOTICS INSTITUTE
CARNEGIE MELLON UNIVERSITY

David Alan Bourne

THE ROBOTICS INSTITUTE
CARNEGIE MELLON UNIVERSITY

ADDISON-WESLEY PUBLISHING COMPANY, INC.
Reading, Massachusetts • Menlo Park, California • New York
Don Mills, Ontario • Wokingham, England
Amsterdam • Bonn • Sydney • Singapore • Tokyo
Madrid • Bogotá • Santiago • San Juan

Library of Congress Cataloging in Publication Data

Wright, Paul Kenneth
Manufacturing Intelligence

Bibliography: p.
Includes index.
1. Robotics. 2. Artificial intelligence--Industrial applications. 3. Production Engineering. I. Bourne, David Alan. II. Title.
TJ211.W75 1988 670.42 87-14457
ISBN 0-201-13576-0

ABCDEFGHIJ-AL-8987

To the children
Sam, Joe, Tom, and Leah

Preface

The goal of the emerging field of "manufacturing intelligence" is to model the skills and expertise of manufacturing craftsmen so that intelligent machines can make small batches of parts without human intervention. We believe that this goal will be accomplished by integrating the results of research in knowledge engineering, manufacturing software systems, robotic vision, and robotic manipulation.

It is the intention of this book to assess actual progress, in research and practice, towards full development and implementation of manufacturing intelligence in the factory. Further, we consider the rich promise that the book holds for bridging the two areas of manufacturing science and computer science. As a result, some general principles are defined for the field. The book is divided into four parts:

- Part One reviews the economic and social influences on the manufacturing industry that are changing its focus from mass production to batch production. In batch production, considerable effort from human craftsmen is still needed to nurture the machines through setup procedures because of the limitations of current machine tools, robots, and sensors. Also, process parameters are set conservatively to make batch runs more predictable. For these reasons, and others, hardware and software that mimic the skills and decision-making process of the manufacturing craftsman must be developed.
- Part Two develops general principles for building intelligent machines. To accomplish this, it considers anthropomorphic systems (i.e., brain, eye, and hand) that must be built in order to physically replace the human craftsman. Each chapter attempts to provide practical design guides that can be used to construct these manufacturing systems. A special effort is made to outline important research problems that remain open.
- Part Three continues the theme of modeling the manufacturing craftsman with a case study of an intelligent machine tool. Initially, the skills exhibited by the human machinist are described, together with details of knowledge engineering experiments. This is followed by a review of the flexible fixtures and cutting sensors that

are being developed to relieve human machinists of their loading and monitoring duties.

- Part Four predicts the manufacturing achievements of the next thirty years by looking at the long-term goals of current research. These predictions take the form of a dialogue between a mechanical engineer and a computer scientist, with the hope of stimulating and provoking similar or divergent research ideas.

The machine tool industry and its users are a vital component of the manufacturing economy, but they are threatened by intense international competition. Innovative ideas, practices, and products are desperately needed if the domestic U.S. industry is to survive. Thus, we hope that this book may help establish some profitable research directions for practicing manufacturing engineers, for graduate manufacturing scientists, and for knowledge engineers interested in the manufacturing domain.

Manufacturing is a multidisciplinary science, and this book is a strong reflection of its diverse problem areas. Therefore, we have tried to develop the material so that a reader who is either familiar with basic manufacturing engineering or who has a good working knowledge of the principles of artificial intelligence will have sufficient background to learn the new material within the confines of this book. We recommend the book not only for use in the university classroom, but for colleagues in industry who are developing automated-manufacturing systems. The open problems are included specifically to stimulate graduate research projects. The book may be of especial interest to industrial and governmental planners who are defining the developments in computer-integrated manufacturing. The material in Chapters 1, 2, and 12 and the glossary will be useful to the lay reader interested in robotics, manufacturing, and international business competition. Suggestions for additional background and in-depth reading are provided in an annotated bibliography.

We are grateful to many individuals from the Flexible Manufacturing Laboratory within the Robotics Institute at Carnegie Mellon University for making this book possible. Several chapters are developed from the thesis work of graduate students. We particularly thank Mark Cutkosky, now at Stanford University, for his contributions to the work on the manufacturing hand, described in Chapter 6; Caroline Hayes for her thesis work, including many of her drawings, on setup planning, described in Chapter 8; Paul Englert for his research work on programmable fixtures in Chapter 9. The software projects in the laboratory have been greatly facilitated by Jeff Baird, Morris Goldstein, Richard Wallstein, and Duane Williams. The smooth operation of the laboratory has been made possible by the efforts of Halil Kulluk and Steve Miketic. Finally, our thanks and respect go to Jim Dillinger and Dan McKeel who not only have built most of the research hardware shown in the book, but also have, in good humor, been the experts for interrogation.

We are indebted to Peggy Martin for the large job of typing and formatting, to Carol Hymowitz for her care and copyediting, and to Geraldine Lockhart for her support. Raj Reddy, Director of the Robotics Institute, is gratefully acknowledged for his encouragement. We are also grateful to our sponsors over the years for supporting this research. They include Westinghouse Electric Corp., Cincinnati Milacron Inc., The National Science Foundation, Kennametal Inc., Chrysler Corp., AT&T Foundation, Speedsteel Inc., and the Commonwealth of Pennsylvania. Westinghouse was responsible for supporting new manufacturing technology and then building a large flexible manufacturing cell with us; we gratefully acknowledge Tom Murrin, Jose Isasi, and Jerry Colyer. Cincinnati Milacron has provided the main impetus for our work on the intelligent machine tool and we thank our friends, Lonnie Burnett, Dick Kegg, and Dick Messinger, there.

The production of this book would not have been possible without the thoughtful comments of reviewers Alice Agogino, Piero Bonissone, David Brown, John Craig, Mark Cutkosky, John Dixon, Robert Kelley, Behrokh Khoshnevis, and Thomas Ward. Mary Jo Dowling and Nicole Vecchi of The Robotics Institute supported us with art and production assistance. Karen Myer and the production staff at Addison-Wesley are gratefully acknowledged for their hard work and attention to detail. Finally, we would like to thank our editor at Addison-Wesley, Barbara Rifkind, for her guiding spirit.

Pittsburgh, Pennsylvania

Paul Kenneth Wright
David Alan Bourne

Contents

Part One

THE MACHINE TOOL INDUSTRY

In very cheap restaurants it is different; there, the same trouble is not taken over the food, and it is just forked out of the pan and flung on to a plate, without handling. Roughly speaking, the more one pays for food, the more sweat and spittle one is obliged to eat with it.
— George Orwell, *Down and Out in Paris and London*[1]

Some (manufacturing operations) indeed have met with success, and are carried on to advantage; but they are generally such as require only a few hands, or wherein great part of the work is performed by machines.
— Benjamin Franklin, *Benjamin Franklin: The Autobiography*[2]

Manufacturing intelligence must be created to produce the customized, high-quality, and yet inexpensive components that are being demanded by today's customer. These demands stress the machine tool setup procedures and the economic goal of "getting a good first part, right the first time." Traditionally, batch manufacturing has relied heavily on the skills of the human craftsman. However, economic factors now require the automation of such batch manufacturing. The next two chapters discuss this background and review the basic techniques in the development of an expert system—tailored toward mimicing the craftsman.

---

[1]From *Down and Out in Paris and London*, copyright 1933 by George Orwell; renewed 1961 by Sonia Pitt-Rivers. Reprinted by permission of Harcourt Brace Jovanovich and A. M. Heath, Ltd., London.

[2]*Benjamin Franklin: The Autobiography*, edited by H.W. Schneider. NY: Macmillan, 1952. Reprinted by permission.

1 Introduction

An untrained observer will see only physical labor and often get the idea that physical labor is mainly what the mechanic does. Actually . . . mechanics don't like it when you talk to them because they are concentrating on mental images, hierarchies, and not really looking at you or the physical motorcycle at all. . . . They are looking at underlying form.[1]

— Robert Pirsig, *Zen and the Art of Motorcycle Maintenance*

HUMAN beings have always enjoyed creating tools that take the muscle and the repetition out of their working environments. Ideally, such tools free the human mind to be more creative in either design or business. Consider, for example, the tools for the construction and the agricultural industries, which have changed dramatically during the last century. One result is that only 3% of the workforce is now needed to produce our agricultural products in comparison with 35% in the early 1900s (Office of Technology Assessment 1986). So, it is a natural progression to exploit a new tool—the computer—in order to build robots and machinery to further reduce the muscle and repetition of shop-floor factory work. With new tools for computer integrated manufacturing (CIM), there is now the opportunity to build fully automated factories.

Even without computers, Henry Ford addressed and solved many of the problems of large-scale factory automation of the same product during the early part of this century. His aphorism—that his customers could have any color of car they wanted as long as it was black—is a testimony to the fact that "mass production" is a huge economic success provided that the product never changes and the initial setup costs are amortized over long batch runs. In Ford's heyday, machine tools were operated by mechanical cams and templates that could be used to make the same component for an automobile ad infinitum. Clever mechanical designs enabled such machines to be loaded and unloaded automati-

[1]From Robert M. Pirsig, *Zen and the Art of Motorcycle Maintenance*, © 1974 Robert M. Pirsig. Reprinted by permission of William Morrow & Company and the Bodley Head.

cally. Conveyor belts were used to connect the various automated parts of the factory. Besides this, human labor was still cheap. In the early years of automobile plants, if the cam-driven automation was not clever enough to do the job automatically, say, fitting a rubber molding around a windshield, then human labor could be used with only a small additional cost.

With the increasing wage rates in recent years, a need has grown to change some of the technology in mass manufacturing to reduce or eliminate manpower. The appearance of programmable logic controllers and then industrial robots capable of simple spot welding and painting operations was thus the natural development of events during the 1970s. These technological advances allowed a shift in industry toward small "batch production."

In the past, Henry Ford understood that society's increasing affluence would create a market in which every American wanted one of his cars. Today, a combination of even more affluence and human vanity has fueled a demand for customized products. Consumers are willing to pay a high price for custom craftsmanship in art work, clothing, housing styles, and even automobiles. The craftsmanship exhibited in these "designer" items is sought after at a price.

The situation in the machine tool user industry (generally the metal products industry) is much the same. The trend in manufacturing is clearly toward production in small batch, or lot, sizes (Miller 1985). As a result, industry needs to develop manufacturing techniques that will allow cost-effective production, even if the batch size is less than ten. Three particular industries come to mind—the aerospace industry, the ship-building industry, and the tool and die making industry. In these areas, small batches are produced with high precision and generally from difficult-to-machine alloys. For example, in 1987 we conducted a survey of aerospace and other high-precision industries and found that size one batches were surprisingly common and relatively small batch sizes were the rule (see Figure 1.1). The small batches are produced in an environment that has been traditionally called a job shop or model shop. In many situations, machinists are asked to accurately fabricate a single, complex component.

In such industries, the availability of traditional craftsmanship is of utmost importance because many unforeseen situations are encountered. For example, the programs for numerically controlled (NC) machine tools may contain errors that need to be corrected at the machine console.

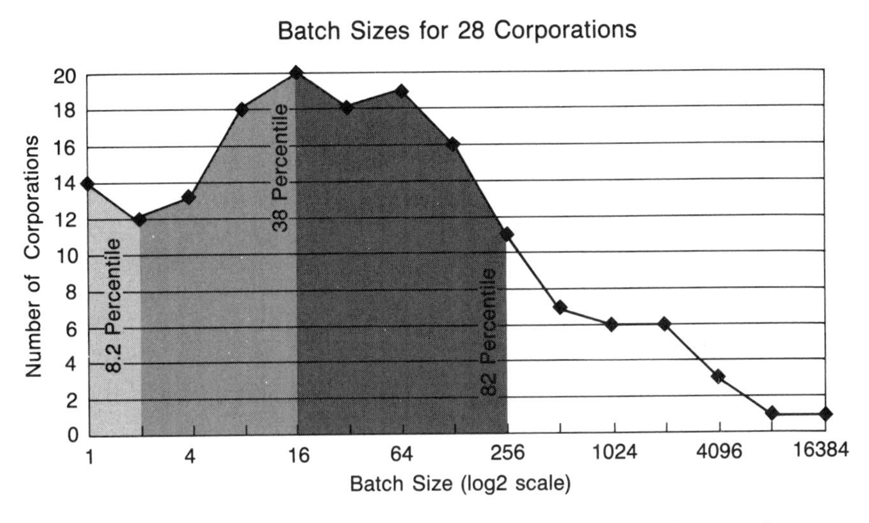

Figure 1.1 Each corporation responded with a range of batch sizes that were produced on a regular basis. These responses were accumulated in this graph. The shaded areas represent the area under the curve up to and not including the data point for that batch size.

Also, there may be unexpected variations in the cutting tool characteristics or in the incoming workmaterial. The skills that the craftsmen employ during the initial planning operations (described in Chapter 8) include spatial reasoning about tools, fixtures, and incoming stock. The craftsmen ponder the correctness of the setup and judge whether changes should be made. During a machining operation, they use a blend of sensory skills that include vision, touch, sensitivity to forces and vibrations, and even smell, to measure whether the process is proceeding along according to plan. After a part is made, the craftsmen measure it and then reevaluate tools, fixtures, and other process control parameters. In doing so, they are relating to the underlying form and dynamics of machining, as Robert Pirsig suggested in his description of a craftsman's state of mind, which was the initial quotation for this chapter. Years of experience are needed to acquire the knowledge of and intuitive feel for some of these activities. Unfortunately, the skilled machinist does not hold the same status as other craftsmen who are employed in the arts or consumer-oriented industries (e.g., high fashion). In recent decades, several factors have led to a severe shortage of skilled machinists in vital industries such as aerospace. The number of apprenticeship programs in large companies has decreased markedly. At the same time, the machine tool industry has not seemed an attractive occupation to high school graduates (although the glamour of computer aided design and robotics has partially reversed this trend), and managers have tended to undervalue the importance of the skilled craftsman's role on the shop floor. As a result, rates of pay have not been attractive, and skilled craftsmen have gone elsewhere to earn a living.

Therefore, we appear to be in a dangerous position in manufacturing history. A variety of societal factors have combined to make these skilled machinist craftsmen in short supply; yet, at the same time, consumer pressures and national defense issues are driving manufacturing to small batch sizes. Here, the role of craftsmanship is vitally important. What are some solutions to this dilemma?

At the front end of any engineering or manufacturing system, the design engineer can address the dilemma. If the designer can simplify either the part description or the overall process description, the resulting manufacturing process will become more predictable. Then there will be fewer surprises on the shop floor that would require a trained craftsman's intervention and intuitive knowledge. The growing interest in the field of "design for manufacturability" (Andreasen, Kahler, and Lund 1983) is of importance in this area, but so far very few universal paradigms are emerging from this research. Ironically, even if the parts could be redesigned to make manufacturing easier, it would still take a long time to switch from current manufacturing methods. The aerospace industry, for example, designs parts and then has them tested and "qualified" by the U.S. Air Force. This qualification can take many years of establishing standards, part designs, and production methods.

So, even if a new design were available tomorrow, it would be several years before that particular part were produced in its qualified form. To change a part design on an existing aircraft would be virtually impossible, and indeed, much of the work in this sector is in the production of spare parts. Realistically, to change component designs, tools, or methods in these traditional, albeit advanced, manufacturing industries is a far slower process than in other high-growth industries such as electronics and computer technology.

We conclude, therefore, that craftsmanship must be automated and that machine tools must be made more intelligent.

1.1 AN AUTONOMOUS BUILDING BLOCK

Figure 1.2 shows the intelligent machine tool as part of a cell and a system that make up a CIM facility. The hierarchy of the automated factory contains four levels of control:

- **A workstation**: one principal machine and a machine controller that, in practice, replaces a single person's station.
- **A cell**: a set of machines and controllers that need to work cooperatively to achieve the desired effect. In practice, the cell replaces several people.
- **A system**: a set of workstations and/or cells that can each operate and be scheduled independently from the others.
- **A factory**: a set of systems that includes all aspects of the factory (i.e., order entry, inventory, and manufacturing).

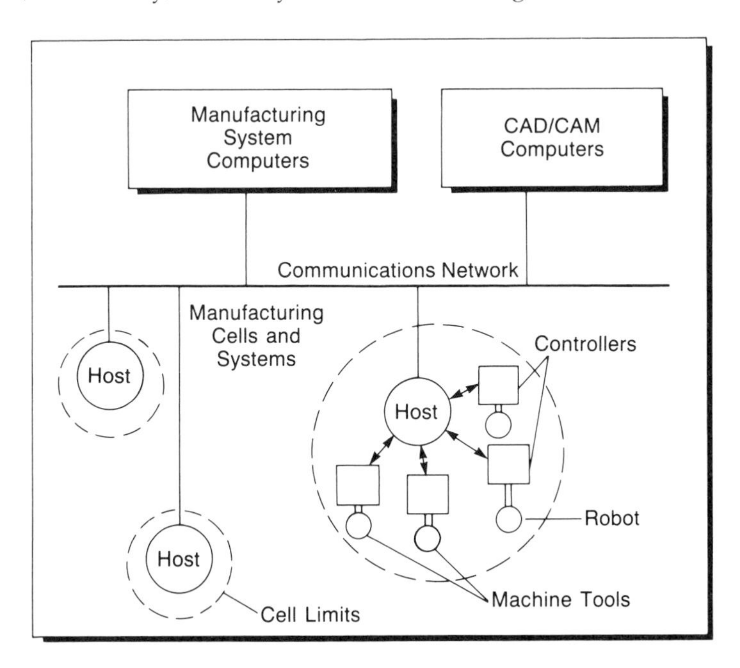

Figure 1.2 The hierarchy of the computer integrated manufacturing (CIM) facility. (Reprinted with permission from The Robotics Institute, Carnegie Mellon University © 1983.)

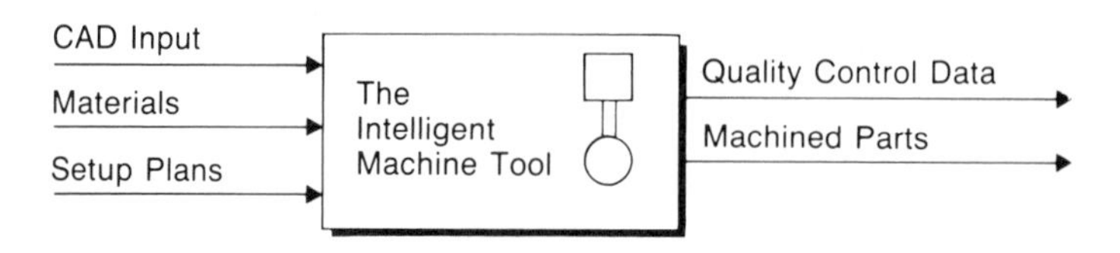

Figure 1.3 Inputs and outputs of the intelligent machine tool.

In the ideal scenario, a designer, working in an office environment, creates a part design and interacts with a software package called the "machinist critic" to ensure that the part can be made successfully. From then on, the part specifications (e.g., geometry and machining parameters) are downloaded through the factory network to an individual intelligent machine tool in a flexible manufacturing cell (see Figure 1.3).

Definition *The intelligent machine tool is defined by comparison with an intelligent human machinist. A higher level scheduler can rely on both of them in the same way. A given input leads to an expected output. Or, the intelligence reports back that the input is beyond the scope of the current system. We must therefore acknowledge that the degree of intelligence can be gauged by the complexity of the input and/or the difficulty of ad hoc in-process problems that get solved during a successful operation. Our unattended, fully matured intelligent machine tool will be able to manufacture accurate aerospace components and "get a good part right the first time."*

At the factory level, the intelligent machine tool, through its appropriate software and hardware, automatically configures itself with fixtures, tools, and stock to manufacture the desired part. Quality control is also part of the manufacturing process plan to guarantee that parts emerging from the intelligent machine tool are accurately produced. The part may then require other operations before it leaves the cell and moves to other areas of the factory.

Since 1960, this computer integrated factory has been the goal of experts in the field such as Merchant (1980). However, the ideal scenario (just described) envisaged by experts is far from complete. In the global view of the world, shown in Figure 1.2 and popularly represented in glossy trade magazines, the machine level of the factory hierarchy is normally represented as a black box with its own controller. From a senior management point of view, there might be the temptation to believe that the machine tool is already an unattended autonomous unit. However, more day-to-day experiences with machine tools quickly show that this is not the case. The myriad of craft skills, discussed in more detail in Part Three, are crucial to making the machines work.

A key reason why the computer integrated factory has not yet been fully realized is this lack of knowledge about the craftsmanship needed at the machine level. Although financial restrictions may prevent companies from investing in new manufacturing technology, this is not the only factor affecting its advancement. Leonardo da Vinci and H. G. Wells both had visions of the future, but no amount of financial resources at those times could have produced flying machines or spacecrafts. The same is true with the factory of the future: large amounts of financial investment cannot make up for the skilled craftsman whose knowledge and experience is needed to compensate for worn tools, faulty fixtures, or errors in NC programs. But like the flying machines and spacecrafts of Leonardo da Vinci and H. G. Wells, time will bring the requisite solutions.

Chapter 2 briefly reviews the industrial trends in flexible manufacturing systems, which at first sight seem flexible enough to make many different parts. However, these systems tend to be efficient only for large batch sizes and produce part accuracies that are in the range of 0.002 to 0.005 inches. Here are some of the restrictions that have been applied vigorously in order to make flexible machining systems unmanned:

- A part is made many times in a manned environment before it is moved to an unmanned environment.
- Part programs are carefully debugged during fully manned operations.
- The incoming parts are carefully constrained to provide predictable stock removal.
- Fixtures and tooling are set up before unmanned operations begin.
- Part geometries that produce critical cutting conditions are avoided.
- Part material is constrained to a limited range of grades and uniformity.
- Feeds and speeds are set very conservatively in an attempt to make the cutting process predictable.

The preceding constraints severely restrict the choice of parts that may be manufactured in an unattended system. Also, even more preparation is needed prior to entering the machining phase (e.g., debugging programs). The result is less flexibility and greater lead times. The production of a "good part that is right the first time" in a rapid prototyping environment is extremely difficult. The full automation of such a process in which only a handful of parts are to be made is currently impossible. In addition, once the required accuracies become as tight as 0.001 inches or less, our current automation abilities for small batch sizes are hopelessly inadequate. If accuracies become as tight as 0.0002 inches, even in large batch manufacturing (e.g., gear manufacturing) a great deal of human overseeing and frequent intervention is necessary to ensure adequate quality control.

A more specific case of the problems involved can be illustrated with a commercial example. Data from the work of Kegg (1985) traces a sampling of 22 different robot parts during their manufacturing operations. This example is summarized in Figure 1.4. In each of the 22 different batches, the first five pieces of each batch were observed closely. It was found that frequent changes of NC tapes, tools, and methods were necessary to obtain good parts. Even after these first five pieces, revisions were still required, and for the next ten pieces, 20% had problems. By this stage, one presumes that difficulties with the NC tapes had been resolved but that tools, fixtures, and other methods still needed adjustment. However, even in the production of the last fifteen pieces in each batch, 10% had some problems.

Evidently, the problems associated with these parts in 22 different batches were resolved at the machine tool by the craftsman. It is a common scene in manufacturing plants to see these skilled craftsmen adjusting tapes, tools, and fixtures in the early part of a manufacturing run. Before the age of computer controls, the machinist stood at the machine tool, carefully observing the part being made and laboring with

First 5 pieces of each part number	Required close observation and frequent change of NC tapes, tools, methods
Next 10 pieces	20% had problems
Next 15 pieces	10% had problems

Figure 1.4 Learning curve for 22 different robot parts (Kegg 1985). (Courtesy of Cincinnati Milacron.)

Figure 1.5 A humorous view of the skills needed by the expert machinist. (From *Machine and Tool Blue Book,* © 1985. Reprinted with permission of Hitchcock Publishing.)

the hand wheels and fixtures of the milling machine. Although machinists today are further removed from the machine, they still labor with the controller, edit NC tapes, and adjust fixtures in order to get the part right. A cartoon from the *Machine Tool Blue Book* summarizes, only too well, the way in which down-to-earth manufacturing engineers feel about NC machine tools and computer control in general (Figure 1.5).

Other data assembled by Kegg show the areas in which errors arise in NC program production. The largest percentages arise from dimensional errors and from programming techniques, perhaps related to misjudgments in interpolation routines (see Figure 1.6). Such errors in NC programs do not materialize until a part is made on the machine tool. To recover, the manufacturing personnel can either fix the error themselves at the machine tool or perhaps refer the program back to the NC programmers. Either way, the skilled craftsmen must notice and analyze the problem before any error recovery can take place.

1.2 GENERAL GOALS AND OBJECTIVES

Automated manufacturing systems will need to embody all the local intelligence of craftsmen, programmers, and manufacturing engineers. Associated with this goal is the aim to fully automate the links between

Reasons for Program Changes
1982

Cause	Total	%	Cause	Total	%
Dimension error	162	19.4	Link error	11	1.3
Prep code error	19	2.3	Tooling: shop	47	5.6
Function error	65	7.8	Tooling: prog	60	7.2
Setup: shop	40	4.8	Defective tape	1	0.1
Setup: prog	30	3.6	Damaged tape	4	0.5
Technique: prog	151	18.1	Machine/control	15	1.8
Excess stock	53	6.4	Comment error	31	3.7
Unknown stock	11	1.3	Other	8	1.0
Engrg revision	29	3.5	New program run	1848	
Routing change	21	2.5	Errors per prog	0.45	
Speeds & feeds	76	9.1			

Figure 1.6 Dimensional errors and mistakes in NC programming techniques account for the largest sources of necessary program changes in a sample of nearly 2,000 programs (Kegg 1985). (Courtesy of Cincinnati Milacron.)

the computer aided design phase and the computer aided manufacturing phase. The production of small batch sizes, even "one-off" components, has been emphasized with high accuracies down to 0.001 inch tolerances. The work is also concerned with rapid prototyping of mechanical parts and the rapid acquisition of repair parts. Table 1.1 lists the commercial needs for the intelligent machine tool.

Along with other robotics research concerned with the interactions between men and machines, the intelligent machine tool should be viewed within a spectrum. It is perhaps worth coining the phrase the man-machine spectrum. Research scientists in the field of autonomous vehicles and mobile robots for hazardous environments (such as mining and nuclear engineering) also acknowledge a spectrum of capabilities. The range goes from situations in which a person is in total control of the situation, to situations in which a teleoperator is used so that the person is removed from the working interface but still oversees the robotic arm, to more autonomous robotic operations in which the machines begin to make their own decisions. We must acknowledge the same kind of spectrum in the machine tool industry and recognize that developments in the intelligent machine tool must, over a period of time, pass through the same transitions. In terms of the near future, we can begin to automate machining processes, but in the early stages it will still be extremely important to have the person in the "loop."

One broad objective of the research is to acquire knowledge about machining. It aims to make an existing skilled machinist even more expert by providing rich information about the manufacturing environ-

Table 1.1 Commercial Needs for the Intelligent Machine Tool

1. Reduce the number of scrap parts following initial setup.
2. Increase the accuracy with which parts are made.
3. Increase the predictability of machine tool operations.
4. Reduce the manned operations in the machine tool environment.
5. Reduce the skill level required for machine setup and operations.
6. Reduce total costs for part fabrication while continuing to make satisfactory parts.
7. Reduce machine downtime.
8. Increase machine throughput.
9. Increase the range of materials that can be both setup and machined.
10. Increase the range of possible geometries of raw stock.
11. Increase the range of possible geometries for the finished part.
12. Reduce tooling through better operation planning.
13. Reduce number of operations required for setup.
14. Reduce setup time by designing parts for ease of setup.
15. Reduce the time between part design and fabrication.
16. Increase the quantity of information between the machine operations and machine control.
17. Increase the quantity and quality of information between the machine control and part design operations.
18. Increase the quantity and quality of information between the human and the machine tool control.

ment, or to give a novice machinist the capabilities of the expert machinist (see Figure 1.7). The other broad objective of the research is to develop sensors for machining. It aims to build flexible and automated mechanical systems, including: robotic part-handling devices, programmable fixtures, and automated tooling systems. The development of computer vision and other sensors for a variety of in-process and post-process quality control checks is also required.

Figure 1.8 combines the two broad objectives in the preceding paragraph. It shows how the research developments in both directions add up and thereby increase the autonomy of the intelligent machine tool. On the x-axis, the added sophistication of the system comes from a growth in knowledge. For example, the first step is to create a machine tool controller that not only contains the cutting tool paths (i.e., the part programs), but also gives the user helpful information on fixture selection, setup, and monitoring. (Experiments in which such knowledge is being acquired from experts are described in Part Three.) As such work continues, the heuristic rules of the craftsmen are unified into multilayer models, which then exhibit complex, problem-solving strategies. As shown on the left of Figure 1.8, the eventual systems will be able to tutor a novice machinist in machining practices.

The y-axis of the graph shows increasingly sophisticated mechanisms (e.g., fixtures, communications, and sensors) that can reduce the number of manned operations. Following the y-axis would result in a teleoperated machine tool system. In this case, a factory engineer could set up and operate several machine tools from a remote terminal. For example, a fixture could be built "on a computer screen," knowing that the icons correspond to particular mechanical clamps. Then, as the icons are assembled around the part drawing, these actions will correspond to physical movements on the machine itself. Actually the ability to do this remotely also depends on a knowledge of machining skills. For ex-

---

Imagine a computer controlled machine tool that is located in a geographically remote land-based facility. Alternatively the machine might be on board an aircraft carrier serving as the resource base for a fleet on maneuvers. Without warning, a telephone call is received, urgently requesting the manufacture of a part that has failed on an aircraft or a ship. If a rich source of information is available, describing both the part and the way it should be manufactured, then it can be made on the spot. From a library in a central land base, such information would be sent over a network to the machine tool at the remote location. It would contain CNC programs and instructions on fixture selection, stock size, initial setup, ordering of CNC subroutines, and machining technology. Then, an engineer who had general skills, but not specific machining skills, could work with this detailed information and produce the part. It would be as if an expert machinist, located many miles away in the central land base, were looking over the engineer's shoulder and describing the actions.

Figure 1.7 A scenario for rapidly producing a prototype or a repair part in a remote location. A rich source of machining information arrives to instruct an inexperienced machinist.

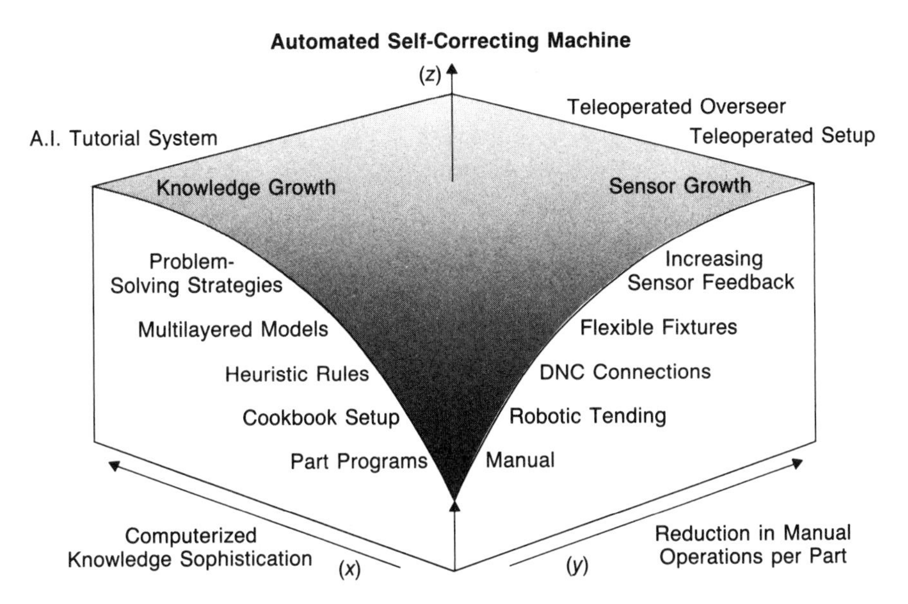

Figure 1.8 The autonomy of intelligent machine tool.

ample, it will be necessary to understand some of the detailed hand-eye coordinations that take place when a craftsman subtly adjusts a part in a fixture to maintain maximum accuracy. Although a robot hand probably will not need to be taught all such subtleties, the automated system will at least need to know the goals behind various adjustments and be able to respond appropriately. Further discussion of the knowledge behind such hand-eye coordinations can be found in Chapter 5 through Chapter 8.

In summary, Figure 1.8 shows that the growth in both knowledge and sensors leads to the automated self-correcting machine, the autonomy of which increases on the z-axis. This combination represents a series of research issues in computer science, engineering, and their combination. We hope that some of the unsolved research problems will provide a catalyst for those who are interested in automating factories.

REFERENCES

Andreasen, M. M., Kahler, S., and Lund, T. 1983. *Design for Assembly*, IFS Publications Ltd., Bedford, England.

Franklin, B. 1949. *The Autobiography*, The American Heritage Series. Edited by H. W. Schneider. Bobbs-Merrill Co., Indianapolis, p. 199.

Kegg, R. L. 1985. Private communication on data accumulated at Cincinnati Milacron Inc., and see article by N. L. Flavell and E. Kincaid, 1983, *American Machinist*, June, pp. 134–136.

Merchant, M. E. 1980. The factory of the future—Technological aspects, *Towards the Factory of the Future*, The American Society of Mechanical Engineers, PED-Vol. 1, pp. 71–82.

Miller, S. M. 1985. Impacts of robotics and flexible manufacturing technologies on manufacturing costs and employment, in *The Management of Productivity and Technology in Manufacturing*. Edited by P. R. Kleindorfer. Plenum Press, New York, pp. 73–110.

Office of Technology Assessment. 1986. *Technology and Structural Unemployment: Reemploying Displaced Adults*, p. 331.

Orwell, G. 1933. *Down and Out in Paris and London*, Harvest/HBJ Books, New York, p. 80.

Pirsig, R. M. 1974. *Zen and the Art of Motorcycle Maintenance: An Inquiry into Values*, Bantam Edition, New York, 1980, p. 96.

2 Evolving Manufacturing Systems

My intention is to tell of bodies changed to different forms; the gods, who made the changes, will help me—or I hope so—with a poem that runs from the world's beginning to our days.[1]

— Ovid, *Ovid's Metamorphoses*

CHAPTER 1 emphasized the economic importance of flexibility in batch manufacturing. The manufacturing facilities must respond rapidly to changing consumer demands and changing economic factors. Indeed, there is considerable evidence that, without flexibility, a manufacturing operation places itself in jeopardy. A combination of events in the late 1970s, including the rising costs of gasoline and the sales competition from Europe and Japan, proved to the American automobile manufacturers that their operations must be more flexible and better able to switch over from one model to another. The tool and die, aerospace, and ship-building industries require even more flexibility than automobile manufacturing because their equipment needs to be rapidly reconfigured to suit very small batch sizes. Until recently in our society, this flexibility could be achieved by adding manpower. Craftsmen were highly skilled and plentiful, and, in the scheme of things, their labor rates were not excessive. This has now changed, and the desired flexibility must be obtained with computers and programmable mechanical devices such as industrial robots and NC machine tools.

Several text books (Armarego and Brown 1969; Pressman and Williams 1977; Childs 1982) review the history of numerically controlled machine tools. Numerical control represents the second major stage in the evolution of machines. The first stage was the ability of a machine to generate its own motor power. The invention of steam engines in the eighteenth century enabled machines to be driven alone for the first time without directly relying on human or animal actuation. The development of numerical controllers between 1952 and 1956 allowed machine tools to position themselves accurately without relying on human judgment to turn handles for machine slideways.

[1] *Ovid's Metamorphoses*, translated 1955 by Rolfe Humphries. (Reprinted with permission of the Indiana University Press.)

At present, we have a closed-loop motor system able to power and coordinate itself, but a "babysitter" is still needed for the machine. This analogy is not meant to be flippant. The current capabilities of an CNC machine may be compared with a child between the ages of 6 and 12. Children have their own motor power, they are well coordinated, and they can carry out quite complicated tasks when "programmed" correctly. But they cannot be left alone for long periods of time without an overseer. Unexpected situations arise, emergencies have to be dealt with, and new plans have to formulated in order to successfully achieve long-term goals. Similarly, the present computer numerically controlled (CNC) machine tool has outstanding capabilities, provided it has been programmed correctly and the environment acts predictably. When things go wrong, the craftsman, acting as the machine tool's babysitter, needs to step in, take action, and make new decisions.

To move beyond these human-dependent machines to a third evolutionary stage, the machines need to learn more about their environment from sensors and evaluative software. From sensor information, the machines must diagnose unexpected situations, and then plan a new sequence of events to ensure that good parts continue to be produced.

This brings us to the domain of artificial intelligence (AI). Learning, diagnosis, and planning have been topics of AI research since the late 1950s. Following a review of some of the pertinent areas of research in AI, this chapter uses case studies to discuss the early applications of artificial intelligence to job-shop scheduling (Section 2.2) to flexible manufacturing cells (Section 2.3), and to programmable mechanical devices (Section 2.4).

2.1 ARTIFICIAL INTELLIGENCE

The intelligent machine tool needs control functions that represent the "brain," the "eyes," the "hands," and other anthropomorphic substitutes for the intelligent craftsman. In the past, the craftsman was responsible for design verification, planning operations, real-time machine monitoring, and inspection procedures. Such skills have called upon the craftsman's knowledge and his abilities to plan complex actions, detect errors, and learn about the environment. At the same time, a craft-related language has enabled communication with other humans and other machines in the factory. To achieve unattended manufacturing, the craftman's intelligence must be captured and used automatically. The next five subsections review some aspects of artificial intelligence that can be useful in manufacturing.

2.1.1 Knowledge Engineering for Expert Systems

Expert systems have proven to be powerful tools for coping with problems that have not yielded to quantitative analysis. For example, it is impossible to compute optimal schedules for job-shop manufacturing because of the exponential nature of scheduling, but it is possible to use sound scheduling principles, known by experts, to develop a workable schedule.

An expert system is a computer program that is developed from extensive interviews with experts in a particular field. It is partly a database of knowledge, together with rules that reflect experience and decision making. Typically the potential user of an expert system then interacts with the program in much the same way as they would have interacted with the expert in the field—explaining problems, performing suggested tests, and asking questions about proposed solutions. The most successful applications in the past have been in chemical and geological data analysis, computer system configuration, structural engineering, and medical diagnosis.

It takes many years for a person to acquire enough information to be called an expert. It usually takes a long time and many interviews to get the information out of the expert person and into the expert system, but, hopefully not as long as it took the expert to acquire the information in the first place. The task can be further complicated if the expert has difficulty describing the process by which he or she arrives at some decisions. Many decisions may be arrived at seemingly by intuitive or "gut feelings." For instance, a machinist may be able to tell when a cutting tool is dull simply by looking at it, but still may be unable to explain how he knows the tool is dull. Of course, the process is not completely intuitive. On further questioning and observation, the machinist might be able to articulate the general shape of the wear scars on the top face, corner, and flank face of the tool, and in addition relate these features to events in the last machining pass. Changes in chip color, vibrations of the machine tool, and the surface finish obtained on the component are the first signs that the tool is becoming worn and should be checked at the end of the machining pass. This suggests that there is a "trend analysis" active in the mind of the machinist. The human trend analysis could take the following form: given that I have been using this tool for so many minutes . . . and the surface finish on the part is beginning to deteriorate . . . but I do not get a rub mark on the bar . . . maybe I could reduce the cutting speed a bit . . . yes, that seems to be stable, now I can get away with things until this part is finished but definitely no more.

As the expert is questioned, a knowledge engineer, who is usually a computer scientist, attempts to fit all the information being gathered into a computational model of the process that the expert is using. This model is built into the expert system in the form of data and rules that manipulate the model. An example of such a rule might be:

> **If** the chip color is dark blue and the vibrations from the machine tool have just increased and the surface is dull, **then** the tool is worn to an unacceptable degree.

The first model that the knowledge engineer creates is probably not complete. After the initial model is finished, it is tested by checking it against known cases. For example, the data might be: a particular chip color, a machining sound level, and the amount of light reflected from the part's surface. If the expert system can identify dull tools from this information as successfully as human experts can, the system passes the test. If the system fails, the knowledge engineer will go back and question the expert again to uncover mistakes in the understanding of the process. Knowledge engineering is this cyclic process of gathering information, building a model, testing the model, and refining it. Figure 2.1 summarizes this activity for machining.

2.1.2 Error Detection and Recovery

The ability to detect and diagnose errors in a manufacturing system is necessary for building autonomous controls for machine tools and factory systems. The problems that an expert system will be able to diagnose in a machine shop include the identification of part programming errors and the location of machine tool failures. Such diagnosis work has already been applied in a flexible manufacturing system in which a robot is used to load components into a hot furnace and then into a forging machine. Table 2.1 illustrates a system that analyzes errors and ultimately infers that there has been an axis error in one of the robots. This system was built in the Cell Management Language (CML). In this example, all the answers have been filled in by a cooperative user, but in fact the robot controller, in conjunction with CML, can also make these decisions.

2.1.3 Planning and Scheduling

At the machine level of a factory, planning and scheduling are required to determine the order of the different cutting operations of a particular

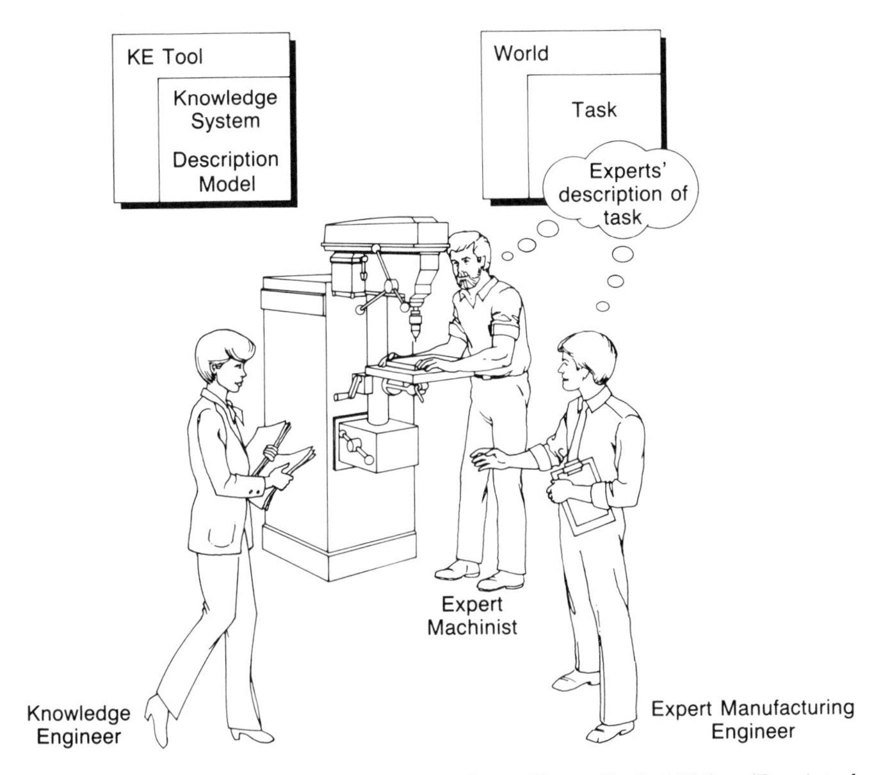

Figure 2.1 Knowledge engineering (based on Hayes-Roth 1984). (Reprinted with permission from F. Hayes-Roth, "The Knowledge Based Expert System," *Computer Magazine*, September 1984, Vol. 17 No. 9. © 1984 IEEE.)

part (e.g., milling followed by drilling followed by tapping). In most cases, there are several possible ways to order the steps involved in producing a piece. Generally, the problem is to find the quickest sequence that minimizes the number of setups in the fixture. However, other factors can be involved in making this decision. For example, the skilled machinist knows that as a component is shaped, it becomes increasingly difficult to hold in a machine tool fixture. Therefore, the machinist may plan to do some of the most difficult cuts first, before the piece becomes awkward to clamp. In other words, it may be wise to concentrate on safe holding and safe machining, rather than trying to minimize the number of setup operations.

At the shop level, planning and scheduling are needed to determine

Table 2.1 Error Diagnosis for a Robot in a Flexible Manufacturing System

Problem: The robot's hydraulic error light is on and the hydraulics have shut down.

CML is sending a message to the robot's computer and waiting for an answer. Was the communication successful? (yes, no) : **yes**

➔ I infer that the problem is either in robot mechanics or software.

CML is downloading a test program which is known to work on the robot. It will automatically receive the answer. This program tests all of the mechanical parts of the cell.
Did the test program work? (yes, no) : **no**

➔ I infer that the problem is in the robot mechanics.

CML is asking the CNC Robot Controller: is the limit switch on? (yes, no) : **no**

➔ I infer that the arm is not fully extended.

CML is looking at the Status indicators on the front of the robot. Are any of the axis lights on? (yes, no) : **yes**

➔ I infer that there has been an axis error.

how to allocate the available machines, materials, and personnel to finish a variety of part orders in the least amount of time or by the scheduled due dates—whichever comes first. Trading off constraints between minimum production time and satisfying a valued customer is an example of job-shop scheduling, and this is an area that systems such as Intelligent Scheduling and Information System (ISIS) are addressing (Fox

1983). Some of the first work in planning was done in the late 1950s by Newell and Simon (1963) with the General Problem Solver (GPS). Planning research has evolved since GPS and has focused on robot planning applications (Fikes and Nilsson 1971; Sacerdoti 1975), chemical experiment planning (Stefik 1981), project management (Tate 1977), and flight management (Robinson and Wilkins 1980).

2.1.4 Knowledge Representation

The knowledge representation is the schematic form of information in an expert system, and this schematic form makes it convenient to solve some kinds of problems and difficult to solve others. Therefore, it is extremely important to choose a representation that matches the requirements of an application; consider Table 2.2.

Table 2.2 in its current form makes it easy to extract an answer to the question: "What is the cost of a motor housing?" By contrast, it is more time consuming to extract an answer to the question: "What parts have a weight between 4 and 8 pounds?" As can be seen from this example, the difficulty arises when the answer to a question is not represented "explicitly." In this case, the information is only one step away from being available: "explicit minus one." However, answers to other questions will be represented "explicitly minus N" or even not at all ("What is the color of the motor housing?"). Of course, not all pertinent information can be explicitly represented. But, if these choices are made wisely for a particular application, the computational requirements will be significantly reduced.

The knowledge representation of geometrical shapes is especially important to manufacturing and to the intelligent machine tool. This knowledge represents the information that characterizes the shape of a part to be manufactured. As it turns out, the choice of a geometrical representation can make one application very easy (e.g., graphics) and another very difficult (e.g., area calculations).

Table 2.2 Parts Order List

MACHINE	COST ($)	SHIPPING WEIGHT (lb.)	SHIPPING CHARGES ($)
Turbine Blade	500	50.9	12
Clamp1	30	8.4	3
Clamp2	125	5.8	3
Motor Housing	50	15.5	4

Figure 2.2 shows a sphere represented with patches, or a number of "tiles," distributed across its surface. This representaion is good for graphics because it can be used to readily generate a realistic picture. In fact, shading is also quite convenient—a constant "shade" can be assigned to each patch with only minimal distortion of the surface. However, if it were ever necessary to calculate the area of the sphere, it would be best to use a completely different representation: a parametric description (e.g., sphere center point and radius). In turn, a parametric description of this sort is virtually worthless for graphics. This dilemma has led researchers to develop translation algorithms between different geometrical representations (Badler and Bajcsy 1978) rather than developing strange new algorithms for awkward representations.

Just deciding how to represent data in the best way is a difficult issue, but it is further exacerbated when the knowledge is partly "control" information (i.e., "how to" information). Unfortunately, an expert's knowledge is almost always a mix of control and data elements. To accommodate control information, "rule-based" programming was developed because this representation makes it possible to "program" small pieces of control knowledge without knowing anything about the final application (Waterman and Hayes-Roth 1978). For example, the rule developed in Section 2.1.1 determines how a rule-based program would behave under very narrow circumstances, but it still may be invoked at any time.

The intelligent machine tool should represent information in a number of different ways and then be able to translate between the representations to take advantage of the merits of each.

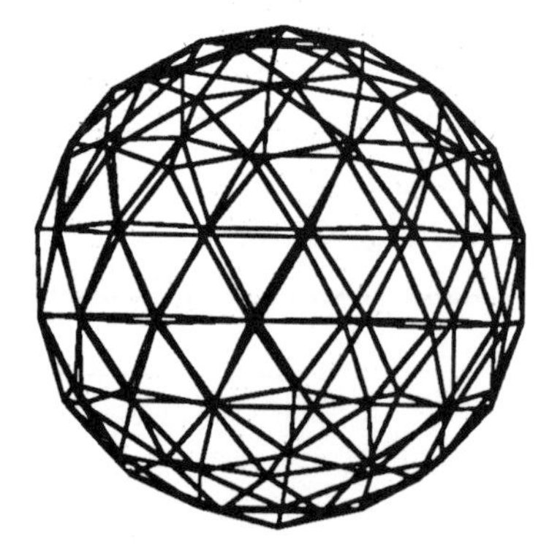

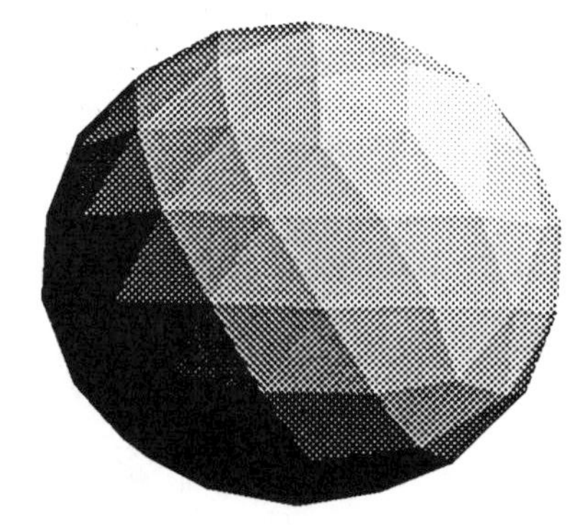

Figure 2.2 A sphere represented with surface patches.

2.1.5 Language Understanding

Artificial intelligence has also concentrated on how to understand and translate between different languages. In Section 2.1.4, we illustrated the need to translate between different knowledge representations. An alternative description of this problem would have been to say that "it is necessary to translate between different descriptive languages that are used to represent knowledge."

However, language understanding and translation have much more to offer manufacturing than translating between knowledge representations. Machine operators need simple and concise languages to communicate part and machine information—these languages must be understood by the new intelligent controller. Engineers need powerful descriptive mechanisms to understand process and diagnostic information—these languages must also be understood. And, at different levels, machines need to send, understand, and translate information between many different machines and people. Some issues involved in language understanding include:

- **Language description**: Before anything general can be done with languages, there must be a way to describe them. Backus Naur Form (BNF) is capable of describing context-free languages, and more complex languages can be described by augmenting BNF in various ways.
- **Language understanding**: To understand a sentence or program, one must first understand its language. Only then can one decompose a sentence into its syntactic parts and attach the semantics (i.e., meanings). Once the sentence is understood, action can be taken to either update a database or a language description.
- **Language translation**: Language translation involves all of the preceding phases of language manipulation. Translation starts with understanding a sentence or program in one language and then generating a new sentence or program in the target language.

When these tools are available and comprehensive, many manufacturing problems can be effectively approached. Part programs can be generated automatically, removing many opportunities for programming errors. At present, using CML, information from one machine (e.g., a vision machine) can be translated into the information suitable for another machine (e.g., a robot). This kind of problem has roots in many other areas of manufacturing, such as the many incompatible databases

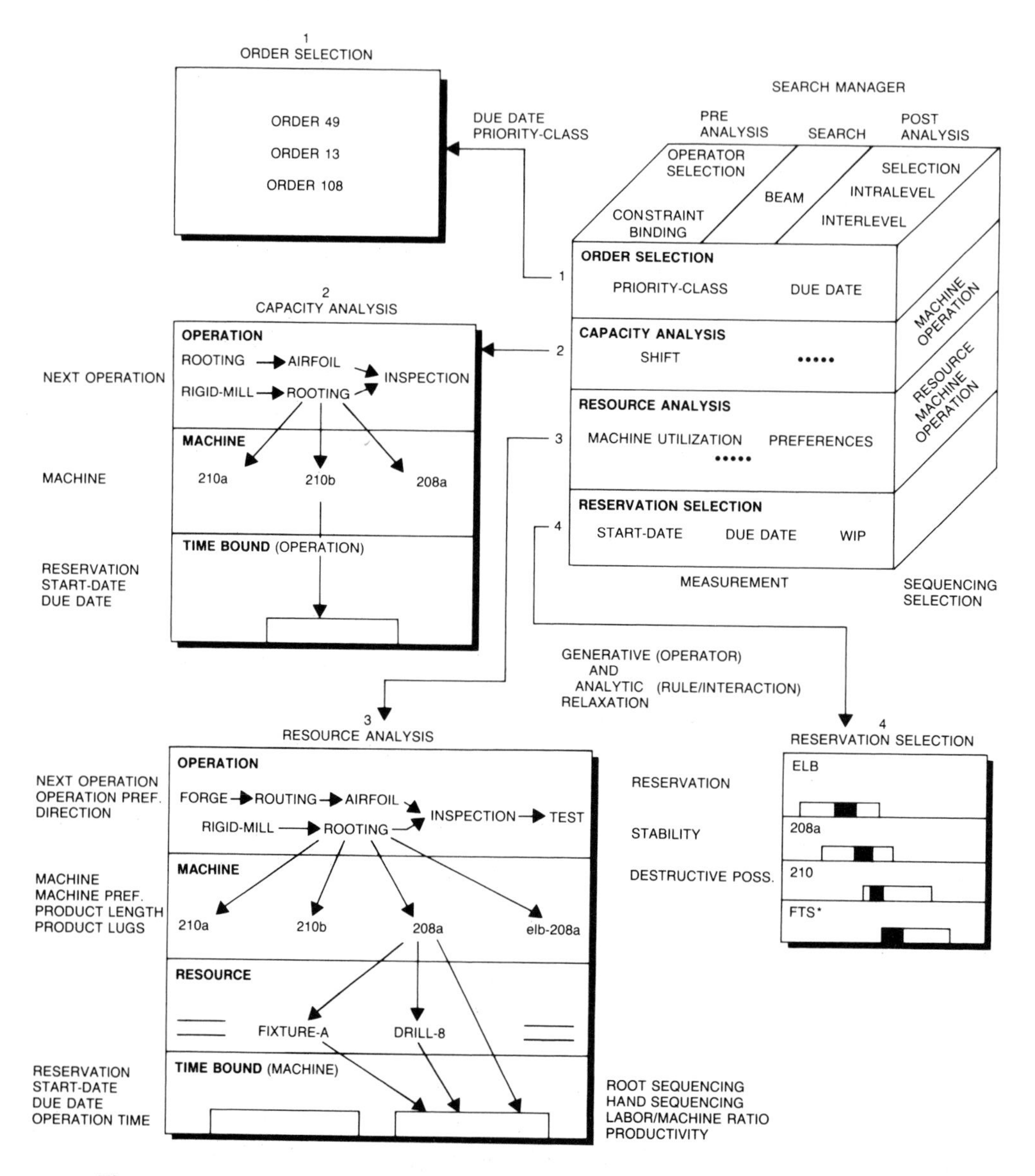

Figure 2.3 ISIS System architecture for scheduling the job shop for which the intelligent machine tool is an integral component (Fox 1983). (Reprinted with permission from Bourne & Fox, "Autonomous Manufacturing: Automating the Job Shop" *Computer Magazine*, Vol. 17, No. 9 © 1983 IEEE.)

in use at corporations all over the world. Automated translation can also make cross database queries feasible.

2.2 THE USE OF AI SYSTEMS AT THE OVERALL FACTORY LEVEL (ISIS)

In recent years, the use of computer systems for shop-level planning has been increasing. This section briefly reviews some of characteristics of ISIS, a job-shop scheduling system that has been developed by Fox (1983). The various workstations such as milling centers, drilling machines, and lathes are defined as operators. The planning system, working with the goal of transforming a raw product description into a final part shape, determines various sequences of the operators. This is difficult to implement without the experience of a skilled production engineer who appreciates the order in which machines must be employed to achieve effective tolerances and surface finishes. In this case, the knowledge of a production engineer is encoded into "constraints" that are used to focus the search for an acceptable machine operation sequence. For this reason, this method is known as "constraint-directed reasoning."

The ISIS system addresses the problems of how to construct accurate, timely, realizable schedules and how to manage their use in the job shop. ISIS carries out a hierarchical, constraint-directed search through the space of possible schedules. The search moves through more and more detailed levels of analysis, with specific types of constraints coming into play at each level. ISIS is capable of incrementally scheduling orders as they are received by the plant as well as reactively rescheduling orders that are affected by machine breakdowns and other dynamic changes in the shop.

The four levels of search conducted by ISIS—order selection, capability analysis, resource analysis, and resource assignment—are depicted in Figure 2.3. Close examination of this figure shows that, in the context of the intelligent machine tool, a great deal of information is needed at levels 2 and 3 in order to establish which operations can be performed in individual workstations and how long these operations take. The quality of the schedules generated by ISIS obviously depends on the accuracy and timeliness of this information. ISIS cannot rely completely on a priori information about machine operations because machine operations are not predictable enough to determine entire factory schedules. To resolve this problem, the intelligent machine tool must close the factorywide feedback loop and start to feed information up the manufacturing chain of command.

2.3 THE USE OF AI SYSTEMS AT THE CELL LEVEL (CML)

During the 1970s a number of flexible manufacturing cells or systems were created, for example, by Popplestone, Amberler, and Bellos (1978); Weck, Zenner, and Tuchelmann (1979); and Ennis and Eastwood (1979). These were created around procedural languages such as Pascal, FORTRAN, and APT. In such systems, despite the size of the program, the operations at any statement critically depended on the previous activities. Generally speaking, the failure of one machine in the sequence required that the system be shut down for fault analysis. These previous systems clearly were not very flexible and were not compliant enough to cope with the contingencies of a typical manufacturing system. The Cell Management Language (CML) has been developed as a new computer language for manufacturing to link complex systems of multivendor equipment together (Bourne 1986). CML combines computational tools from rule-based systems, object-oriented languages, and new tools that facilitate language processing. These language tools make it convenient to build new interpreters for interfacing and understanding a range of computer and natural languages; for example, in CML it is easy to build an interpreter with machine tool languages that can understand and generate new part programs. Once constructed, such interpreters can be linked together and used to make intelligent decisions for systemwide action planning and diagnostic error recovery.

Between 1984 and 1987, CML was used to build two flexible manufacturing cells. The first cell focused on machining operations (see Figure 2.4). The major equipment includes two CNC machine tools, an industrial robot, vision-based inspection equipment, and an intense environment of sensors and computers.

The second cell (Wright, Bourne, Colyer, Schatz, and Isasi 1982; Bourne 1984) manufactures preforms for turbine blades (see Figure 2.5). This cell operation is the first step in manufacturing at this plant. It initially takes a cylindrical billet of stainless steel and makes a preform by open-die forging. In later parts of the manufacturing process, these open-die forgings are closed-die forged to obtain a complex airfoil section. Finish machining and grinding operations are then done on some of the blade features in order to obtain the final part. Metaphorically, the forging cell is an automated blacksmith. It is composed of ten machines, and in the

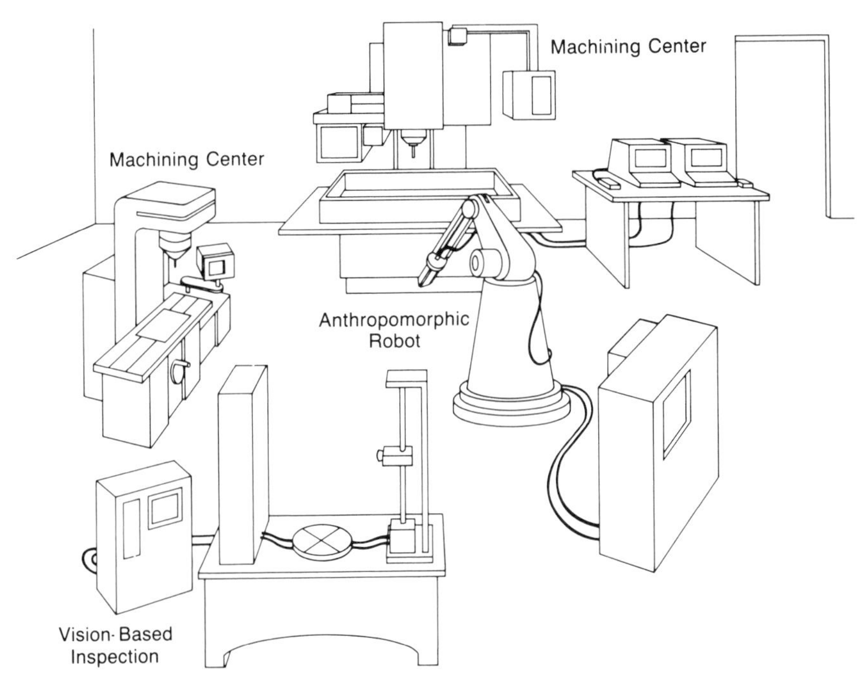

Figure 2.4 An overview of a flexible machining cell.

bottom left-hand corner of Figure 2.5 a vision system locates the incoming billets. The first robot guided by the vision system picks up billets from the rack and then transfers them to the rotary furnace that operates at 1200°C. The robot has been designed to withstand the hostile environment of the hot furnace. From the furnace, the robot loads the open-die forge that changes the cylindrical shape into an elongated shape, which approximates the final turbine blade geometry. After the part formation, a second robot moves the part to a gauge station that reconstructs a three-dimensional model of the part using a second vision system. (This is described in more detail in Chapter 5.) The second robot then moves the part to a cropper that trims the excess material from the preform, then to a stamper that punches a batch code onto to the surface of the part, and finally to a rack that is destined for the next area of the factory. The forging cell is controlled by a star-shaped computer network with a host at the center and four different kinds of machine controllers at the tips of the star, one per machine. The cell is monitored by 150 sensors, which are strategically located to detect error conditions in equipment.

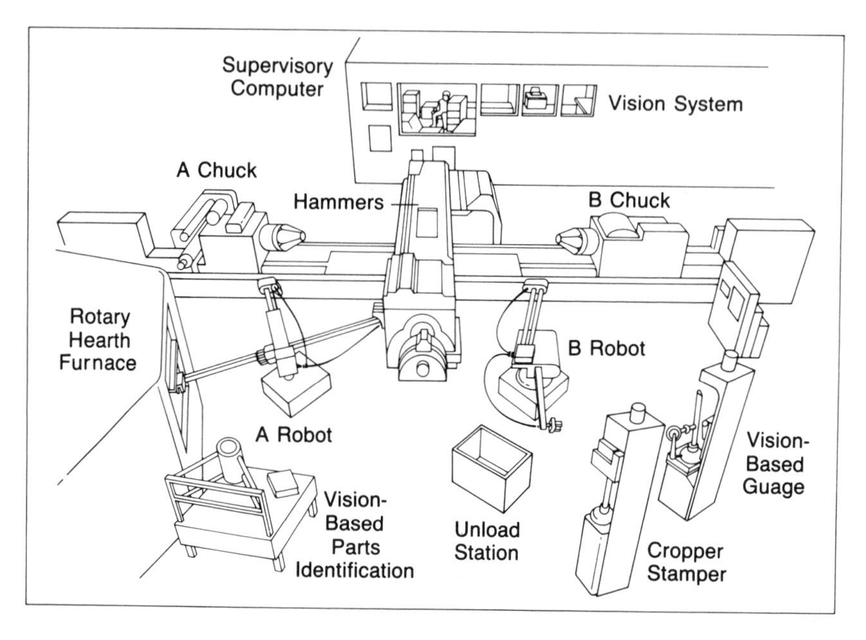

Figure 2.5 The flexible manufacturing cell for open die forging installed in the Westinghouse Corporation's plant in Winston-Salem. (Reprinted with permission from Wright, "A Manufacturing Hand" in *Robotics and Computer Integrated Manufacturing* Vol. 2, No. 1, 1985, Pergamon Journals Ltd.)

We have not been able to achieve complete autonomy in this cell because of the extreme dangers inherent in its failure. However, the system does effectively reduce setup times, increase machine utilization, and minimize the number of bad parts produced. An advantage of trying to build a completely autonomous system is that it becomes a scientific laboratory for performing experiments in manufacturing and process physics.

This facility was ready for experimentation in late summer 1982 and ready for production in early 1983. By late 1984, the first version of the Cell Management Language ran the cell. In contrast to other flexible manufacturing systems, CML was deliberately designed to be a rule-based language that operated on a complete and ever-changing model of the cell. In this application, CML's primary objective was to dynamically reschedule the machines in the cell. At any given point in time, the system decides what to do next on the basis on information from the machines and their sensors. In other words, the system does not decide in advance what schedule to follow; at each step of the process, it looks at everything that is happening in the cell environment and decides what to do next on the basis of the current situation. If something unexpected happens in the middle of a process (e.g., the machinery jams or communication to a robot fails), the system can detect the error through machine-to-machine communications and sensor readings. Then, the system can schedule the appropriate corrective operations.

Also, CML incorporates a language translation scheme; it constructs and understands messages sent in a multivendor machine system such as that shown in Figure 2.5. Thus, CML not only contains information about the machine tools and their scheduling requirements, but also information that describes the idiosyncrasies of the various machine tool controllers and their control languages. CML can then be used to perform a translation service for each of the machine tools. In this respect, it can be compared with the language translation service at the United Nations. To accomplish this task, CML must undertake many functions. In particular, CML receives a message, parses it, attaches semantics to the parsed form, generates internal structures, updates the internal model of the manufacturing cell, fires new rules as a result of the state change, and finally automatically writes and sends new programs to whatever new machines are affected by the change (Bourne 1986).

2.4 THE USE OF AI SYSTEMS FOR PROGRAMMABLE MECHANICAL DEVICES

Beyond knowledge engineering in a software sense, there is now an additional need to develop mechanical fixtures that can duplicate the skills of a craftsman. This mechanical hardware should be compatible with the new computer languages such as CML and possess a degree of flexibility not yet fully realized in manufacturing. The key point is that while the human machinist is there to provide the dexterity and the human flexibility, it is not particularly necessary to have a clamping device that is dexterous or sensor based. But, as operations such as machine tool loading, parts inspection, and tool changing become automated, the manual dexterity of human machinists must also be duplicated.

Knowledge engineering has so far been regarded as the capturing of mental or "book" knowledge. However, in manufacturing, many skills are manual skills, which require sensations and intuition. In the future, it will be necessary to ensure that the mechanical devices are intelligent and that they begin to duplicate these human sensations. One is tempted to coin a new phrase to mirror knowledge engineering, namely, "sensation engineering."

If we conceive of the mechanical devices as being intelligent in their own right, they can become for us a source of valuable information. For example, the machine tool clamps shown in Chapter 9 can be used to both digitize the shape of a part as well as clamp it into position.

2.5 THE STATUS OF MANUFACTURING SYSTEMS AND CNC MACHINING CENTERS

2.5.1 Flexible Manufacturing Systems

A flexible manufacturing system or cell is an "orchestra" of machine tools. Just as each musician follows his or her own musical score, each CNC machine tool follows its own NC program. The coordination and scheduling of these musical and engineering systems is exceedingly important; without it, even a group of gifted soloists is downgraded to a chaotic assemblage. To continue the musical simile, the best conductors have a feel for the nuances of individual instruments; they can read and send signals to and from each musician during a live performance; and most important, they can hold a complete work together. All of this is a rather tall order for a computer hosting the performance of a live manufacturing system, but for reliable autonomous manufacturing, many of these qualities are in fact needed.

For a general survey of flexible manufacturing systems (FMS) throughout the world, the proceedings of the International Conferences on Flexible Manufacturing Systems (1982 through 1987) provide many case studies (see Rathmill 1983). In one conservative sense, an FMS is just an efficiently organized group of machine tools for batch manufacturing. Such systems were first discussed by Williamson (1968), and the early applications consisted of only one or two machines with simple conveyor systems and, surprisingly to us now, no computer (Dempsey 1983). During the period from 1968 to the early 1980s, the concept of flexible manufacturing systems and cells became increasingly popular. Computer control, with flexible part routings through the FMS, made it possible to design larger systems with more machines and with a variety of processes (e.g. machining, grinding, measurement, and cleaning) going on simultaneously.

Recently, manufacturing systems have been scaled down to respond to costs and technical difficulties that are associated with orchestrating large systems. A recent economic review by Zygmont (1986) collates the war wounds of many FMS users. Although remaining enthusiastic about the design philosophy, the users are now limiting the size of new installations to about four or five machines and from the outset are not expecting great flexibility. All the users realize that individual machine tools must be equipped with reliable, rugged sensors and that the host software must be more responsive to unexpected problems. It is then hoped that large autonomous systems will be able to operate without human intervention for extended periods.

Despite this air of realism and mild cynicism, the market trends for the next few years are positive. Based on an article by Krouse (1986) that reports on data from the Yankee Group (Boston), the number of FMSs operating in the United States will grow from the 50 that were operating in 1985 to a projected 284 systems in 1990. This represents an increase in revenue from $143 million dollars in 1985 to $832 million dollars in 1990. This would be shared among several companies notably Kearney and Trecker (Bloomfield Hills, Michigan) who claim 28% of the current market; Cincinnati Milacron (Cincinnati, Ohio) who claim 23% of the market; Giddings and Lewis (Ford du Lac, Wisconsin) with 14% of the market and White Sundstrand (Belvidere, Illinois) with 11%. The user group includes companies such as Boeing, Deere & Co., General Electric, Hughes Aircraft, IBM, and Westinghouse.

2.5.2 CNC Machining Centers

To reiterate, the user group is presently reducing the size of their flexible manufacturing systems because of the unpredictable behavior of individual CNC machine tools. To behave predictably, the CNC machine tool and its controller will need to contain detailed information in a broad range of areas: how to set up the machine tool, how to monitor machining activities, and how to detect and then recover from errors, to name a few.

Machine tool controllers can be used to effectively operate unattended for periods of time, provided that painstaking engineering is carried out prior to the machining run. Seifried (1985) has proved this point, but only after having invested the requisite effort. He used a horizontal machining center equipped with unusual options that provided enhanced capabilities for unattended machining. The principal options were a mechanical touch probe, automatic work changer, and a sensor for torque-controlled machining. The mechanical probe was mounted in the tool carousel and handled as a tool. An electrical signal indicated probe deflection with a high degree of precision. The probe was used in a variety of ways, for example to align the machine to the part, to detect excess stock and increase the number of rough milling passes, to check part dimensions, and to detect missing machining operations and, by inference, broken tools. The automatic work changer consisted of an oval track with eight pallets that held the parts prior to automatic loading. Torque-controlled machining was used to permit more aggressive machining strategies, provide wreck protection, and to avoid shutdown when hard spots or dull tools were encountered.

2.6 SUMMARY

Before autonomous manufacturing systems can be built, two challenges must be overcome:

- To integrate machine actions into a cohesive flexible system.
- To eliminate the dependency on human experts for setup, monitoring, and evaluation.

Software systems such as CML provide a partial solution that makes systemwide action planning and diagnostic error recovery possible. However, a general-purpose tight coupling between heterogeneous systems remains an open problem and is considered in Part Two. The elimination of human experts will be possible when the experience of skilled machine operators, assembly workers, and repair personnel can be captured in software systems; the beginnings of these problems are considered in Part Three.

REFERENCES

Armarego, E. J. A., and Brown R. H. 1969. *The Machining of Metals.* Prentice-Hall Inc., Englewood Cliffs, N.J.

Badler, N., and Bajcsy, R. 1978. 3-D representations of computer graphics and computer vision, *Computer Graphics,* Vol. 12, pp. 153–160.

Bourne, D. A. 1984. A multi-lingual database bridges the communication gap in manufacturing. In *Robotics and Factories of the Future,* edited by S. N. Dwivedi. Springer-Verlag, New York, pp. 545-552.

Bourne, D. A. 1986. CML: A meta-interpreter for manufacturing, *AI Magazine,* Vol. 7, No. 4, pp. 86–96.

Childs, J. J. 1982. *Principles of Numerical Control.* Industrial Press Inc., New York.

Dempsey, P. A. 1983. New corporate perceptives in FMS, *Proceedings of the 2nd International Conference on Flexible Manufacturing Systems,* Vol. 1, pp. 3–18.

Ennis, G. E., and Eastwood, M. A. 1979. Robot systems for aerospace batch manufacturing, Interim Technical Report, IR-812-8, McDonnell-Douglas Corp.

Fikes, R. E., and Nilsson, N. J. 1971. Strips: A new approach to the application of theorem proving to problem solving, *Artificial Intelligence,* Vol. 2.

Fox, M. S. 1983. The intelligent management system: An overview. In *Processes and Tools for Decision Support,* edited by H. G. Sol. North-Holland Pub. Co., New York.

Hayes-Roth, F. 1984. The knowledge-based expert system: A tutorial, *Computer Magazine,* Vol. 17, No. 9, p. 11–28.

Humphries, R. 1955. *Ovid's Metamorphoses.* Indiana University Press, Bloomington, Ind., pp. 1-3.

Krouse, J. 1986. Flexible manufacturing systems begin to take hold, *High Technology,* Vol. 6, No. 10, p. 26.

Newell, A., and Simon, H. A. 1963. A program that simulates human thought. *In Computers and Thought,* edited by E. Feigenbaum and J. Feldman. McGraw-Hill, New York, p. 3.

Popplestone, R. J., Amberler, A. B., and Bellos, I. 1978. RAPT: A language for describing assemblies, *The Industrial Robot,* pp. 131–137.

Pressman, R. S., and Williams J. E. 1977. *Numerical Control and Computer Aided Manufacturing.* John Wiley and Sons, New York.

Rathmill, K. 1983. *The 2nd International Conference on Flexible Manufacturing Systems.* International Fluidics Services Limited and North-Holland Pub. Co., Bedford, England.

Robinson, A. E., and Wilkins, D. E. 1980. Representing knowledge in interactive planning, *Proceedings of the First Annual National Conference on Artificial Intelligence,* American Association for Artificial Intelligence , pp. 148–150.

Sacerdoti, E. D. 1975. A structure for plans as behavior, Ph.D. Thesis, Computer Science Dept., Stanford University.

Seifried, R. 1985. Unattended machining, SME Technical Paper, MR85-468.

Stefik, M. 1981. Planning with constraints (MOLGEN: Part 1), *Artificial Intelligence,* Vol. 16.

Tate, A. 1977. Generating project networks, *Proceedings of the Fifth International Joint Conference on Artificial Intelligence,* Cambridge, Mass., pp. 888–893.

Waterman, D. A., and Hayes-Roth, F. 1978. An overview of pattern-directed inference systems. *In Pattern Directed Inference Systems,* edited by D. A. Waterman and F. Hayes-Roth. Academic Press, New York, pp. 3–24.

Weck, M., Zenner, K., and Tuchelmann, Y. 1979. New developments of data processing in computer-controlled manufacturing systems, Technical Paper, MS-79-161. Soc. of Manufacturing Engineers, Detroit.

Williamson, D. T. N. 1968. The pattern of batch manufacture and its influence on machine tool design, *Proceedings of the Institution of Mechanical Engineers,* Vol. 182, No. 1, p. 870.

Wright, P. K., Bourne, D. A., Colyer, J. P., Schatz, G. C., and Isasi, J. 1982. A flexible manufacturing cell for swaging, *Mechanical Engineering,* Vol. 104, No. 10, pp. 76–83.

Zygmont, J. 1986. Flexible manufacturing systems, *High Technology,* Vol. 6, No. 10, pp. 22–31.

Part Two

BUILDING INTELLIGENT MACHINES

Physics has a history of synthesizing many phenomena into a few theories. For instance, in the early days there were phenomena of motion and phenomena of heat; there were phenomena of sound, of light and of gravity. But it was soon discovered, after Sir Isaac Newton explained the laws of motion, that some of these apparently different things were aspects of the same thing. For example, the phenomena of sound could be completely understood as the motion of atoms in air. So sound was no longer considered something in addition to motion. It was also discovered that heat phenomena are easily understandable from the laws of motion. In this way, great globs of physics theory were synthesized into a simplified theory.
— Richard P. Feynman in *QED*.[1]

The craftsman brings personal experience and hand-eye coordination to any manufacturing, assembly, or repair task in the factory. An autonomous factory must duplicate this experience and skill. To accomplish this, we give practical guidelines for building generic components that are used to make intelligent machines. These components make it possible to duplicate a craftsman's thinking patterns, visual abilities for inspection and monitoring, and manipulative abilities for part handling.

[1]Richard P. Feynman, *QED: The Strange Theory of Light and Matter*. © 1985 by Princeton University Press. Reprinted with permission.

3 Unifying Intelligence for Manufacturing Tools

We are thus introduced to a new principle of "relativity," which holds that all observers are not led by the same physical evidence to the same picture of the universe, unless their linguistic backgrounds are similar, or in some way be "calibrated."[1]

— Benjamin Lee Whorf in *Science and Linguistics.*

TODAY most factories are a science museum of manufacturing technology. The plant floor exhibits manufacturing equipment from the early 1900s to the latest robotic technology, but it does not function well together. Teamwork can maximize the efficiency of both human and automated systems; however, this is difficult to achieve when the team members have different skills and competencies. In particular, the computer controls in manufacturing environments all possess different capabilities and limitations. A major goal in manufacturing is to create unified environments in which every person and every machine can take full advantage of each team member's special skills. This leads us to develop a new "relativity" theory of robotics in which each control system can operate in its own unique way, and the global system can "calibrate" and translate between independent manufacturing controllers.

3.1 EVOLUTION OF AUTOMATED SYSTEMS

Every engineer who builds automated systems must decide how to effectively integrate system components into one system. Currently, it is only possible to build systems when the machine-to-machine interface problems have been "designed out," which typically involves buying all equipment from the same vendor. Unfortunately, automated systems are usually constructed, out of necessity, from components provided by multiple vendors, and the components rarely conform to unifying stan-

[1] Benjamin Lee Whorf, *Science and Linguistics in Language Thought and Reality* © 1956 MIT Press. Reprinted with permission.

dards. This same dilemma is also encountered in strictly software systems.

Because system integration is difficult, many manufacturing groups settle for buying one piece of equipment into which they plug vendor-supported options. For example, a vision system company may sell a robot as an option that would be plugged into the vision system's central processor. Alternatively, a robot company may sell a vision system as an option that would be plugged into the robot's central processor. These "technology groups" are very difficult to manage in a factory environment because they all operate differently. Usually, different engineers are assigned to each technology group, and each engineer is compelled to advocate a system's special merits. This is not the best way to spend engineering efforts in today's factories.

General Motors has proposed an important first step to building integratable systems. GM engineers reason that if they design a "universal" protocol for all machines, publish it, and then only buy equipment that conforms to these rules, they will be able to jump the system integration hurdle. GM's protocol is called the Manufacturing Automation Protocol, or MAP (Adler 1984). This effort has received remarkable support from both computer and machine tool manufacturers. Although the project is running slightly behind schedule, the manufacturing community is seeing a steady flow of results, and some of the world's largest corporations are deeply committed. When this first step reaches its potential, a new factory will be able to be built with new machines, purchased from the cooperating companies. Once these machines have arrived, they will be able to be assembled into a functional network so that the machines can easily pass information. This would be similar to the telephone company taking over the automation business and using their telephone network technology to build systems. With this analogy, one machine tool could pick up a phone and call another machine tool no matter where it resides in the factory.

However, making a connection between machines and successfully passing information is only one aspect of integrating a system. If a machine receives information and cannot decode it, the information is virtually worthless. This would be similar to the experience of an English speaker answering a phone and discovering that the other party speaks only French.

MAP attempts to breach this language barrier by providing an application layer on top of the message formats and exchange protocols. This application layer is a collection of standardized binary codes that are designed to eliminate the need for a real understanding between machines. Essentially it is the same as a third party setting up a secret code between machines (e.g., 1 means download, 2 means upload, 3 means start robot program). This may suffice for situations in which an all powerful third party provides each of the machines with a goal before the "conversation" begins, but it does not help much if each machine has its own concerns that exceed the limits of the secret code. The teamwork in an automated system involves much more than simple answers to prearranged questions. Teamwork involves problem solving and creative questions and solutions from each of the participants in a manufacturing task. This is true in the factory, no matter whether equipment is new or old, simple or complex.

A science fiction writer might envisage a special earpiece that would be attached to our respective telephones. The earpiece would translate incoming foreign language messages into messages in our own native language. In factory terms, this would be like a machine tool sending a part program to another machine and then having the receiving machine be able to extract the basic meaning of a program. This earpiece technology would make it possible for each machine to understand the purpose of other machines and to cooperate with the goals of the whole system.

A software version of such a telephone earpiece has been implemented and tested in an industrial setting. The Cell Management Language (CML) has been designed to include the tools that are necessary to build language-understanding and language-generation mechanisms necessary for manufacturing tasks (see Chapter 4).

3.1.1 Relating Manufacturing Terms Across Manufacturing Modalities

Manufacturing tools are programmed differently, not only because vendors choose to make them different, but also because they must refer to different modalities in manufacturing. Vision tools refer to two-dimensional pictures, whereas controls for robot arms refer to velocities and six degrees of freedom. How can these ideas be unified? Hand and eye information must be intertranslatable to accomplish hand-eye coordination tasks. On the surface, this seems like an impossible task because a robot manipulator cannot analyze pictures.

There must be a control strategy that can act as an intermediary between different modalities and the external world. To accomplish this goal, the current concepts in robotics will be made generic enough that one high-level control concept can combine several domain-dependent ideas. At first, it may seem that we have taken some common characteristics of robotic systems and turned them upside down, because robot problems will be rephrased into vision terminology, and vision problems rephrased into robot terminology. But by rephrasing the domain-dependent terminology, we hope to amplify the underlying generic mechanisms. Some examples of our rephrasings are now given in Table 3.1 before the unification is more fully developed in Sections 3.2.1 and 3.3. These rephrasings make it possible to unify manufacturing control mechanisms, thus minimizing the interpretation and translation that is necessary to switch between them.

This unification of manufacturing terms makes it convenient to pursue a range of different strategies to attain a final goal. For example, imagine a robot and a stationary vision system that are located alongside an automobile assembly line. The robot is responsible for welding spot joints as the car is moving, and the vision system is partially responsible for locating the spot-weld locations. As long as the car is properly illuminated and there are no major obstructions, the visual sensor is capable of tracking the car and finding gross spot-weld positions. For the moment, we are glossing over technical problems that are very difficult within one sensory domain so that the interactions between multiple senses can be seen in a relatively simple context. However, when the car moves into heavy shadows or behind obstructing equipment, the robot must continue the chase by pursuing the car only with its arm by maintaining tactile contact. In this situation, despite the temporary loss of visual guidance, the robot arm will be able to follow the car for a short period by obtaining tactile reference checks from datum points on the chassis of the vehicle.

Physically, there is only one difference between these two modes of operation, namely, the sensing medium used to track the object. In the visual case, light is being reflected off the car and is interacting with the sensor. In the tactile case, as the robot touches the car, an electromagnetic field is being sensed. These two modes of pursuit are so similar that, for reasons of theoretical simplicity and practical transition time, there should only be one control algorithm that mediates between them.

Although at first it may seem as though the robot arm has a much more difficult job because it has to "search around" for contact, this is the same problem that vision must solve. The vision processing involves searching within a picture or waiting for the visual circumstances to improve, for example, for the automobile to emerge from the shaded area.

Table 3.1 Examples of How Control Concepts in One Domain Rephrased Into Control Concepts of Another Domain to Ease Problems Inherent in Translation

- **A vision system "positions" its cameral hand on the center of an object.** A key part of vision is the selection of a visual subject followed by the effective capture of the subject in an image. This is not unlike a hand grasping an object.
- **A robot arm recognizes an object and begins to "zoom" in on its location.** Grasping an object is a two-step process: (1) the hand moves to and surrounds an object and (2) the hand grasp is conformed, or "zoomed" into the shape of the object. This is similar to vision because zooming requires that the subject be identified and centered before it is possible to zoom.
- **A vision system uses "compliance" to locate objects.** Compliance is usually associated with the mechanical flexibility of a robotic system. It allows a robot to adapt to the uncertainties and variations in the environment. For example, compliance in a robot wrist enables a peg being held in the gripper to "track" the position of the hole during the assembly process. This control concept may now be extended to vision. If a vision system subtracts successive images, a perfect match between two images results in a zero image. If a visual subject moves, the vision system can "comply" with it by moving to maintain a zero image of the subject, thus tracking it.
- **A robot arm adjusts its "focus" to locate parts.** Many robot applications, such as a peg in a hole, can be made considerably easier by intentionally introducing positional noise (e.g., vibration) into the insertion process (Sanderson and Perry 1983). Similarly, vision can subtract pictures in and out of focus to find object edges (see Chapter 5).

3.2 TRADEOFFS IN SYSTEM COMPONENT DESIGN

The design for each subsystem of the intelligent craftsman is faced with basic design tradeoffs that are extensively analyzed in the following three chapters. These tradeoffs occur when there are two desirable characteristics of a system that are difficult to achieve at the same time. For example, a robotic hand may have to be very powerful to lift the

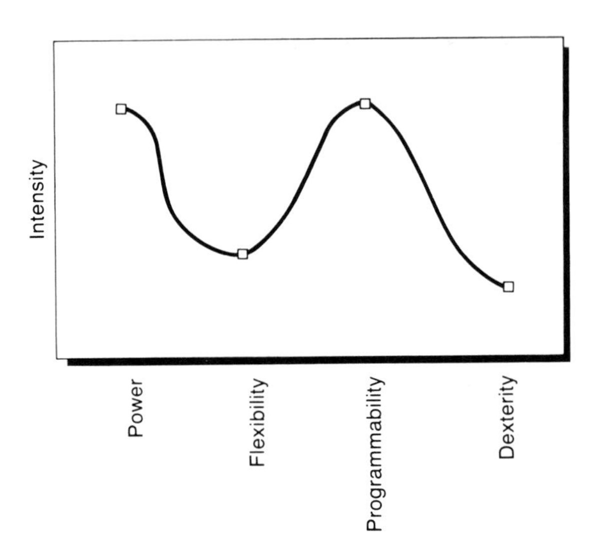

Figure 3.1 Design intensities.

parts that are used in an application. However, this is difficult to achieve if a high degree of dexterity is also needed in an assembly task, because the two characteristics are negatively correlated: as one design intensity goes up, the other goes down. For a given set of tasks that we call the "use spectrum," there are required "use intensities" as shown in Figure 3.1. Then, Figure 3.2 shows a generic version of the tradeoffs between key design characteristics.

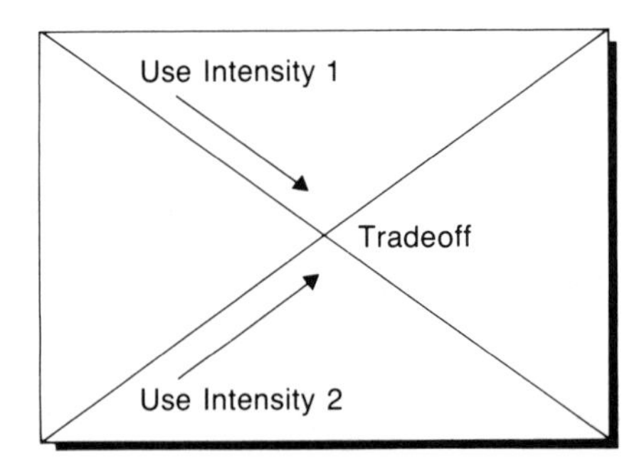

Figure 3.2 Use intensity 1 vs. use intensity 2.

These design characteristics may vary, but the choice of the use spectrum can completely determine the difficulty of the tradeoffs. For example, the manufacturing hand could be designed around many possible features, including power, accuracy, and flexibility. However, if the use spectrum covers a wide range of lifting tasks, the power of the system is probably the dominating design feature, and minor fine-motor tasks should be eliminated from the use spectrum. Thus, by concentrating on the balance between the scope of the use spectrum and the difficulty of the key design tradeoffs, it is possible to build control systems that are robust enough to cope with industrial applications.

3.2.1 Designing Controls

Every system (e.g., brain, eye and hand) in the intelligent craftsman is controlled by an algorithm that inherently tries to minimize effort while still finding a good solution to a domain-dependent problem. Figure 3.3 shows an algorithm that will be used as a template for implementing control systems in the next three chapters.

The control systems are always active, but they change their performance whenever the goal of the system is changed. For example, a user presents an intelligent robot with a goal, which in turn causes the robotic system to form a plan of attack. This first plan is then executed, as the robot goes through a series of moves, and the final state is matched with the desired goal. If it does not match the goal, the control algorithm manipulates the robot in ways that are known to be safe and yet effective. For example, the fingers of the robot may adjust their grasp and try

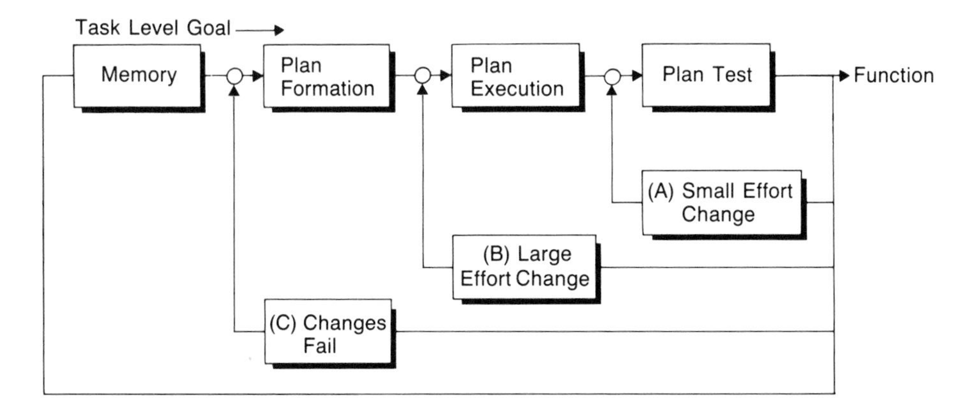

Figure 3.3 A least-effort-first control system.

to make up for small inaccuracies. If this does not work, more drastic actions are taken; for example, the whole grasp geometry may change.

Imagine the task of unscrewing a bottle top. At first expecting an easy operation, the first plan might be to twist the cap with the tips of one thumb and a finger; then, encountering resistance, more fingers would be brought into use (see feedback element A in Figure 3.3). If the resistance is much higher than expected, the finger grasps will be abandoned in favor of a wraparound palm grasp where the whole hand is brought to bear on the cap (see feedback element B in Figure 3.3). Ninety-nine times out of one hundred, these adjustments will make it possible to remove the cap. Ultimately, if these attempts fail, the system forms a completely new plan. In our bottle top example, the system could choose to use a pair of pliers, bang the bottle top on the edge of a bench, or simply choose a different bottle (see feedback element C in Figure 3.3). Each level of adjustment to the plan requires a successively greater effort and involves a higher risk of finally running out of options.

A least-effort system like the one in Figure 3.3 is a goal-directed system because it would not do anything without an initial goal. But intelligent systems require more than simple goal-directed reasoning. The real world does not know or care about the system's goals, and it is unlikely that it will always cooperate. For example, a welding robot with a camera is looking for a seam, but the system cannot see the seam through the smoke caused by the welding process.

Such circumstances have motivated many researchers to build "data-driven" systems, but in these systems it is often difficult to recognize an object without some preconceived notion of what it might look like. For example, an air conditioner on the top of a building and an automobile have exactly the same visual appearance from an aerial view, except that the cars are usually on roads, and air conditioners are usually on buildings (Rosenthal 1978). With this knowledge, it is possible to make the correct interpretation of the aerial image. Hence, the dilemma: should an intelligent system be driven strictly by its goals, by incoming data, or by some combination of these extremes?

Mixed control strategies have also proven difficult because it is never clear when to give priority to the data or the preconceived goal. The most novel AI systems have struggled with this dichotomy: mobile robot navigation (Brooks 1985), scene understanding (Hanson and Riseman 1978), speech understanding (Erman, Hayes-Roth, Lesser, and Reddy 1980), natural language understanding (Marcus 1980), and gaming (Berliner 1977). However, none of these efforts has reached a consensus on this design issue. They all use their own unique approach to this balancing act between incoming data and preconceived goals.

1. **Fine:** Set "small adjustment necessary" when measured parameters do not fit within preset bounds of precision $\{\pi_1, \pi_2, ..., \pi_n\}$.
2. **Coarse:** Set "large adjustment necessary" after trying all small adjustments and "small adjustment necessary" is set.
3. **Timeout:** Set "solution not found" after trying large adjustments for time τ.
4. **Reset:** Return to a home position when a gradient threshold is exceeded in domains $\{\delta_1, \delta_2, ..., \delta_m\}$.

Figure 3.4 One possible basis plan.

One way to achieve a mix of strategies in the least-effort control algorithm in Figure 3.3 is to preload it with a basis plan for all other plans. Note that symbolic data is expected to pass between the control elements in Figure 3.3. A basis plan is the plan that is resident in the system before there has been any opportunity for a particular strategy to be chosen and passed through the control system. However, this basis plan must be prescribed so as not to obstruct the goal directions that are flowing in from the top.

The first three steps of the plan in Figure 3.4 are the default controls for the feedback elements in the control algorithm. The fourth step is a "safety" loophole that lets the system reset itself whenever extreme conditions such as extreme heat or a fast approaching object are encountered. Later chapters will show that individual manufacturing control systems are essentially an elaboration of this single basis plan.

One purpose of developing a generic control algorithm is to centralize important control features into the algorithm. For example, most controls are expected to converge on an accurate solution in a reasonable period of time. There are also other more exotic features that have proven valuable in our studies. Perhaps the most useful of these is dexterity. When a basis plan and a goal-directed plan work together to solve a problem, "dexterity" is a term that describes the grace and competence of the system.

Dexterity encompasses every element of the control from the opening desire to the final function, as well as all the intervening mid-course corrections. It is most commonly used to refer to hand-eye manipulations. Adroitness, skill, and neatness of handling are the dictionary definitions of dexterity. Thus, for robotic hands and manufacturing, dexterity is associated with (1) a large number of degrees of freedom for

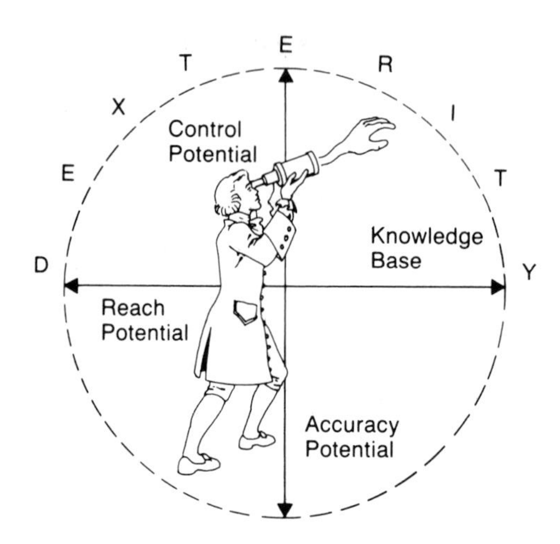

Figure 3.5 The breakdown of dexterity.

sensor-based, fine manipulations of an object, (2) the ability to achieve fine resolution with high accuracy, (3) the ability to attain these resolutions at high speeds, and (4) the ability to achieve the preceding qualities within a large working volume so that many parts of the environment can be accessed by the manipulator.

However, these concepts can be easily and naturally extended to other domains, such as problem solving with the manufacturing brain and scene analysis with the manufacturing eye. For example, a "dexterous" vision system (1) will be able to operate and provide information in a large number of degrees of freedom for understanding, with extensive image analysis; (2) it will be able to achieve fine resolution with high accuracy and with zoom and focus; (3) it will be high speed, and finally, (4) it will achieve the above in a large working volume with pan, tilt, yaw, and translational control.

An awareness that the concept of dexterity has generic attributes for all subelements of a manufacturing system, guides the design and orchestration of manufacturing software, vision systems, and manipulators.

Figure 3.5 breaks dexterity down into subcomponents that can be used in harmony to expand the level of a system's dexterity. The "knowledge base" is the reservoir from which skills can be extracted and used to solve the problems posed by a particular task. Once the knowledge of a task is available, it can be used to precondition the "control logic," so that a solution can be quickly forged and executed. However, the physical system must have the potential to "reach" an object so that the object can be manipulated to satisfy the task. For example, if a hand does not fit between two vertical bars, it does not matter how well control logic is able to manipulate an object such as a key on the other side of the bars. This idea of "reachability" becomes "visibility" in the vision domain. Finally, the "accuracy potential" of the mechanical system limits the tasks that can be accomplished. All these components determine the grace and competence that can be achieved by the overall system. As implied in the center of Figure 3.5, both the hand and the eye have similar requirements for dexterity.

3.3 UNIFICATION

The final control algorithms that operate the manufacturing eye, the manufacturing brain, and the manufacturing hand all run independently of each other. This gives each subsystem as much self-control as possible. However, to satisfy the global goals of the system, the subcomponents must be integrated as shown in Figure 3.6. Two design issues are now raised and discussed in the following subsections:

1. How are the results from one subsystem exchanged and used by one another?

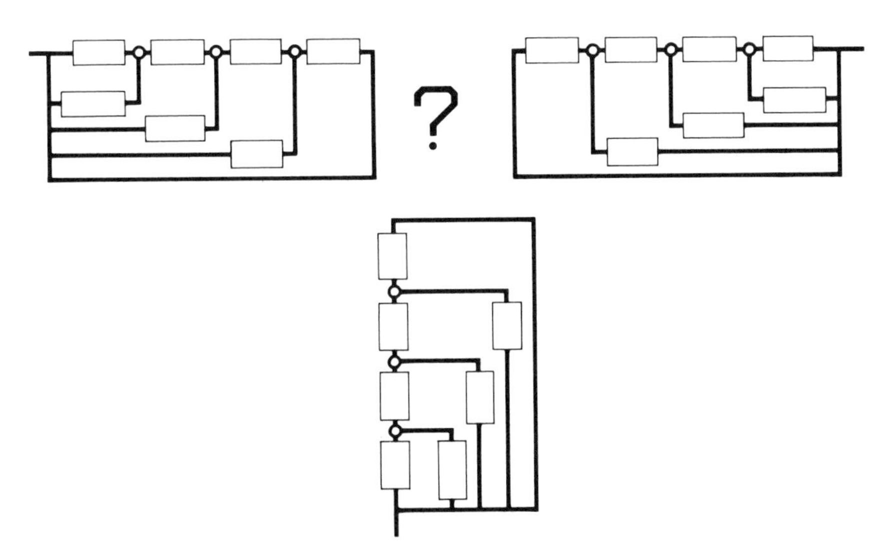

Figure 3.6 The unification of control systems.

2. How can the results that are localized in each subsystem be orchestrated with each other so that the total system satisfies it goals within an appropriate time frame?

3.3.1 Exchange of Control Information

As discussed earlier, each of the subsystems refer to different domains that may have direct counterparts in adjacent systems as shown in Table 3.1. The exchange of control information between the three systems is like translating between the English, French, and German languages. A straightforward translation from one language into another is, minimally, a mapping of the objects presented in one language to the corresponding objects in another (e.g., a book = un livre = das Buch). Unfortunately, this mapping is not always possible in practice, and it is often necessary to resort to mapping objects into object descriptions and vice versa. For example, there is no Chinese word for "computer," so a translator has to describe in Chinese terms what a "computer" looks like, how it operates, and its overall purpose. To clarify these difficult translation issues, we present a number of situations that are designed to illustrate different aspects of the problem.

The story in Figure 3.7 illustrates which cognitive objects, as opposed to physical objects, people use to help them sort out the sensations received from the world. Why does one person invent one set of objects in his head, and another person invents a different set? Where do these objects come from? The story is used to argue that objects, as we know them, are a direct by-product of our personal, historical, and cultural experiences. The African object problem in Figure 3.7 illustrates this effect by juxtaposing two radically different world views. These different views highlight the role of cognitive objects in complex interpretation tasks. The cognitive objects that an automated system uses will also determine how manufacturing systems can reason about their environment.

Object boundaries are as unique to individuals as is their thinking. Some similarities in object boundaries are based on culture; these make communications possible. Once the mechanisms for creating and using objects have been established, it is also possible to invent hypothetical objects. These hypothetical objects can be used to solve problems and to organize tasks.

For example, consider a wooden pallet filled with parts. When a fork-lift manipulates the pallet, there is no need to consider the contents of the pallet. Thus, the fork-lift needs to only represent a complete pallet as a

Imagine a strange native from an unknown tribe in Africa. There is nothing biologically peculiar about this native, but some aspects of his daily culture are completely distinct from our western traditions. One of the differences between our culture and his is that in Livingston-Ville everyone eats alone. In fact, it is quite embarrassing to be seen eating in Livingston-Ville. Sex on the other hand is a chance for everyone to get together and celebrate the day's events.

One day I actually got to meet a Livingstonian and discovered that he was quite amiable and communicative—so much so that I invited him home. My house is quite conventional by most standards: a living room, dining room, and the rest; and by habit I gave the usual house tour. When we walked into the dining room, I saw, as usual, an old but charming dining room set that was inherited from my grandparents. Our guest on the other hand saw four chairs and a large table, bigger than any he had ever seen before. He could not imagine why all this furniture would all be in the same room, not to mention in the middle of the house! And to him, the second floor of the house was just as puzzling as the first. While I saw three bedrooms, all neatly made up, our guest wondered why he was invited when the house had an appearance of an unfriendly hospital ward. The bed set was not even ready for the night's festivities.

Figure 3.7 Why are objects constructed: The African object problem?

cognitive object. Subsequently, the manufacturing cell can also use this pallet, cognitive object to define a batch of parts. But, when a robot unloads the pallet, only the contents matter. The robot needs meaningful cognitive objects for the inside boundaries of the pallet as well as for the billets stacked inside it. The fork-lift and robot each has a completely different conception of the objects in the world and how they are structured.

Nature presents both humans and automated systems with the open challenge of making sense of the prevailing universe. We have demonstrated that the universe can be broken up into different objects, so that it is possible to carry out certain tasks of which communication is one. What objects should be constructed to help reason about an environment? The relatively new field of qualitative physics also addresses this basic question of choosing the right ontology (Bobrow 1985, Hobbs and Moore 1985).

Imagine shopping and selecting fruit from a basket filled to the brim with ripe apples. The senses would be overwhelmed with the rich smell and a patchwork of colors. Also, a kinematic sense of position would be felt as an apple was grasped. Such information from the individual senses in Figure 3.8 can stand on its own, and it can also contribute to the

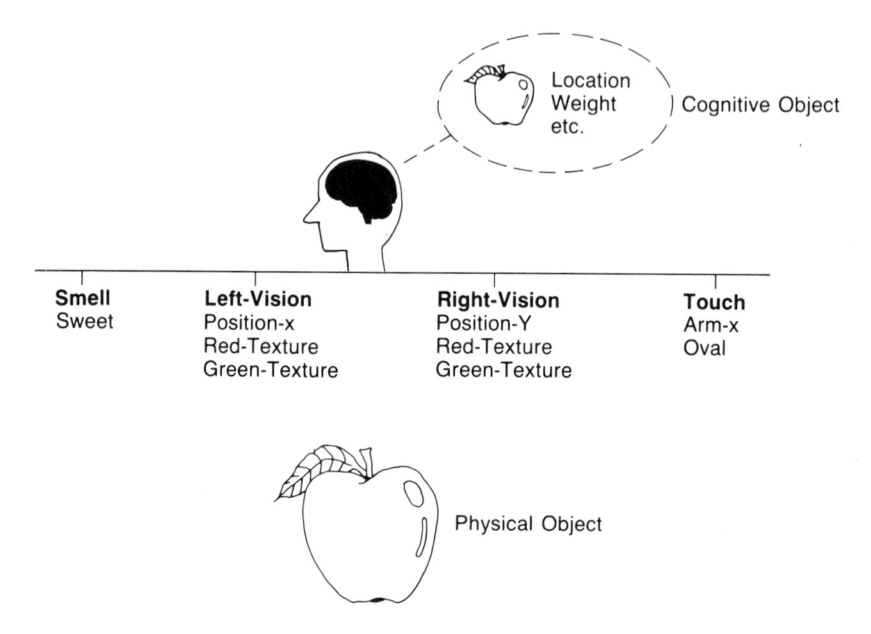

Figure 3.8 What objects are constructed: The cognitive apple?

construction of cognitive objects. For example, it is possible for a system to determine that there is a global source of a sweet smell, a certain red patch, a particular hard feel, and a corresponding kinematic position. This process of coping with the different systems is an attempt to calibrate (or translate) the information between them. But in fact this is not strictly possible; scents and colors do not directly match up with specific positions in the kinematic domain. Instead, an artificial bridge must be built to explain the connections, and this bridge is essentially an internalized version of the apple. It is a place in which abstract properties, perhaps not represented by any one sensor, can be freely associated with the "software object," which is the cognitive apple. The requirement of making sense of input across modalities helps to explain what cognitive objects must be constructed for task-based reasoning.

The question of how to apply this in manufacturing automation demands further practical examples. A simple case is a program that describes the actions that are necessary for a robot to pick up a part. The software controller of the system may initially translate the robot program into a vision subprogram to answer the question: Where do I look? But it is only obvious how to translate a very small part of the robot program—namely, the part that matches with semantic ideas in vision, such as location. What should the system do, once the object has been successfully located? The visual images give no information about weight, necessary grip force, nor acceleration, but these features must still be specified in the robot domain. The gripper perhaps needs some trial-and-error manipulations to learn about this new application. If the gripper force in a robot program is very high, it is possible to deduce that the object is heavy or even metallic. This, in turn, may provide additional clues about what visual properties to look for in new images, such as looking for points of high contrast. High contrast features are a good heuristic for finding specular reflections and a corresponding metallic object. As successive objects in a batch are viewed, located, handled, and put down, the various subcomponents of the system effectively learn from each other.

Figure 3.9 summarizes the following scenario:

1. Data about an object arrives at the visual system and is processed.
2. The results of this processing are then used as a key into an existing database of objects.
3. The database returns a host of potential candidates that would possibly correspond to the image.
4. To disambiguate the input, differences between the potential candidates are calculated, and a plan is formed to drive another sensor

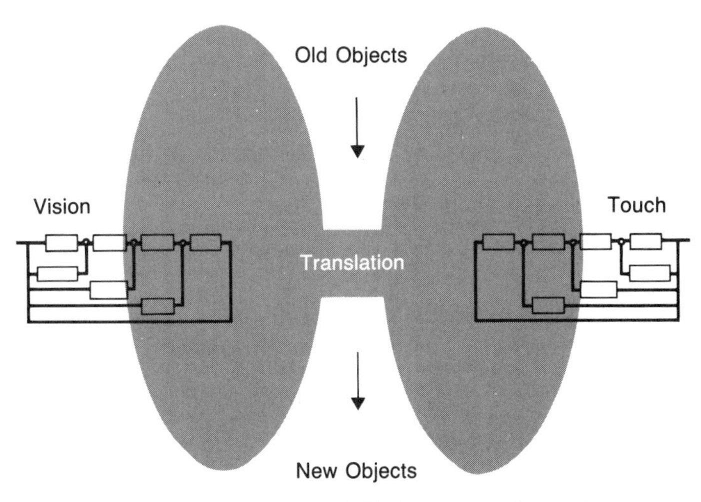

Figure 3.9 How objects are constructed: Visionese to robotese?

in such a way as to highlight the differences. For example, one plan might be to touch the object at point X, and if there is a force encountered then it is object A, otherwise, it is object B. This plan would discriminate between an actual hole in a mechanical part and a discoloration in the part that visually looks like a hole.

5. Execute the "discrimination function," and report on which object is present.

This scenario simply recognizes objects that have already been constructed. However, it is inevitable that new objects will be encountered and not recognized. In this case, new objects will be the result of the interaction.

3.3.2 Orchestration of Subsystems for Global Control

Table 3.2 summarizes some of the interactions that occur in a "generic" task between the manufacturing brain, the manufacturing eye, and the manufacturing hand. Each step in the generic task is implemented by a closely correlated operation in the various domains, so that a unifying translation is simplified (see again Table 3.1). As the generic task proceeds through its successive steps, each subsystem tries to contribute a successful result to aid in the next step across the tabular board. As a successful conclusion, cooperating subsystems create concrete results, which can be used in a manufacturing process. These results "drop out" at the bottom of Table 3.2.

Throughout the book, we provide various concrete examples of how the subsystems are organized to achieve global goals. Some applications in robotics and automation are more concerned with planning or general monitoring. In these cases, the various subsystems can work at different rates. But, in other applications, the real-time coordination of subsystems is crucial. Here, sensors, actuators, and software must cooperate and exchange information at compatible rates. Referring to Figure 3.6, the three subsystems should cooperate so that the results drop out simultaneously from Table 3.2.

We preview some applications in other chapters that highlight the real-time unification problems found in intelligent systems.

1. In Chapter 5, a vision system is used to measure a forging that has been produced in a manufacturing cell. The purpose of this system is to inspect the part, decide if it is correct, and feedback information to the forge and furnace before the next component is

made. The brain and eye columns in Table 3.2 must work with individual efficiency, but also they must work at a rate that delivers information at the bottom of the table in a timely way. Simply put, the brain managing the schedule of the cell cannot

Table 3.2 Domain Specific Control Functions that are Related to "Generic" Task Steps

GENERIC TASK	BRAIN	EYE	HAND
Locate Object	Search Database Looking for Object Description	Scan Scene Looking for Object Description	Feel Environment Looking for Object Description
Zoom-in on Object	Put Object Attributes in Working Memory	Acquire Centered Closeup of Object	Grasp Object Using Best Estimate Grasp Position
Interrogate Object	Collect Associated Objects and Object Uses from Database	Collect 3-D Alternative Views of Object	Collect Alternative Grasps of Object
Analyze Object	Extract Derivative Information: Deductions	Extract Key Visual Attributes: Color, Shape, etc.	Extract Key Kinematic Attributes: Weight, Shape, etc.
Move Object	Set Goal for New Object Position in World	Track Gross Object Movements	Execute Gross Movement
Work Object	Plan Sequence of Actions to Transform Object	Collect Accurate Dimensional Information	Execute Fine Motor Movements to Effect Transformation
	↓	↓	↓
	Implicit World Modeling	Implicit Inspection	Implicit External Changes: Machining Assembly, etc.

"wait around" for the eye to do its inspection task. While this inspection station is in production, its sluggish response time has greatly diminished its utility for controlling run-time parameters in the manufacturing cell.

2. In Chapter 6, a robot arm with a position-sensing wrist is used to perform a grinding operation along the edge of a metal plate. The wrist locates the uncertain position of the plate and then instructs the robot controller on a suitable position trajectory. In this case, the brain and arm columns of Table 3.2 must work and cooperate in a timely way. Actually, we have found that with the limitations of today's robot controllers the robot's brain is slow to react to the information arriving from the hand (i.e., the wrist plus grinder in this application). As a consequence, we await improved robot controllers before robotic edge grinding can be effectively automated in industry.

3. In Chapters 9 and 10, sensor-based machine tool fixtures and instrumented cutting tools are used to monitor a variety of in-process conditions during machining. These new devices could be added to Table 3.2 to show that each device has to keep pace with the brain and vice versa. A successful application is shown in Figures 10.13 and 10.14. Here, the sensors report that the drill is about to fail, while still leaving the managing brain enough time to stop the machine tool. This avoids a catastrophic drill failure.

3.4 DISCUSSION

The chief objective behind Table 3.2 is to explore how to build robust connections among a number of subsystems. The first step is to establish an abstract control design that can be used for each subsystem. When the subsystems must be connected, a translation scheme is used whenever possible to transfer the ideas from one modality to the other. However, this translation mechanism may initially fail because concepts in one domain are found to be missing in the other domain. To resolve this deficit, an internal object is constructed, so that the representations from each modality can be collated. The software that is required to mediate between the control systems is (*1*) a database of objects and object properties, and (*2*) a translation scheme between different representations. These software tools are discussed in much greater detail in the next chapter.

The resulting architecture gives each subsystem as much autonomy as possible. In this way, the best source of information is given the dominant role in the ensuing operations. The control software for craftsmanship cannot rely on a fixed strategy to make decisions, but must be flexible to determine which sense is giving the most reliable information. Other senses, in turn, can be used to confirm or disprove the initial conclusions.

3.5 SAMPLE OPEN PROBLEMS

In the next three chapters, we offer open problems that stress the issues that cross between the domains of the manufacturing brain, the manufacturing eye, and the manufacturing hand. This chapter starts the tradition with a handful of problems that concentrate on the connections between the subsystems.

1. Develop a software system where programs can be built from a library of routines without previously knowing argument requirements. As a start, it would be useful to consider object-oriented systems that send and receive messages as a parameter-passing scheme. Further, it may be necessary to build enough logic into the message-passing scheme to save state, so that dialogues between modules can automatically query for and deliver parameters. This problem involves inventing a way not to "hard code" particular parameter names, although there must be some agreement built into the message formats.
2. Build a control system for a manufacturing cell and base it on the object-oriented paradigm described in the first problem.
3. Design and implement a small system that can translate between subsets of two natural languages of your choice. This system should concentrate on automatically constructing new objects that are attached to a description derived from the source text.
4. For a group of sensors, design a set of experiments to determine which sensors should be dominant for a particular range of tasks. For example, when should a touch sensor be dominant over a heat sensor, and when should a sonar sensor be dominant over a stereo depth map? Build a control system that makes it easy to change these dominance relationships.
5. Analyze the design checks (in Chapter 5) and the design guides (in Chapter 6) and develop a unifying set of design principles.

REFERENCES

Adler, M. B. 1984. GM manufacturing protocol. *Proceedings from CAM Symposium*, University of Cincinnati, October, pp. 159–170.

Berliner, H. J. 1977. Search and knowledge. *International Joint Conference on Artificial Intelligence, Vol. 5*, pp. 975–979.

Bobrow, D. G. 1985 (editor). *Qualitative Reasoning About Physical Systems.* MIT Press, Cambridge, Mass.

Brooks, R. A. 1985. A robust layered control system for a mobile robot. *A.I. Memo* 864, MIT, Sept.

Erman, L. D., Hayes-Roth, F., Lesser, V. R., and Reddy, D. R. 1980. The HEARSAY-II speech understanding system: Integrating knowledge to resolve uncertainty. *Computing Surveys,* Vol. 12, No. 2, pp. 213–253.

Feynman, R. P. 1985. *QED - The Strange Theory of Light and Matter.* University of Princeton Press, Princeton, N.J., p. 4.

Hanson, A. R., and Riseman, E. M. 1978. VISIONS: A computer system for interpreting scenes. In *Computer Vision Systems,* edited by A. R. Hanson and E. M. Riseman. Academic Press, New York, pp. 129–164.

Hobbs, J. R., and Moore, R. C. 1985 (editors). *Formal Theories of the Commonsense World.* Ablex Publishing Co., Norwood, N.J.

Marcus, M. P. 1980. *A Theory of Syntactic Recognition for Natural Language.* MIT Press, Cambridge, Mass.

Rosenthal, D. 1978. An inquiry driven computer vision system based on visual and conceptual hierarchies. Ph.D. thesis, University of Pennsylvania.

Sanderson, A. C., and Perry, G. 1983. Sensor-based robotic assembly systems: Research and applications in electronic manufacturing, *Proceedings of the IEEE,* Vol. 7, No. 7, pp. 856–871.

Whorf, B. L. 1956. Science and linguistics. In *Language, Thought and Reality; Selected Writings,* edited by J. G. Carroll, MIT Press, Cambridge, Mass.

4 The Manufacturing Brain

Is It Possible for Computing Machines to Think?

No —if one defines thinking as an activity peculiarly and exclusively ***human****. Any such behavior in machines, therefore, would have to be called thinking-like behavior.*

No —if one postulates that there is something in the essence of thinking which is ***inscrutable, mysterious, mystical****.*

Yes—if one admits that the question is to be answered by ***experiment*** *and* ***observation****, comparing the behavior of the computer with that behavior of human beings to which the term "thinking" is generally applied.*[1]

—E. Feigenbaum and J. Feldman in *Computers and Thought.*

THE manufacturing brain is the control center for all the intelligent systems that are needed to automate the factory. This control center must mediate between all the systems and provide a deep computational resource (see Figure 4.1) from which all the subsystems can draw. For example, strategic goals are plotted for each subsystem so that effective solutions are within their limited computational reach.

This chapter discusses different architectures for an intelligent control center. The best architecture is flexible enough to manage a wide variety of normal situations, and it is robust enough to handle an unknown situation without serious error. To build the best control system, the system designer must have a range of software tools with which to solve manufacturing problems. This chapter also develops these important tools. Once these tools are available, it is also possible to build systems semiautomatically.

---

[1]Feigenbaum & Feldman, *Computers and Thought*, © 1963 McGraw-Hill Inc. Reprinted with permission.

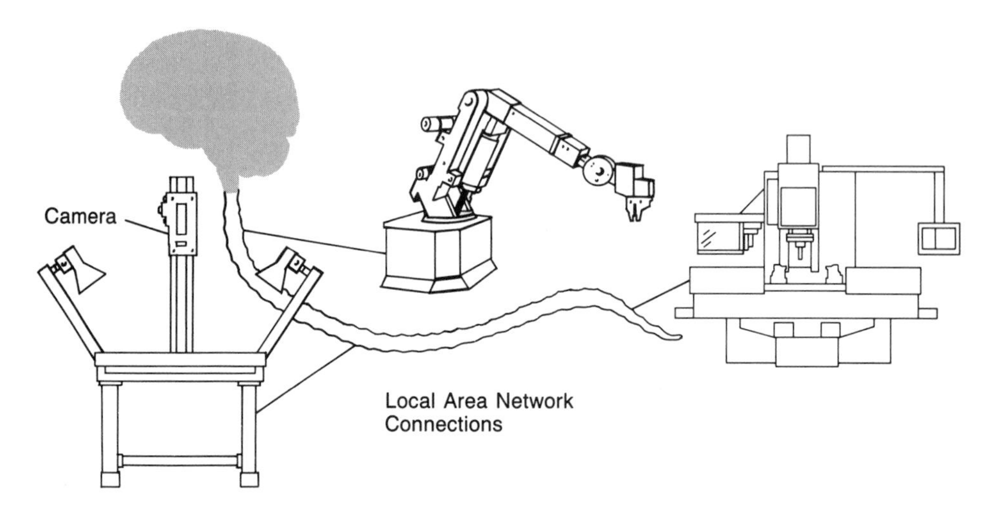

Figure 4.1 Macro anatomy of a manufacturing control center and brain.

4.1 EVOLUTION OF MANUFACTURING CONTROLS AND BRAINS

Control systems were built from mechanical relays for almost 30 years (approximately 1940–1970). These relay systems were stored in large cabinets that emitted distinct clicking noises as the relays opened and closed. To change the function of the controller, it was necessary to make physical changes in the arrangement of relays. This is when "hard wired" meant that the relays were actually wired together.

The advent of the microcomputer chip made building a substitute for relay networks economically feasible. Ironically, rather than redesigning the controller, control suppliers simply used the microchip to simulate the old relay systems. The additional power that the general-purpose computer provided was instead used to graphically interact with the manufacturing engineer.

In the late 1970s, control suppliers finally started to develop controllers that took advantage of commercial chip architectures. These computer chip controllers implemented simple control organizations based on a serial flow of information from sensors to actuators (see Figure 4.2). For example, a limit switch triggers processing that looks up a cause (e.g., lathe chuck open), plans an action, and finally activates an event (e.g., a robot load of a billet).

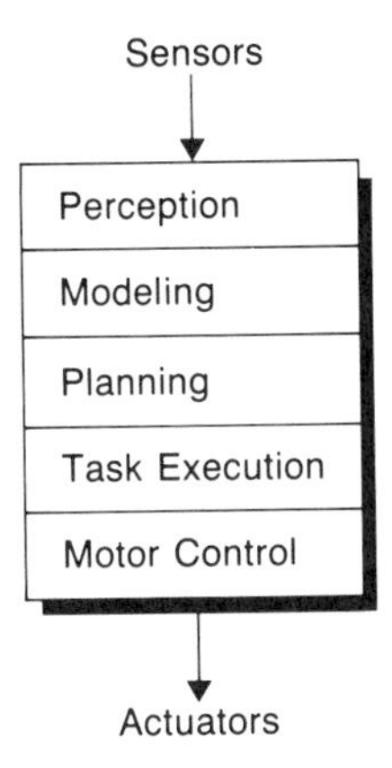

Figure 4.2 A traditional decomposition of a serial control system; presented in Brooks's work on mobile robots (Brooks 1985). (Reprinted with permission from Rodney A. Brooks, " A Robust Layered Control System For a Mobile Robot" AI Memo 864, MIT 1985.)

As an alternative, Brooks has proposed parallel control systems for mobile robots that are based on natural behavior, rather than on organizing the processing into a series of steps (Brooks 1985; Brooks 1986; Moravec 1986). We have transposed his mobile robot behaviors into machine tool behaviors (see Figure 4.3). The behaviors are organized into a parallel structure so that each independent control function is responsible for only one behavioral task (e.g., chip removal). At any

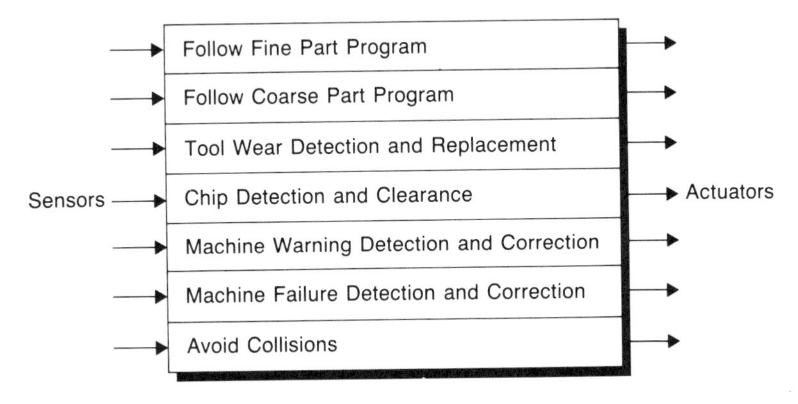

Figure 4.3 A behavioral decomposition of a machine tool's parallel control system; based on Brooks (1985). (Reprinted with permission from Rodney A. Brooks. "A Robust Layered Control System For a Mobile Robot" AI Memo 864, MIT 1985.)

time, the behavior of the system can be governed by one of the control functions suspending the action of the others. For example, while a part program is executing, chips could be detected and cleared, interrupting the part program. While the chips are being cleared, a potential collision between the machine spindle and the part might be detected, interrupting the chip removal. For this combination of events, a plan would be constructed to first avoid the collision, then resume the chip removal, and finally resume the part program.

How do serial and parallel control strategies compare? A serial controller is plagued by potential bottlenecks, i.e., every processing step has to work reliably in order for the whole system to advance. By decomposing the control into parallel units, such bottlenecks can be avoided. However, both system types still have many problems in common.

1. A serial control can usually process only one task at a time. This makes it susceptible to unexpected events that cast the system out of its intended domain of operation. Parallel systems also must handle unexpected events graciously. However, it is easy to forget important connections between the independent control units, leaving the parallel controller in the same situation as the serial controller.
2. Error checking and correction is not always maintained in all modes of operation, no matter whether the independent "control stages" are in series or in parallel.
3. Planning strategies are computationally expensive, and they are the first to be cut from the control roster.
4. Simple sensor checking (e.g., limit switches) can sometimes be triggered by unexpected events (e.g., trapped dirt or physical failure). Without adequate planning facilities, it is difficult to correct for this.

In addition, Brooks's decomposition by behavior faces a basic problem: the number of desirable behaviors can become cumbersome. The connection between each behavioral task must be elaborated, and the priority of each task must be determined. Otherwise, the control system can result in a deadlock situation where one task (e.g., object avoidance) overrides another task (e.g., chip removal), only to be overridden by the first, ad infinitum.

To build a controller, a basic design tradeoff must be optimized (see Figure 4.4). This tradeoff should balance the number of control elements with the organizational complexity that the final system can tolerate.

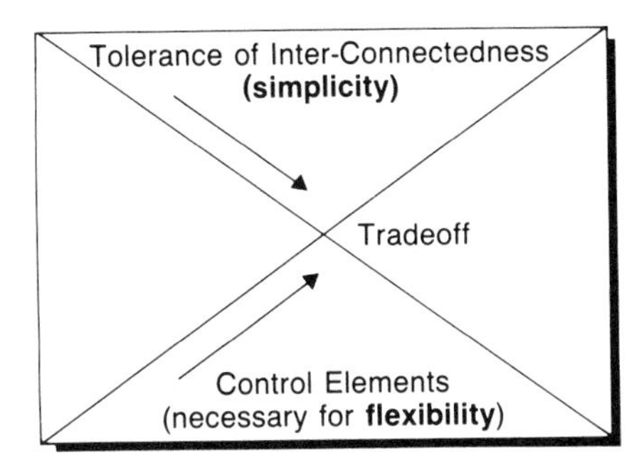

Figure 4.4 Key control tradeoff.

4.2 CENTRAL CONTROL SYSTEM

To balance the connection/complexity tradeoff, we advocate a compromise between the serial and behavioral decompositions. Chapter 3 already has tipped our hand, showing that we intend to decompose an intelligent machine tool's control into its major subsystems. Our view is to let each major subsystem (e.g., the manufacturing eye, the manufacturing hand, and the manufacturing brain) run virtually autonomously. The advantage of this decomposition is that it breaks the serial bottleneck and fixes the complexity of the connections between the parallel components.

By decomposing the control into major subsystems, each system can operate near full capacity, even if adjacent subsystems were to fail completely. However, even a system with a small fixed number of parallel components must concentrate on the system interconnections. These interconnections involve message exchanges that must be translated appropriately. The resulting dialogue makes it possible for a group of systems to both follow a previously written script and establish a new script in a surprise situation. In this case, the manufacturing eye could determine an appropriate new script and communicate it to the manufacturing hand, at which time the two would cooperate in its execution (O'Donnell 1985; Bourne 1986b).

The decomposition of the total manufacturing brain includes the central control as one of the possible subsystems (compare the control schemes in Chapters 3 through 6). Here the same control logic is used to search for a solution, find a good hypothesis, test it, and, if the test fails,

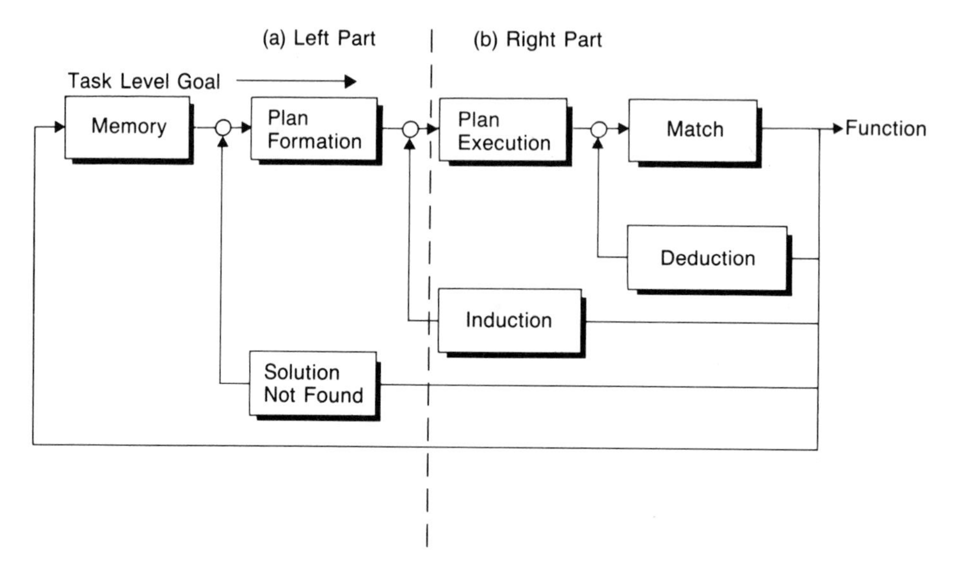

Figure 4.5 Central control system

find an alternative (see Figure 4.5). The deductive adjustment in Figure 4.5 is intended to be a variation on the solution that is both easy to find and bound to be valid. If deductive inferences fail to produce a workable solution, stronger methods of changing the solution such as inductive inferences must be used.

Our goal for the central control system is to make most common problems manufacturing fit under one generalized umbrella. This umbrella would spell out the macrostructure of a problem to ease the job of an autonomous system. Therefore, the macrostructure must be flexible enough to admit many different control strategies. As in Chapter 3, it is assumed that a number of these control systems will be connected by a translation network. To test this hypothesis of control, we analyze six problems common in the manufacturing environment (see Figure 4.6).

The first impact of the generalized control scheme is that problems tend to be decomposed into two problem parts (see left and right parts of control diagram in Figure 4.5, which correspond to part a and part b in Figure 4.6, respectively).

The first is typically characterized by general-purpose symbolic planning, called "linguistic" reasoning in Chapter 7, and the second is characterized by the task's execution, called "non-linquistic" reasoning in Chapter 7. Essentially, the rest of the control scheme integrates these split parts with the rest of the manufacturing brain and ties together the stages into a closed-loop control.

PROBLEM 1

a. Sensor Understanding—To understand a sensor reading, a model of the sensor must be employed to relate the sensor readings to sensor meanings.

b. Modification—Once this correlation has been established, it is possible to determine whether the sensor is operating successfully for a given task, so that minor adjustments can be made to optimize the readings. If this fails, the sensor position may have to be adjusted in order to collect useful information tion (e.g., a volt meter placed correctly on a wire or a vision system pointed at the correct target).

PROBLEM 2

a. Grasp Planning—To grasp an object, it is useful to approach the object with a plan already formulated for how to grasp it.

b. Grasping—In the grasp attempt, small adjustments to finger positions and stiffness may have to be asserted. If all else fails, this grasp plan may have to be abandoned in favor of another.

PROBLEM 3

a. Process Planning—A process engineer typically uses information about the part design, material properties, manufacturing constraints, and economics in order to develop a manufacturing plan.

b. Manufacturing—Once a process plan is formulated, it can be tested and adjusted to maximize part tolerances and overall part costs. If the process plan fails to make the part, a new plan must be developed.

Figure 4.6 Six major problems in manufacturing (continued).

Figure 4.6 (continued).

PROBLEM 4

a. Scheduling—In order to maximize the utilization of machines in a factory, careful consideration must be given to part orders, part priorities, machine availability, tooling availability, inventory availability, and personnel. Only then can a reasonably good schedule for parts and machines be formulated.

b. Dispatch—Once a schedule is in hand, a dispatcher carries it out. As problems in manufacturing disturb the schedule, the dispatcher can make small adjustments to the schedule to avoid building a completely new schedule.

PROBLEM 5

a. Diagnosis—When a machine breaks down, a manufacturing engineer uses visible symptoms, knowledge of the physical workings of the machine, and test equipment to determine the source of the problem.

b. Repair—From this, a repair plan is prepared and implemented to correct the problem. Small adjustments in the repair plan can usually be attempted before a new repair plan must be formulated.

PROBLEM 6

a. Part Design for Function—A part designer uses a number of physical, geometrical, and functional constraints to develop a part design.

b. Part Design for Manufacturing—The result of this design must then be tested against the capabilities of the manufacturing facility to verify that it is feasible to produce the part. Small adjustments can usually be made before a design must be scrapped in favor of another design more suitable for manufacturing.

As new problems arise, a new control framework is constructed automatically in software, and the control boxes are filled in with task parts. Of course, this automatic construction of a new control is itself a task that must be put in this framework (e.g., (a) split the task and (b) test the control).

After a control is constructed, at this very coarse level, the details of the task must begin to flesh out. This involves trial and error, searching in other domains for related information and, more easily, collecting knowledge from outside sources. The rest of the chapter concentrates on automating this last area. For example, cell control is used to illustrate the levels of a task that must be filled out for any manufacturing problem. In summary, the resulting manufacturing brain will consist of a collection of these controls for different tasks, together with detailed task information.

To fill out the task information, it is practical to use a pyramid of software tools that also reflects the necessary levels in a manufacturing task (see Figure 4.7). At the bottom of the pyramid, there are tools that make up a programming environment and are convenient for solving real-time, sensor-intensive problems that are common in manufacturing. At the next level, there are generic programs that can be reused for any number of applications. At the loftier levels, there are systems that write their own programs and systems that learn from experience. We must prove that this pyramid is more than a software Tower of Babel. To accomplish this proof, we show how this pyramid of tools has already been used to construct the control program for the forging cell (see Figure 2.5).

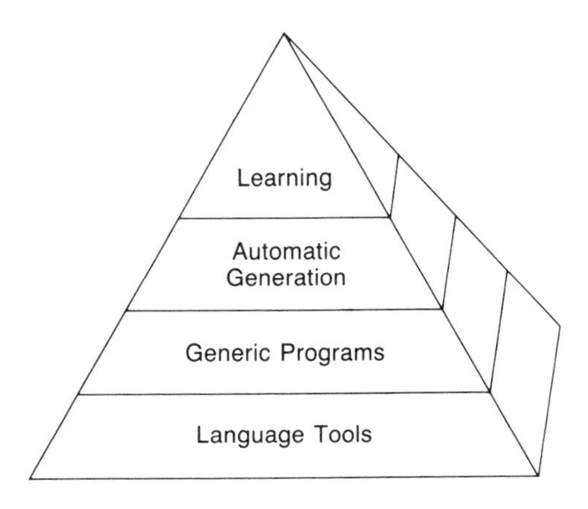

Figure 4.7 A pyramid of tools for building intelligent task systems.

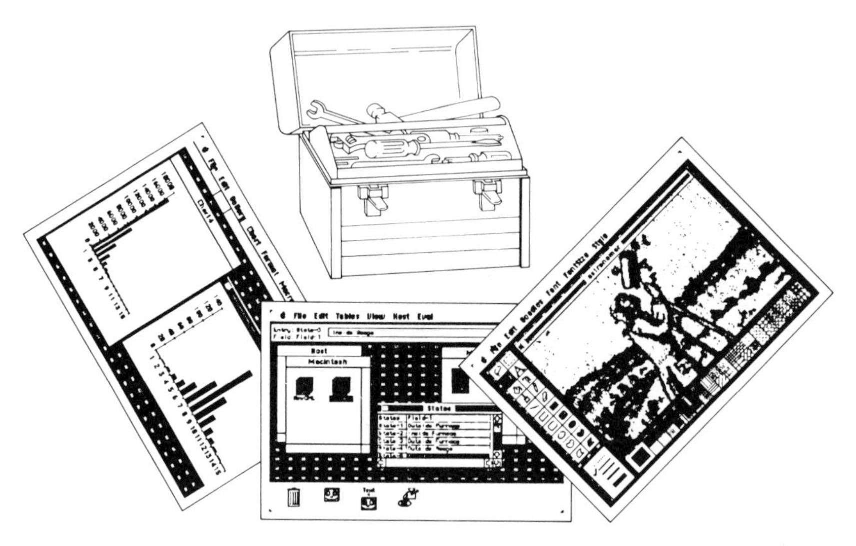

Figure 4.8 A programmer's software toolbox.

4.3 THE EVOLVING TOOLBOX

The faster programs can be constructed, tested, debugged, and put into production, the easier it will be to automate the construction of manufacturing systems and, of course, it will also be easier to build them "by hand." In every sense of the word, software "production" has the same properties as any other kind of capital production. But, while we build automated factories to make soft drinks and automobiles, software has remained in the realm of the ad hoc job shop. Before a software factory can be made profitable, we must invent generic tools that can help to create the finished products, in this case, computer programs. Conventional factories as we know them could not exist without the basic tools of production (e.g., simple hand tools and machine tools). So, what tools are needed to make computer programming more generic and hence more productive?

Our experience building control software for manufacturing cells has shown that only a few generic tools must be added to the programmer's "toolbox" (Bourne 1984). Like the toolbox used to repair a car, there are many sizes and shapes of each type of tool (e.g., ring wrenches, crescent wrenches or context-free parsers, and context-sensitive parsers). These tools (see Figure 4.8) make up the bottom of the software pyramid.

Many kinds of programming tasks are involved in building factory systems. Like the programmer's tools, these tasks can be categorized into "program families" so that major simplifications can be devised. Figure 4.9 illustrates generic manufacturing tasks and their approximate interrelationship. The bottom half of the figure illustrates a range of tools that is required to intelligently communicate with machines, people, and other software packages.

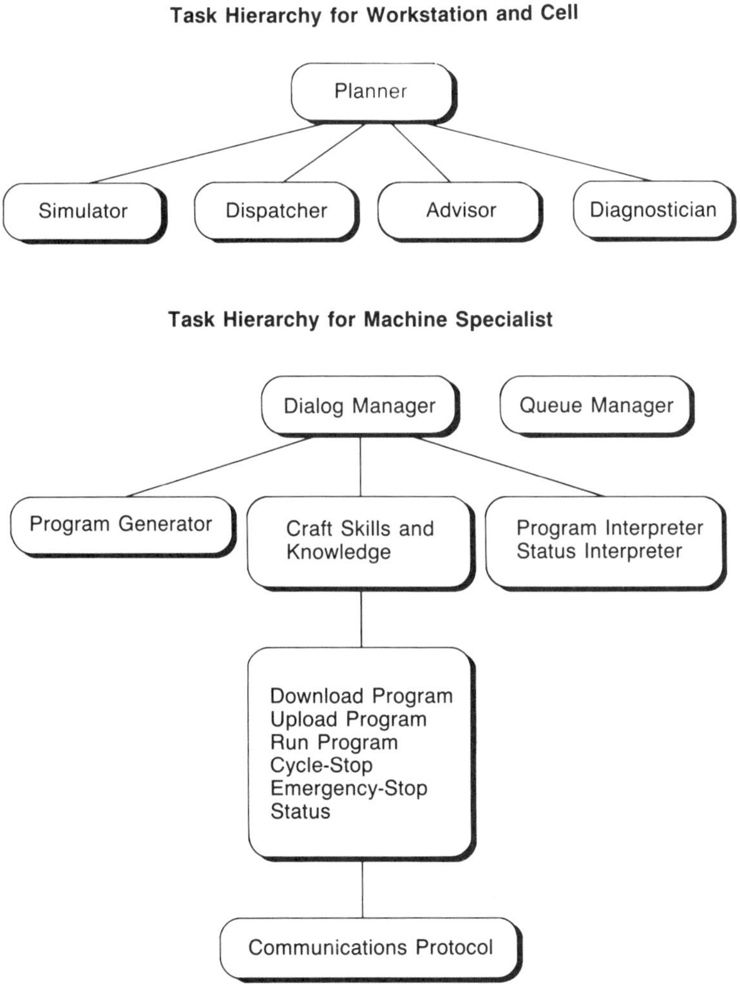

Figure 4.9 Common tasks in factory systems.

An analysis of the activities of on-site factory programmers would quickly reveal that the majority of the programmers' time is spent developing pre-processors and post-processors. A pre-processor is the software that can intelligently format data before an operation begins, whereas a post-processor is the software that can intelligently format data after an operation finishes. For example, it is often necessary to write a pre-processor that converts APT (a high-level language for describing how to machine parts) into MCL (a low-level language that is understood by most machine tool controllers). Another example is the problem of moving computer aided design (CAD) information from one CAD station to another. This is typically accomplished by writing a post-processor that translates a part description language into a common exchange format such as the International Graphics Exchange Standard (IGES). Once this CAD information is in the exchange format, it can be moved to another CAD station that supports the same standard. At this point, the exchange format on the second CAD station is converted to an internal and usually proprietary representation. Distributed manufacturing systems involve a large amount of format conversion and language translation because there are always dissimilar systems in the network.

The translation phases of a project must be finished before the main project can even begin. How can a flexible manufacturing system (FMS) be built if it is not possible to communicate between the machines? How can a production planner in a company ask questions when the answers are in more than one database system? Unfortunately, everyone who works in manufacturing comes to realize that these problems represent the status quo rather than a rare programming difficulty.

To alleviate these problems, we start with a set of tools that are designed to solve this family of problems (the bottom half of Figure 4.10). By combining the use of these tools into a complex program that uses application data, it is possible to build novel translation systems. As Chapter 3 suggests, a system that flexibly translates data from one domain to another bears the mark of intelligence.

Figure 4.10 lays out a range of tools that are needed to build these flexible translation systems: (1) generic I/O routines, (2) generic expert system routines, (3) generic parsing routines, (4) generic language generation routines, and (5) a control structure to carry out dialogues with outside sources.

Suppose, we are building a program to intelligently communicate with a robotic arm. In this case, it is reasonable to start with a generic program that has already been designed for communication tasks. Ap-

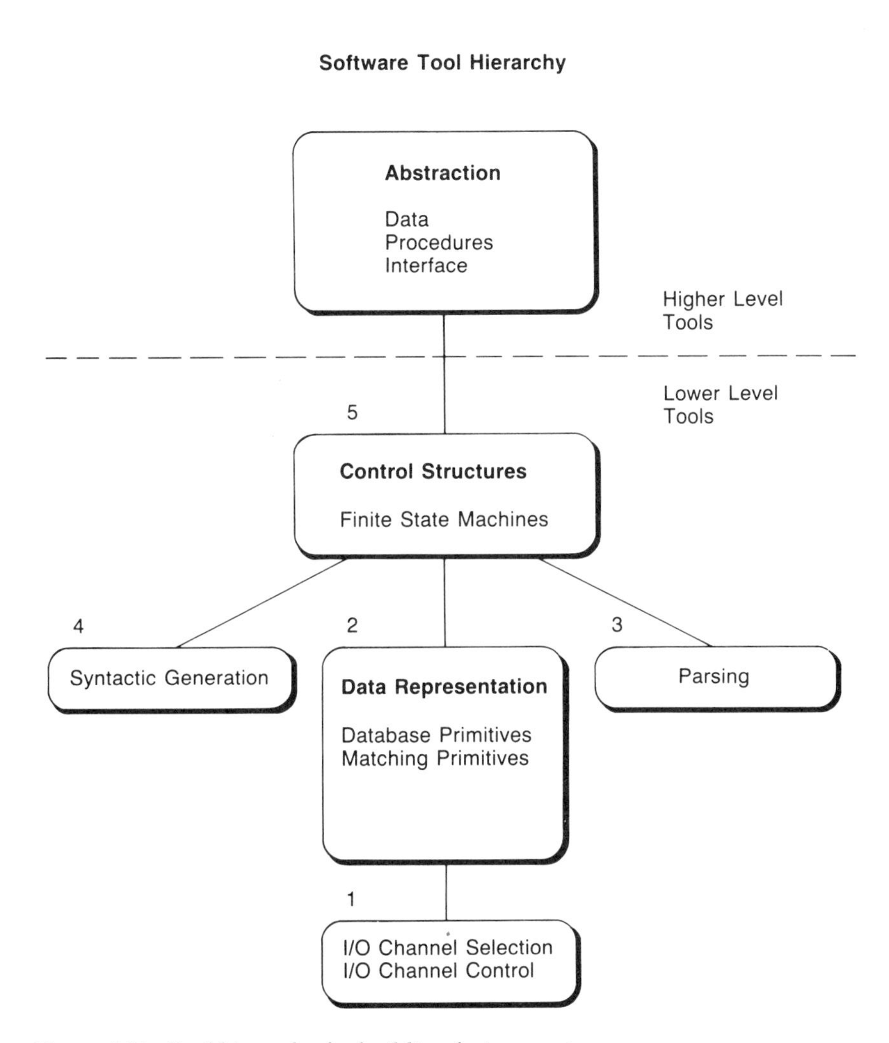

Figure 4.10 Tool hierarchy for building factory systems.

plication data must then be added to the generic program to determine the name and type of the I/O channel, to describe the robot command language (or even the programming language for the robot controller), to describe the tasks for the robot, and finally to describe the dialogues that are meaningful to the robot. This information allows the final program, which we call a "robot machine specialist," to carry out the message exchanges seen in Figure 4.11.

- **Action**: Request location of part from vision specialist.
- **Action receipt**: Location of part is: x,y,z.

- **Action**: Build robot program for part pickup and manipulation.

- **Action**: Request download of this new robot program.
- **Action Receipt**: Download failed, another part program already exists in the robot controller.

- **Action**: Request robot to delete the old part program.
- **Action Receipt**: Old robot program deleted.

- **Action**: Request download of new robot program.
- **Action Receipt**: New robot program download was successful.

- **Action**: Request run of new robot program.
- **Action Receipt**: Robot program is running.

Figure 4.11 Typical message exchange between central controller, vision controller, and robot controller.

This scenario makes heavy use of both the language translation tools and the finite state control that synchronizes the message exchange. In addition, database primitives must be invoked to extract the information necessary to construct the robot program and the robot commands that the dialogue uses. The responses from the dialogue participants must also be understood using language tools, and database retrieval tools must be used to find the appropriate reaction to a message.

Once individual programs can be built in the machine specialist family, they can be combined into complex, multispecialist networks. But, before this can be undertaken, each machine specialist must be packaged so that the final system need not consider the internal details of each specialist. The specialist on the outside of its package becomes just a name, and there must be a valid way of communicating with it based on a previously arranged message format.

The software tool hierarchy is not complex at the top (Figure 4.10) because the same tools that are used to solve the machine specialist problems can be used to solve other problems at the top of the task hierarchy (see Figure 4.9). The language used to communicate with the object is standardized so that all the basic commands can be understood by each machine specialist, and only one message format needs to be prepared. To complete the module, the sequence of events is controlled by the dialogue engine, and the responses returned from each machine specialist use the database primitives and the matching primitives to determine the next state. The result is a "higher" level module that is virtually identical to all of the "lower" level modules. The simulator, advisor, and diagnostician in Figure 4.9 are the same as the machine specialists in every respect, but they are specialists in other tasks besides part production.

There are many advantages for devising artifact families, such as part families used in manufacturing and program families used in rapid software prototyping. One advantage of grouping artifacts into appropriate families is a drastically reduced machine setup time that can include every facet of manufacturing: fixtures, tooling, part programs, material transfer methods and control, and so on. This analogy carries through to programming so that this task does not appear to be mysterious, but simply another aspect of the setup process. The generic programs that provide the basis of a program family make up the second level of the software pyramid (see Figure 4.7).

4.3.1 Languages and CML

In computer systems, there are three good ways to make programming tools available to the programmer. The first way is to build a new programming environment, which may have access to traditional programming languages. For example, the UNIX operating system has a number of tools (e.g., pipes) that make programming easier (Ritchie and Thompson 1974). The second possibility is to design a new language that makes it easier to express solutions in a particular domain. This has been a popular way of introducing new tools. For example, the Programming Language In The Sky (PLITS) was designed to program distributed systems (Feldman 1979); AUTOPASS and RAPT were designed to program mechanical assembly tasks (Lieberman and Wesley 1977; Popplestone, Ambler, and Bellos 1978; Lozano-Perez 1979); and AL was designed to control robots (Mujtaba, Goldman, and Binford 1982). Finally, the most conservative approach is to provide a subroutine package that is integrated into an existing language. For example, Paul gives a detailed outline of a subroutine package to control robots (Paul 1981).

The Cell Management Language (CML) is a new language that is designed to simplify the programming of manufacturing systems. CML is especially useful when machine components are used that were not previously designed to work together (Bourne 1986a). This task poses a

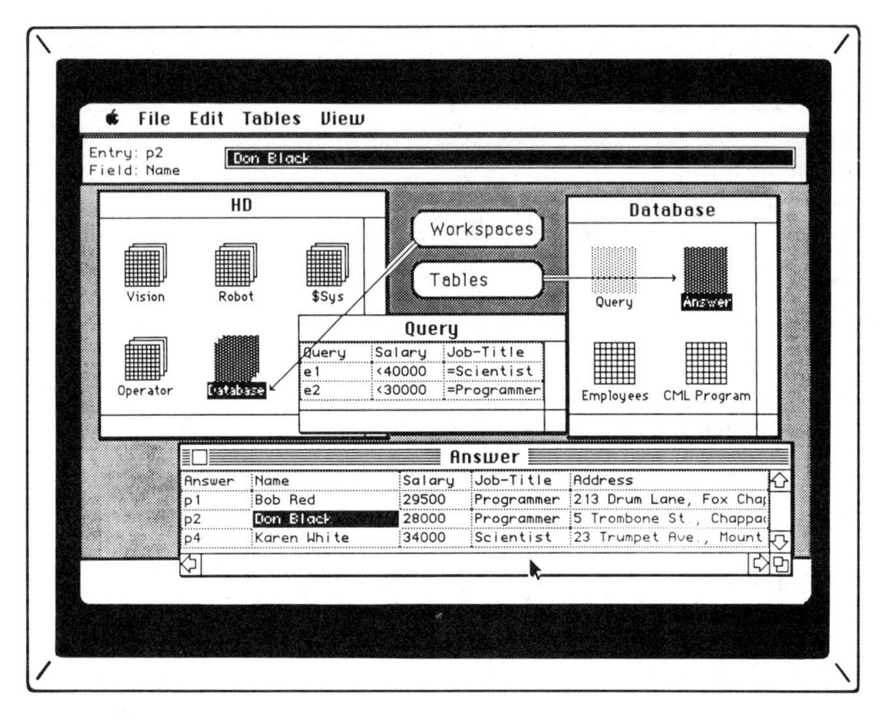

Figure 4.12 Managing data in CML.

number of difficult problems requiring software tools that have not been brought together in the past. For example, manufacturing machines (e.g., robots, milling machines, and grinders) are often programmed in completely different languages, so tools are necessary to build programs that can "understand" and "generate" messages in these languages. In our test application (Figure 2.5), four different programming languages were used to control the ten machines.

CML is founded on a basic table structure, where both data and programs are stored exclusively in tables. This means that traditional database operations can be performed for database-like tasks (Figure 4.12) as well as for tasks that are more unusual, such as automatic programming. It is this ability to swiftly update programs (e.g., internal CML programs, external robot programs, and external machine tool programs) with a single database operation that makes CML so useful in heterogeneous system environments.

To organize a large CML program, groups of tables are put into a workspace (see Figure 4.12). In this figure, the window labeled "HD" shows one workspace, called "Database." The window labeled "Database" shows four tables that are inside the Database workspace, and two of these tables (Query and Answer) are opened at the bottom of the display. This example uses the Query table as a guide to copy the appropriate database entries into another table. One CML command applies a query to a database, building a simple system that resembles IBM's Query By Example language (Zloof 1977). Again, programs can be manipulated in the same way in order to implement an automatic programming scheme.

The workspaces function not only as an abstraction, but also as an "object" similar to the objects in SMALLTALK (Krasner 1983) and objects in other systems (Laff and Haipern 1985). Workspaces can send messages to each other to transfer control. To manage this interchange, there is a central system workspace called "$Sys" that has dispatch information for messages (see Figure 4.13). If a message is sent to "Robot," the system looks up the dispatch parameters in the "$Lang" table. There are three

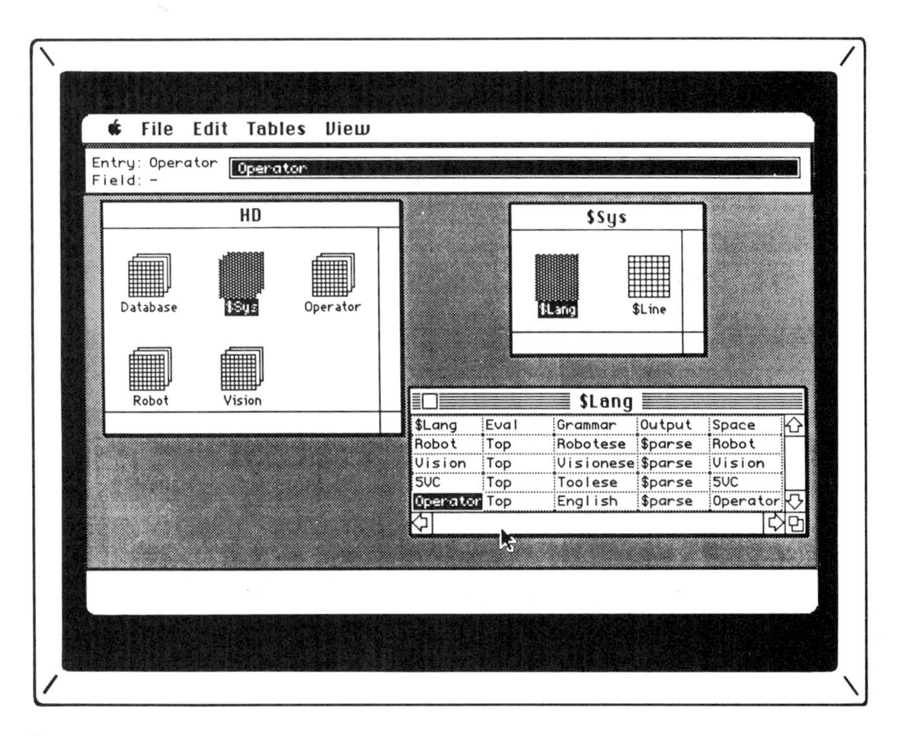

Figure 4.13 $Sys message dispatch.

parameters for workspace-to-workspace communications:

1. Grammar: a table that specifies the language syntax for the parser.
2. Output: a table that lists the parse results.
3. Space: gives name of the workspace that will process the message.

When a message is received, it is automatically forwarded to the appropriate workspace. Here, the CML parser is automatically invoked and uses a grammar description to break down the message into labeled tokens (see the $Parse table in Figure 4.14). In this example, the "grammar" guides the parser while it breaks up the input string: Program Error 33. Information about the word separators is hidden in the grammar so that the lexical analysis can be done at the same time as the syntactic analysis. In this case, the separator is "any number of blanks." Finally, a function that processes the message is automatically

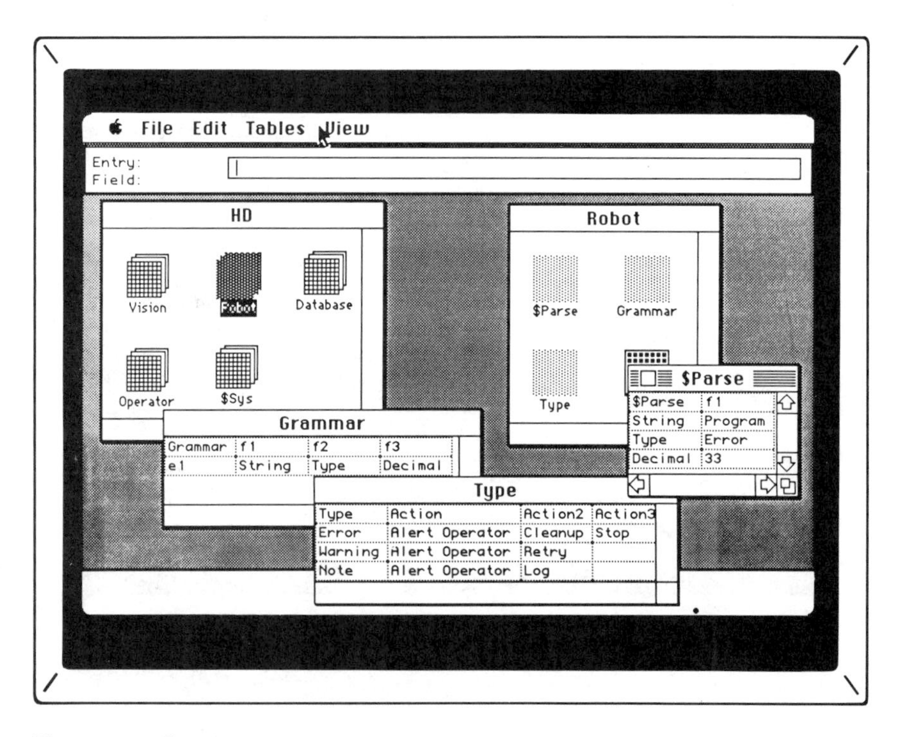

Figure 4.14 Parsing messages in CML.

invoked. After the function completes, the system retrieves another message and the cycle starts over again. This same style of message processing is used to interpret messages from external machines and users.

The messages that are sent between workspaces can also be organized into a dialogue that is designed to carry out a task. For example, every request to a machine tool may involve several contingent messages to accomplish a single task (see again Figure 4.11). CML offers a number of tools that facilitate the construction and management of complex message exchanges. Together, these tools have been used to build control systems for flexible manufacturing. The next section goes beyond the new language tools and concentrates on how to build systems, assuming this initial software base.

4.4 SETUP PROGRAMS FOR MANUFACTURING

Manufacturing engineers are accustomed to setting up their machine tools before a new part style is produced. However, they are not accustomed to writing complex computer programs, not to mention writing computer programs for ten different machine tools and the associated machine-machine interactions. Therefore, if there is to be any hope of generating programs quickly and reliably on a manufacturing site, the programming task must become a much more systematic exercise. For this purpose, we introduce the idea of "setup programs."

Definition *A setup program is a program that is completely designed and implemented from drawings and other graphical interactions. In particular, typing on a keyboard is limited to supplying simple names that are meaningful for the application. This is possible because a setup program starts with a generic base and adds only the application details.*

This graphic programming permits manufacturing engineers to avoid learning many complex programming skills. Even if these engineers are willing to learn new skills, the manufacturing environment is not conducive to this task. The environment is often noisy and dirty, with pressured schedules as the daily routine; the plant floor is not a place to think calmly about the complexities of programming.

One alternative to engineers writing their own programs is to contract out the plant's software projects. An advantage of this strategy is that contracts usually can be structured so that the outside company takes most of the risk. This strategy also means that projects can be made more predictable, both in terms of cost and final results, than if they were implemented in house. However, a premium must be paid to have someone else do the work, and this premium must be repaid for each new application.

Although, the first in-house project is almost always late, underpriced, and overstated, the next project can essentially start where the first project finished. But in-house programming is desirable only if the technology of the problem is well within the employees' capabilities, and this is usually limited at a single manufacturing facility. Hence, we see the importance of setup programs.

The remainder of this section describes a system that has been constructed to set up programs for flexible manufacturing cells. However, this general methodology can be used to set up programs for any well-defined task domain.

4.4.1 The Nature of Setup Programs

Setup programs can almost always be recognized as belonging to a certain program family. Just like a screw is recognizable even if it is short or long, few threads or many, one or two slotted, steel or aluminum, programs also can be quite different and yet remain identifiable. Without these program families, the programmer's task would be much more difficult because there would be many more choices to make in the implementation.

Figure 4.15 graphs the kind of tradeoff that can be expected when a typical range of programming strategies is used. As more information is built into a programming system, the generality of the possible results may decrease, but so does the total programming effort. The quality of the programming tools usually has the effect of flattening, but not eliminating, these curves.

The tools labeled across the abscissa in Figure 4.15 start with a general-purpose language such as: LISP, Pascal, or FORTRAN. Starting from this basic language, it is possible to use a subroutine package that does much of the difficult work within the scope of a commonly available language. Some old packages include: math subroutine packages, graphics routines for particular monitors, and database interface functions. To write programs that use these packages, it is necessary to both know how to program as well as how to use the relevant subroutines in the package. In some cases, the actual source of the subroutines is provided to the user; in other cases, only the subroutine names, arguments, and functional descriptions are provided. Learning how to use one of these sub-

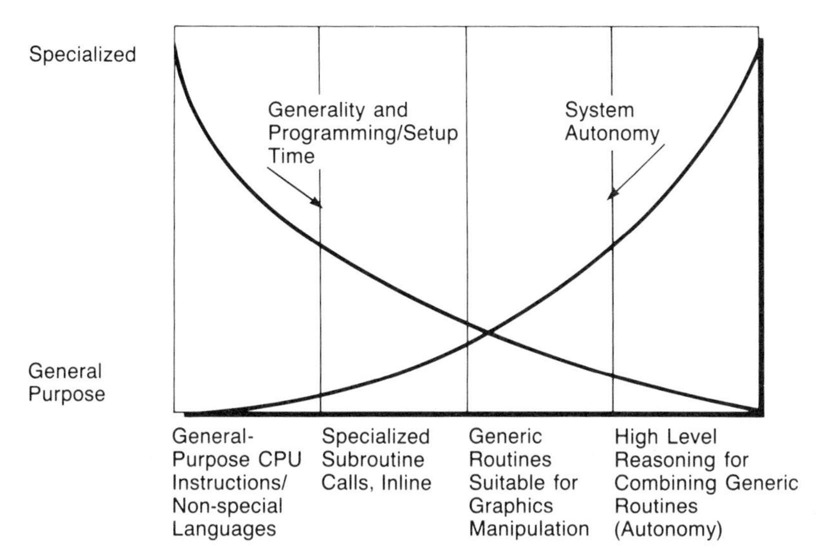

Figure 4.15 The Programming time–generality tradeoff.

routine packages can be as difficult as learning the basic language in the first place. But, once that knowledge is attained, these packages can be used very effectively to build programs.

The next step beyond subroutine packages, in Figure 4.15, is object-oriented programming. These objects can be both routines and data; their features can be inherited from their superior; and the interface to them is usually in the form of messages rather than argument lists. This abstraction makes it appropriate for a programmer to think of these things actual objects, which are represented by icons (i.e., easily identifiable pictures). The routines can then be executed by pointing at the pictures, and they can be combined into complex programs by logging a series of user actions. That is, a programmer can write a program by showing the system once how to do a task and then having the system remember the sequence under a new picture.

Finally, at the extreme right of Figure 4.15, the system can automatically generate its own plan of action and then autonomously carry it out. As it turns out, this automatic programming problem is tractable only if the level of actions available to the system are at the task level, that is, directly reflect the application. It is no accident that these are just the tasks that are quite easy to label with meaningful icons; they are designed with this in mind. If the primitives are below the level of a task description, the complexities of the planning problem are beyond most automatic systems.

4.4.2 Building Control Software Semi-Automatically

Our task is to automatically build control software for manufacturing systems (see third level of Figure 4.7) in such a way that the program primitives mirror the cognitive objects (see Chapter 3) that a human engineer would typically use to configure the machines in this sort of system. Of course, the primitives change depending on whether the system is a workstation, manufacturing cell, or flexible manufacturing system. But, the basic mechanisms for setting up the control software are still the same, just with different objects.

A complete control system has a number of different elements, each of which must be carefully designed and implemented. This is true regardless of whether or not the system is built with automatic programming tools. The breakdown of a control program follows:

1. **Statics**: The static structure of a control program is a group of modules that make up the final program. However, this group of modules has no goals and no reason to do any work. Therefore, the statics of a program are the "passive" components of a program that include procedure and data for general situations, but nothing that covers specific situations.
2. **Operator Interface**: The operator interface is the most visible part of a control system and, in the minds of most people, it is the control system. This admits that the graphics front-end is given more importance than it may deserve, but this importance does have some foundation. The operator is part of the control system, and the quality of the control system can be strictly enhanced by making the interface easy to learn, easy to understand, fast to use, error resilient, and visually provocative.
3. **External Programs**: A complete control system often requires the use of multiple systems. It may be necessary to partially program key routines in each external system (e.g., a robot program that moves the robot in front of a furnace).
4. **Stable Dynamics**: The stable dynamics of a control program are defined in terms of the static program structure. This definition results in data and logic that determine what an application does and what the sequence of events must be in order to achieve the

system's goals. The stable dynamics of the control programs have no information about how the system should respond to error situations, unless it is embedded in the static structure.

5. **Error Dynamics**: The error dynamics of a control program are defined in terms of the stable dynamics. This definition results in logic and data that are added to the already compiled stable dynamics. They are responsible for managing situations that occur whenever the stable dynamics no longer apply because the manufacturing environment has changed in some unexpected way. For example, what should the system do when an automatic guided vehicle never arrives at its final destination?

By using the terms "statics" and "dynamics," we are suggesting that a comparison be made between software objects and physical objects. In this way, the mystery is taken out of software, and the software is reduced to the level of a physical piece of steel so that a similar analysis can be performed on both types of objects. This analogy continues in the section on open problems by raising the possibility of higher order program properties of control systems (e.g., stress, strain, and yield strength).

4.4.3 Statics: Program Structure

The first step is to lay out the system components' basic configuration and, at the same time, piece together generic software modules into the shape of the final system. These generic modules define all the properties that are unique to the object as well as establish a standard way of interfacing the module to a system of similarly constructed packages. For example, a module that represents a robot contains descriptions of the robot's programming language, command language, general purpose, and basic limitations. The module also contains procedures that manage the connections with the other modules in the final system, for example, an interface language for message passing and a dialogue manager for coordinating the messages.

A "configuration system"—a database of these generic machine-description modules together with a collection of icons that represent them—has been built. Then, by picking a series of these icons and by demonstrating their interconnection, it is possible to build the structure of the basic control system. As icons are selected, corresponding machine specialists are automatically included in the control system being constructed.

Building a system with this flexibility makes it possible to experiment with a number of different system architectures. For example, the most popular architecture for factory control system has been hierarchical (Simpson, Hocken, and Albus 1984). In this architecture, each control node is responsible for a set of tasks at a given level of abstraction. A robot controller is responsible only for robot actions, and a dispatcher at a higher level is responsible for the coordination between a group of machines without having to know the details of any one of them (see Figure 4.16). Unfortunately, this scheme often involves extraneous message passing because, if two machines need to communicate, they must first pass the messages all the way up the hierarchy to the "lowest common boss." However, the final control system for many manufacturing cells is only two levels deep, and so the message overhead is not prohibitive. This was, in fact, the architecture used in the forging cell shown in Figure 2.5).

However, there is also a growing interest in heterarchical control be-

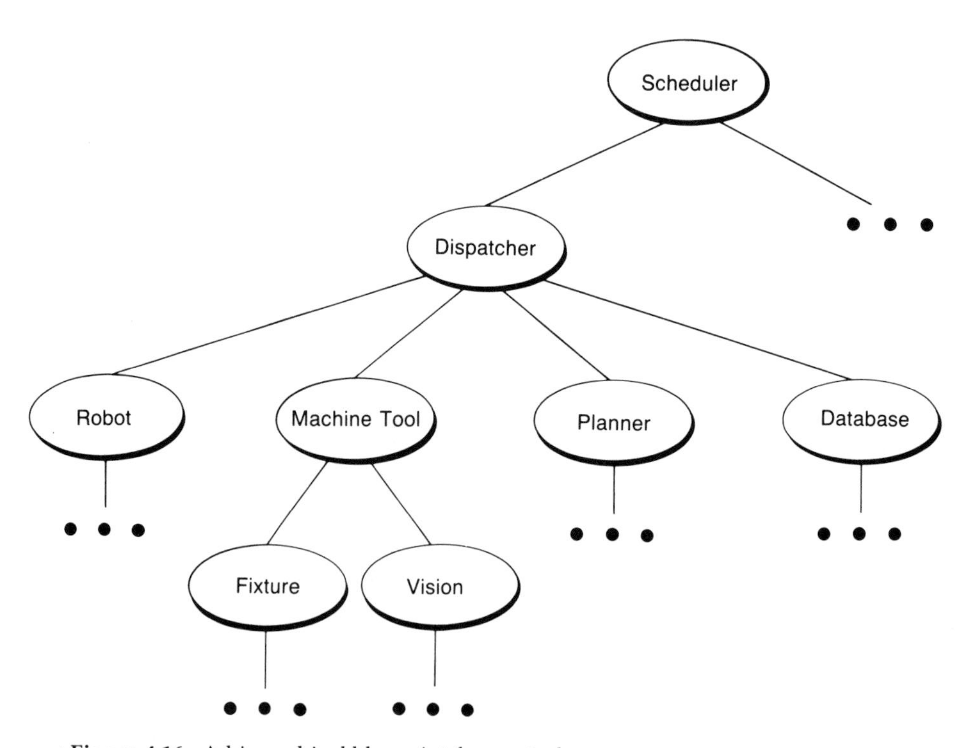

Figure 4.16 A hierarchical blueprint for control.

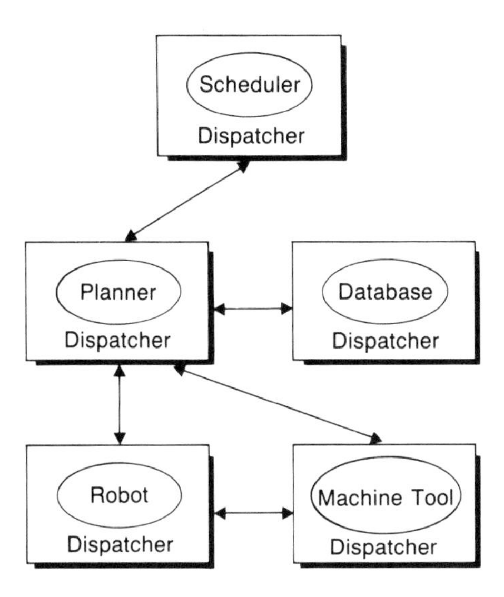

Figure 4.17 A heterarchical blueprint for control.

cause it avoids the unnecessary messages in hierarchies and because it is similar in concept to object-oriented programming (see Figure 4.17). In this architecture, control nodes can communicate on an as-needed basis. This communication makes it possible for each node to have a higher degree of self-determination, thus requiring more intelligence at each node. The distribution of intelligence makes the final system much more resilient when a control node fails. In a hierarchical system, if one intermediate control node fails, all the control nodes below it no longer have enough information to operate. In the heterarchical scheme, the system is only damaged when a system that performs a vital function fails, and this can be safeguarded against by providing redundant control systems. The trend in man-operated factories is to balance these two architectures. The top level of the hierarchy tries to retain some measure of control over the entire system, while still attempting to provide the bottom levels with a greater measure of autonomy.

The least popular control system, from a manufacturing user's perspective, is anarchical. However, these control elements are perhaps the most commonly built by manufacturing suppliers. In an anarchical system, each control component would be able to carry out all the functions of a full system (see Figure 4.18), which also means that they are probably not designed to work as a system or team (see Chapter 11). In fact, the ten machines in the forging cell (see Figure 2.5) were supplied by ven-

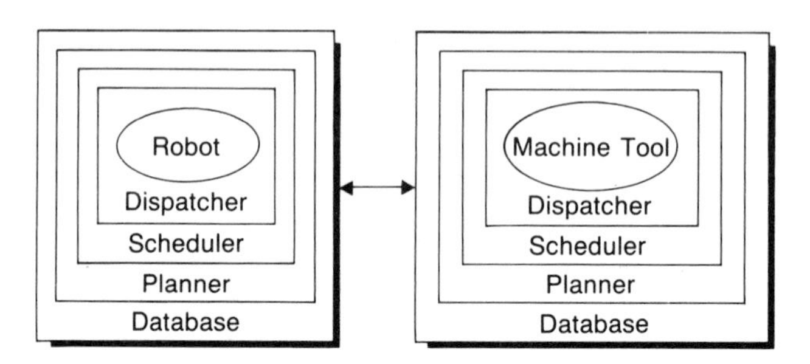

Figure 4.18 An anarchical blueprint for control.

dors as "standalone" machines, and they had to be converted into usable system components. For example, they all needed complete communications support so that both the functions of the front control panel and the part storage facilities could be duplicated remotely.

Which architecture is best for a manufacturing system? This is still an open question. In fact, even the international corporate structure has not been able to make a definitive architectural choice. The result has been that the structure of a corporation usually reflects the personality and the aesthetics of the founder or chief executive. We should not be surprised that the shape of a control system equally reflects the prejudices of the designer.

By having a fast way to configure a system, it is possible to determine an optimal control arrangement for a given application and to develop general-purpose design guides. No matter what the choice is for the final system architecture, there must be a powerful "glue" that holds the modules together. In CML, this infrastructure has the following mechanisms:

1. **A Message-Passing Mechanism**: This can be accomplished in a number of ways, for example, with dedicated memory for each connection (a kind of "soft" serial line) or with one central mailbox facility that processes the messages one by one. The message order in the mailbox can be determined by arrival time, priority, or a combination of both. CML uses the central mailbox scheme.

2. **Logging Mechanism**: In case the system fails, it is important to be able to reconstitute the state of the system from a log of exchanged messages (Liskov and Scheiffler 1982). Another strategy used in some sophisticated manufacturing control systems runs two machines in case one of them breaks down. In this event, the other

machine knows the complete state of the system, and it can take over the system's physical control.

3. **Distribution Mechanism**: Each control node must be movable to separate processes or even separate computers. This ability allows the system to be scaled up beyond the powers of a single machine and the limitations imposed by a single process. This is a feature of CML.
4. **Dialogue Mechanism**: Each control node must be able to keep track of the state of each dialogue that is in progress. This is also a feature of CML.

4.4.4 Operator Interface

The next step is to design the graphics so that the operator can monitor and interact with the final system. We have built a CAD-style program that allows the system implementor to draw the operator interface and, from that drawing, to construct the final program that the operator will use. Figure 4.19 shows an operator display being constructed that will eventually be used by an operator to control the system.

A system like CML that separates the user interface design from the system design (see also Smith, Lafue, Schoen, and Vestal 1984) has several advantages:

1. **Uniformity:** User interfaces are mostly repetitive from application to application: the command language, options, display, and help facilities. However, almost every program has an interface that must be completely learned from the beginning. One advantage of a CAD system for building an operator interface is that it offers a set of tools that can be combined in ways that are different for each task, but bear a strong similarity. Thus a user who is familiar with the family of interfaces can use any of the interfaces with minimal learning.
2. **Development Time:** System designers no longer need to spend the time designing and implementing an operator interface from scratch. Most projects begin with a very simple operator interface, because of the expense of building a complete and sophisticated operator interface. As a result, most user criticisms focus on the ease of use and capabilities of the interface. The designer and the user place a completely different value on the quality of an interface.

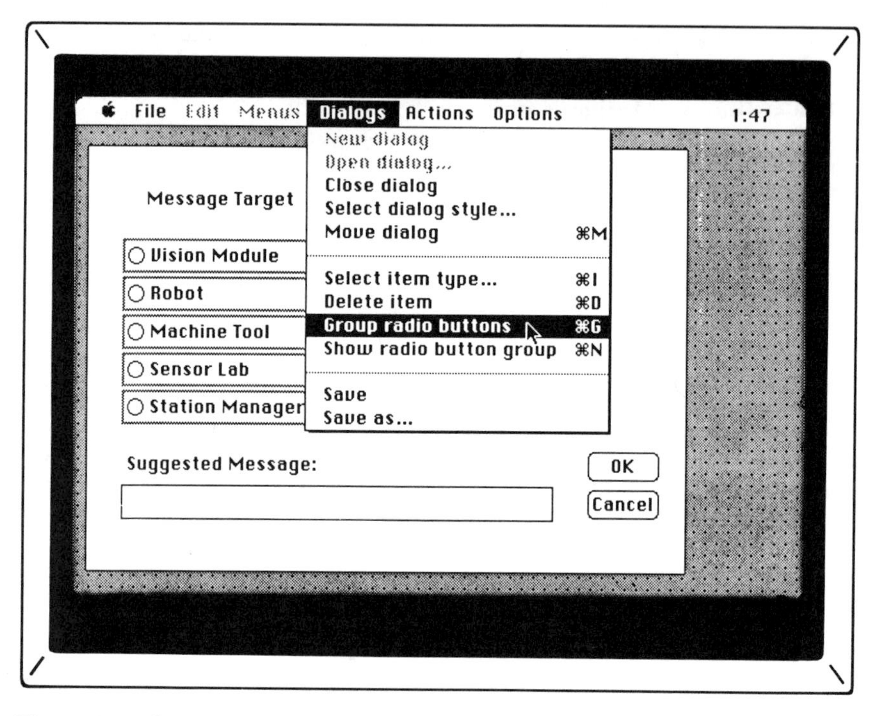

Figure 4.19 Operator interface being constructed for teaching dialogues.

3. **Sophistication:** If one CAD family of interfaces is designed once for a whole range of applications, it is worth the time and effort to build features that would never be built in one-of-a-kind systems. For example, many features are enjoyable for users but are rarely included in applications: windows, menus, graphics input (e.g., a mouse or a joy-stick), and graphics output.
4. **Computational Resource:** It is also advantageous to put the user interface on a completely separate machine, so that the sophisticated interface does not compete for computational resources with the basic application. An operator can also learn about the underlying system without it operating.

4.4.5 Stable Dynamics: Program Actions

Once the operator interface is built in CML and a skeletal form of the control software has been prepared, the system can begin execution. At this stage, the control system does not have any task programmed because there is no logic or dialogue to guide the succession of events.

We have built a system that automatically constructs rules to control what should happen next in a manufacturing cell. At any given time, the cell (see Figure 2.5) can be characterized by a collection of machine states. For example, the robots in the forging cell could be in their "home position" and the forge could be "executing part program 27;" these are the states. Now some logic is needed to trigger a robot to unload the forge after the part program is complete. The operator can effectively write this logic by pointing at an icon, for robot, on a terminal screen that was prepared in the system's static configuration. The system constructs the rule by putting the current state of every machine in the condition part of a rule, and the operator's command in the action part of the rule. Therefore, in this case, the resulting rule would read:

If (robot A is home) and (robot B is home) and (forge running program 27)
then (robot B unload forge)

This rule is then added to an accumlating list of rules. When a rule matches the current situation, the action part of the rule is initiated, which triggers a message to the "robot machine specialist." This module then handles all the low-level details that make up the command. When no rules match, the operator can issue a new command and hence build a new rule.

As this process proceeds, the number of situations that need rules to activate events goes up exponentially, until a majority of cases have already been handled. After all the machines in the cell have been in operation for a while, the number of messages not seen before dramatically decreases. By using this style of programming in the forging cell with its ten machines, no new situations were encountered under normal operating conditions after only four to five cycles of the cell (Bourne 1986a).

In some situations, a rule may be required to initiate an event even though it seems like an old situation. This happens because a machine not involved in a transaction, in this case, an inspection station, may be changing state and thus preventing rule matches. To avoid this situation, machines can be grouped by the operator or programmer into critical units. With a few exceptions in the forging cell, only the robots and a "target machine" had to be checked in the conditional part of the rule. Therefore, the state of the inspection station would never be considered by a rule that was designed to unload the forge.

4.4.6 Error Dynamics

If an error occurs in the manufacturing cell for forging, it is important to take three steps:

1. Manage the error so that it does not propagate beyond the immediate situation.
2. Diagnose the cause of the error.
3. Take corrective action to resolve the error in order to resume production.

The first step is the most important one and is probably the least understood. We have seen evidence for this lack of "error management" in both the Three Mile Island and Chernobyl nuclear accidents. A stuck valve and a faulty indicator at Three Mile Island started a whole chain reaction of both automatic and human errors, which finally resulted in a serious nuclear event (Lombardo 1981). This kind of chain reaction also can occur in manufacturing.

For example, on one occasion, the automated cell for forging had a chain reaction that started with a robot gripper welding itself to a red hot part. The robot gripper failed to open even though the chuck of the forge was tightly gripping the other end of the billet. As a result of this deadlock, the robot failed to back away from the CNC forge and the heavy machinery started into its process. After a brief tug of war, the robot arm broke away from its base, which propelled the connecting bolts into the ceiling 60 feet above. Finally, a human was able to reach the emergency stop buttons and bring everything to a halt. At this point, the emergency stop caused the furnace door to be left open for an extended period of time. This meant that a number of hydraulic machines were exposed to high temperatures, which could have started a factory fire. Fortunately, the production engineer recognized the situation before a new chain reaction could begin.

It is almost always true that, although the first error may not be very damaging, the error becomes successively worse (Bourne and Fox 1984). If the first error can always be contained, the problem of managing errors is effectively solved.

The second step in error management is to automatically diagnose the cause of an error. Several successful expert systems have been developed to diagnose a situation (Barr and Feigenbaum 1982). Although some of these systems work well, it is quite tedious to build and test them to verify the results. Although only few systems aid in the construction of diagnosis systems (Pan and Tenenbaum 1986), some expert system shells provide the appropriate environment as a working start.

The third step in error management is to automatically take actions that will correct the error. This step is similar to the topic discussed in Chapter 10, and some of the diagnosis systems can also suggest ways to correct an error. However, the confidence in these systems is currently not high enough to commit to the machine recommendations without first consulting the human analyst.

4.4.7 The Complete System

After all five program parts have been constructed, the control system is complete. Many simulation systems for cell control and one actual system have been constructed in this way. The result is that it can take as little as a day to build a complete software control system.

4.5 PROBLEMS THAT FACE THE MANUFACTURING ENGINEER

It is difficult to install any advanced technology in the factory today unless at least one trained specialist is on call. These advanced technologies pose problems just because they are advanced and rarely, if ever, can be purchased as turnkey systems. This has prompted us to build an array of tools for CML that is designed to make the manufacturing engineer's job much easier (see Figure 4.20). The long-term goal is to allow nonprogrammers to manage the construction of large software probjects (e.g., building intelligent control systems).

The x-axis of Figure 4.20 describes a range of tools that can be used to automatically generate and understand programs or text that fits within a predefined family. The y-axis of Figure 4.20 describes tools that range from advisory systems to systems that run automatically.

One tool that has not been previously discussed is the CML expert (upper right corner of Figure 4.20). This system takes a CML program as input and then writes a letter to the programmer. The letter gives recommendations for program optimization, comments on program style, warnings about obsolete functions, and advice about how to structure CML programs. Such a tool makes it possible to quickly learn about the idiosyncrasies of a new programming language environment.

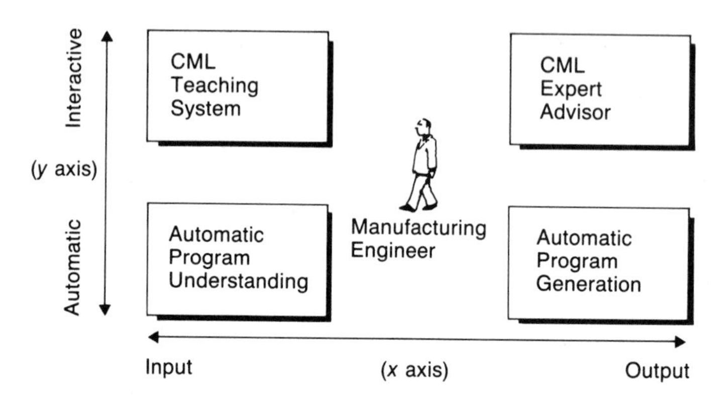

Figure 4.20 AI tools for the manufacturing engineer. (Reprinted with permission from the *AI Magazine* Vol. 7, No. 4, pages 86–96, published by the American Association for Artificial Intelligence.)

Figure 4.21 is an example of a letter written by the CML expert. The letter is made from a series of pre-programmed paragraphs. The paragraphs are invoked using a rule-based system in CML and are accumulated as an outline of the letter. Once the first selection of paragraphs is made, the outline is scanned and modified so that redundant paragraphs and appropriate concluding paragraphs are added. The programmer then can use the letter as a guide to improve the CML program.

A system that can easily analyze textual material in one language and then generate an analysis in another can solve many of the remaining problems in automated manufacturing. For example, a CAD description can be analyzed, information added from the user where needed, and finally a part program generated to make the part. Alternatively, a part program that runs on one machine tool can be interpreted and rewritten to run on another machine tool, with different language capabilities and without any reprogramming. This same problem repeats itself in every corner of the factory.

4.6 DISCUSSION

Truly intelligent man-made machines are still an unborn species. However, we have shown that by breaking down a system into the right components and then putting the components back together with the right kind of "glue" (i.e., translation software), the intelligence barrier that separates man and machine could possibly be overcome.

Dear David:

I have spent some time analyzing the functions you have used in CML and have come to a number of conclusions. Nothing I say here should be taken as absolute, since the analysis is only performed at a surface level.

A huge portion of your program is devoted to input and output operations. Perhaps you should try to centralize some of these activities in one module.

You have used quite a few of the control features, which indicates to me that you have spent quite a bit of time refining the program.

You have used a number of database oriented operations, which tells me that you have appreciated at least this aspect of CML.

I have noticed that you tend to use quite a few "begin" conditions. It is usually quite easy to bring these together into a set of rules that will show the logical structure of a program at a glance. The other advantages of using rules is that they tend to run faster than their "begin" counterparts, it is often easier to change the program to manage more situations and to include parameters that allow the overall function to be much smaller.

Input statements are often convenient to use if you are writing a "test" program, but they should not be part of any production software. The problem with using "inputs" is that they do not verify that the input-string is in fact in the range of appropriate responses. Try to use the connect statement instead.

I noticed that you are using a "print" command. These are often convenient in test programs, just as a quick way to write out some sort of message. However you should seriously consider switching these to "generate" commands, since the output of a "generate" command can be easily redirected to external prosesses and files.

Thank you for trying the CML expert system, I hope my comments about your program have proven to be helpful.

Sincerely,

The CML Wizard

Figure 4.21 Letter automatically written to suggest improvements in a program. (Reprinted with permission from the *AI Magazine* Vol. 7, No. 4, pages 86–96, published by the American Association for Artificial Intelligence.)

In the short term, we have concentrated on software tools that make building increasingly intelligent systems much easier. These tools range from simple database primitives to more sophisticated letter-generation programs. As a result of using these tools, we have successfully built a system based on the software pyramid presented in Figure 4.7. The resulting cohesive structure and the daily production of parts in the forging cell are proof that this pyramid of tools is not the feared software Tower of Babel.

After the success of the forging application, it has become apparent that many applications are composed of a multiplicity of dissimilar systems. To satisfy the demand of these applications, CML is being commercially developed (Westinghouse Electric Corp. 1987). The next section looks beyond the state-of-the-art into a set of unresolved problems for constructing intelligent control programs.

4.7 SAMPLE OPEN PROBLEMS IN THE CONTROL OF AUTONOMOUS SYSTEMS

Although the following two chapters also contain open problems that concern control, this section stresses general software control problems that are not directly related to the control of the manufacturing eye or the manufacturing hand.

1. Develop an intelligence test that extends to lower life forms and mechanical systems. An initial effort might be made to rate systems by the number of different types of real-time inputs that generate a desirable response (e.g., a shadow detector, a thermostat, and a keyboard interrupt). It would be beneficial to normalize this IQ test to an accepted test for humans, thereby providing more significance in the lower ranges.
2. Use general-purpose translation tools to solve one of the following manufacturing problems.
    a. Develop a uniform interface to two or more different CAD systems.
    b. Develop a uniform interface to two or more different machine tools.
    c. Develop a translation procedure between two traditional languages (e.g., FORTRAN to Pascal). It may be necessary to make this system interactive so that missing information can be added during the translation process.

3. Decompose the programming problems discussed in Section 4.4.6 on error dynamics. Isolate a group of tasks that lend themselves to teaching strategies and develop the computational tools that are necessary to accomplish it.

4. What happens to a system when it goes outside of its intended domain of operation (e.g., an unexpected error or event)? Develop a simple expert system that remains in control despite being outside of its intended domain. It must achieve this not by giving false information, but by continuing to give valid information whenever possible. Extend the solution to operate a real-time controller.

5. Develop a theory of higher order properties for programmed control systems. The idea would be to apply concepts that are used in mechanical systems (e.g., stress, strain, and fatigue) to programs. These properties would be used to quantify the strengths and weaknesses of a program, just as the strengths and weaknesses of a metal part can be quantified. Today, virtually no quantitative analysis considers the "strength" of programs. Instead, researchers have relied on sophistry (e.g., "robust," "fail-soft," and "graceful degradation") to describe the qualities of programs that they are seeking.

6. Develop a system that can plan and spawn a new control process for new tasks, such as the control system in Figure 4.5 applied to the tasks in Figure 4.6. At this point, a pyramid of tools appropriate for the task domain should be identified and used to flesh out the specialized task information.

7. Database theorists have developed the idea that one fact in a database belongs in just one place. Without this constraint, difficulties arise when the database is being updated because a single fact has to be tracked down in multiple places and then changed. Extend this notion to programs, so that only one "program fact" occurs in one place. CML is a good model for this exercise because the programs are already represented in database format.

8. Develop a programming model and language where time is an essential component. In this language, it should be possible to "set" the time of a program point (relative to the program starting time) and have the compiler organize the program to guarantee that all the time constraints are satisfied. Some research has considered

time constraints, but none has actually reorganized the program to satisfy them.

9. Outline the methods of machine learning that have been employed in research systems. Why have these combined methods failed to scale up to the demands of new applications? Refer to the top level of Figure 4.7 and its interconnections to the lower levels.

10. As an extension to Problem 1 in Chapter 3, develop a computational model where both internal and external software objects are connected by the same of kind of mechanism. In particular, the system should interface to external objects by taking two steps. First, the system should identify the object as either an old recognizable object or as a new unknown object. Then, based on this identification, initiate a dialogue that takes advantage of previous information about the object. The message exchanges between both internal and external objects should be used to cooperatively solve problems that are posed to the system from inside and out.

REFERENCES

Barr, A., and Feigenbaum, E. A. 1982. *The Handbook of Artificial Intelligence, Volume 2.* Heuristech Press, Stanford, Calif., Ch.8.

Bourne, D. A. 1984. A multi-lingual database bridges communication gap. In Manufacturing. In *Robotics and Factories of the Future,* edited by S. N. Dwivedi. Springer-Verlag, New York, pp. 543–552.

Bourne, D. A. 1986a. CML: A meta-interpreter for manufacturing. *AI Magazine,* Vol. 7, No. 4, Fall, pp. 86–96.

Bourne, D. A. 1986b. Manufacturing: Acquiring craft skills through dialogues. *Intelligent Robots and Computer Vision: Fifth in a Series.* SPIE, October.

Bourne, D. A., and Fox, M. S. 1984. Autonomous manufacturing: Automating the job-shop. *Computer,* IEEE Computer Society, Sept.,pp. 77–88.

Brooks, R. A. 1985. A robust layered control system for a mobile robot. *A.I. Memo* 864, MIT, Cambridge, Mass., Sept.

Brooks, R. A. 1986. Achieving artificial intelligence through building robots. *A.I. Memo* 899, MIT, Cambridge, Mass., May.

Feigenbaum, E. A., and Feldman, J. 1963. *Computers and Thought.* McGraw-Hill, New York, p. 3.

Feldman, J. A. 1979. High level programming for distributed systems. *Communications of the ACM,* Vol. 22, No. 6, pp. 353–368.

Krasner G. 1983. *Smalltalk-80: Bits of History and Words of Advice.* Addison-Wesley, Reading, Mass., pp. 10–28.

Laff, M. R., and Haipern. B. 1985. SW2—An object based programming environment. *ACM Symposium on Language Issues in Programming Environments*, Seattle, Wash., pp. 1-9.

Lieberman, L. I., and Wesley, M. A. 1977. AUTOPASS: An automatic programming system for computer controlled mechanical assembly. *IBM Journal of Research and Development*, Vol. 21, No. 4, July, pp. 321–333.

Liskov, B., and Scheiffler, R. 1982. Guardians and support for robust distributed programs. *ACM Symposium on Principles of Programming Languages*, pp. 7–19.

Lombardo, T. G. 1981. TMI Plus 2. *IEEE Spectrum*, Vol. 18, No. 4, pp.28–44.

Lozano-Perez T. 1979. A language for automatic mechanical assembly. In *Artificial Intelligence: An MIT Perspective*, edited by P. H. Winston and R. H. Brown. MIT Press, Cambridge, Mass. pp. 245–271.

Moravec, H. 1986. Personal correspondence based on his unpublished book *Mind Children* to appear in Harvard Press.

Mujtaba, M. S., Goldman, R., and Binford, T. 1982. The AL robot programming language. In *Computers in Engineering*, edited by G. D. Gupta. American Society for Mechanical Engineers, New York, pp. 77–86.

O'Donnell, J. T. 1985. Dialogues: A basis for constructing programming environments, *ACM Symposium on Language Issues in Programming Environments.* June.

Pan, J. Y., and Tenenbaum, J. M. 1986. PIES: An engineer's do-it-yourself knowledge system for interpretation of parametric test data. *AI Magazine*, Vol. 7, No. 4, Fall, pp. 62–71.

Paul, R. P. 1981. *Robot manipulators: Mathematics, programming and control.* MIT Press, Cambridge, Mass.

Popplestone, R. J., Ambler, A. P., and Bellos, I. 1978. RAPT: A language for describing assemblies. *The Industrial Robot*, Vol. 4, No. 1, March, pp. 10–17.

Ritchie, D. M., and Thompson, K. 1974. The UNIX time sharing system. *Communications of the ACM*, Vol. 17, No. 7, pp. 365–375.

Simpson, J. A., Hocken, R. K., and Albus, J. S. 1984. The Automated Manufacturing Research Facility of the National Bureau of Standards. *Journal of Manufacturing Systems*, Vol. 1, No. 1, pp. 17–32.

Smith, R. G., Lafue, M. E., Schoen, E., and Vestal, S. C. 1984. Declarative task description as a user-interface structuring mechanism. *Computer*, IEEE Computer Society, Sept., pp. 29–38.

Westinghouse Electric Corp. 1987. CML, commercially managed by C. Wingert, Westinghouse, Box 160, Pittsburgh, Pa. 15230.

Zloof, M. M. 1977. Query by example: A data base language. *IBM Systems Journal*, Vol. 16, No. 4.

5 The Manufacturing Eye

"Yes, I have a pair of eyes and that's just it. If they wos a pair o' patent double million magnifyin' gas microscopes of hextra power, p'raps I might be able to see through a flight o' stairs and a deal door; but bein' only eyes, you see, my wision's limited."
— Charles Dickens in *The Posthumous Papers of The Pickwick Club.*

THE manufacturing eye is the richest sensor at the craftsman's disposal. It can be used to solve a wide range of problems that are present in the factory. For example, in machining work, the machine's control can rely on a vision system to verify that there are no chips in the cutter path, to measure parts, to check tools, and to assure that the part is firmly seated on the fixture.

This chapter outlines a new approach to machine vision. This approach stresses the importance of selecting images to confirm or reject hypotheses posed by system goals, rather than concentrating on single images. In addition, we have accumulated a checklist to help the reader determine whether vision is the most appropriate solution for a particular industrial application.

5.1 EVOLUTION OF MANUFACTURING CAMERAS AND EYES

Computer vision first appeared in university laboratories in the early 1960s (Roberts 1965), but it did not become practical enough to be considered a serious manufacturing tool until the late 1970s (Gleason and Agin 1979). The increased interest in this technology was due to the development of special-purpose computers that could process binary pictures in reasonable times. In essence, these "vision modules" were the first CNC controllers for machine vision. Paradoxically, computer vision researchers were making the most significant advances in the science of machine sight (Marr 1982) at the same time that engineers were making severe compromises to make vision work in the factory. Figure 5.1 shows the evolving machine eye from both the academic and the industrial perspectives.

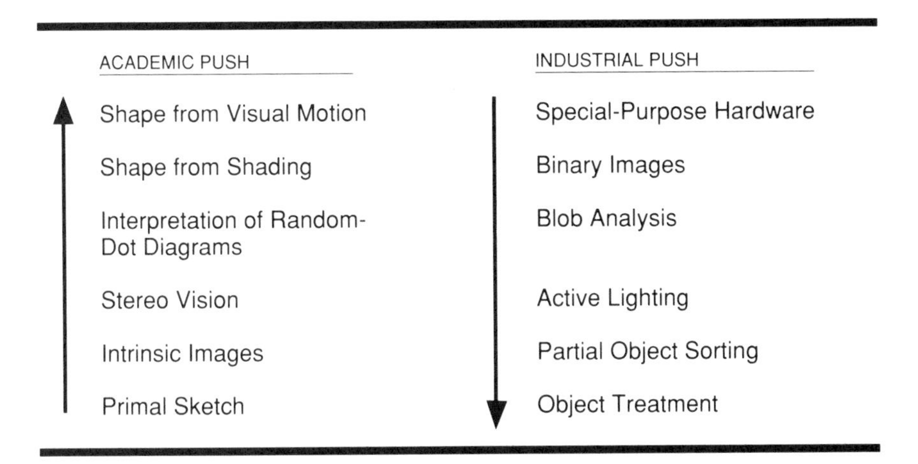

Figure 5.1 Academic advances and industrial compromises.

The academic side clearly realized that an image is a composite of many sources of information; these sources are called "intrinsic images." For example, the intensities in an image are formed from many external properties such as the reflectance properties of objects, illumination levels, shadows, and distance (Barrow and Tenenbaum 1978). This realization started a flurry of activity, first, in the extraction of the intrinsic images, and second, in their separate interpretation (Horn 1986). At the same time, Marr's group at MIT was developing computer models of vision based on the psycho-physical evidence available from humans. This work concentrated on developing vision algorithms that would exhibit properties similar to human perception on small but difficult problems. Before Marr's work, Julesz (1971) showed that humans could detect systematic shifts between stereo pairs of random dots by fusing the images into one three-dimensional representation. Then, Marr and Poggio (1976) developed a computer algorithm that would compute the same results as humans perceive. This approach was used to understand the fundamentals of matching objects in stereo pairs (Marr and Poggio 1979) and to understand the perceptual effects of moving objects (Hildreth 1983).

At the other end of the community, industrial engineers were struggling to make the vision systems easier to use in the factory (Bullock 1978). This started by identifying vision algorithms that had been first used successfully in the 1960s, and then by making them run very fast by putting them directly into special-purpose hardware, or microcode. This practical approach was biased toward using binary images that were run-length encoded. From this encoding, it was easy to extract "blob" descriptions in real time. In addition, active lighting was one way to produce a binary image, thus eliminating all but one of the intrinsic images. This intrinsic image is the structured light information, and that is only one derivative step away from surface orientation. Needless to say, this made active lighting one of the most important vision tools for the industrial engineer. If these simplifications were not enough, mechanical sorters were used to isolate one object from a batch so that the image processing could not get confused about the object boundaries. And, if all else failed, the objects were treated by painting them in special way or by putting a barcode on them with the identification information already encoded in it. Villers (1984) reviews a number of industrial systems that each use a combination of methods to make vision systems practical.

The primary goal for a manufacturing eye is to resolve a tradeoff that both the academics and industrialists are struggling with in their own ways. Figure 5.2 shows the struggle between building a system that can truly develop a global understanding of a scene as opposed to building one that can very accurately measure particular properties of objects. To understand the full scope of a scene, many sources of information (i.e., intrinsic images) must be discovered and fully used. Although using vision for accurate measurements requires that the nonessential information be kept out of the images, the field of view must be decreased, the lighting must be optimized, and an optical table should be considered to reduce vibrations caused by the environment. These global and local properties have never been jointly optimized in any machine vision system.

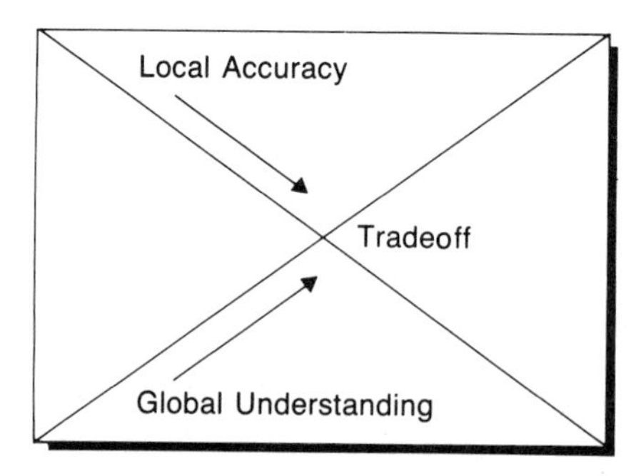

Figure 5.2 Key vision tradeoff.

5.2 VISION PREHENSION AND IMAGE ACQUISITION

Most computer vision projects have not addressed the issue of how to acquire images. The projects start with a picture, and the entire problem is in its interpretation. For example, the following questions might be addressed.

- Here is a picture. What can be found in it?
- Here is a picture. Measure the dimensions of the only object in it.
- Here is a picture. Where are the boundaries of the objects?
- Here is a picture to match with a model of what the picture should look like. Does it match and how well?

All these questions focus on a single image, and this does not address the broader issues of how to get the image in the first place. By starting with a single image, the ensuing vision problems are either more difficult (because there is no contextual information) or easier (because the contextual information can be "hardcoded" into an interpretation procedure).

Instead of starting with a single picture, we first start with the problem of choosing which picture to get. This makes it possible to solve more practical problems. For example, if a machine is emitting a high-pitched noise, the intelligent craftsman has to decide where to look to confirm or reject possible causes for the noise. In this case, a high-pitched noise might be caused by a vibrating part, or a dull tool, cutting fluid passing through a nozzle, or the machine hydraulics. Where does the vision system look; which picture should it get? This is the reason why we have adapted the metaphor of prehension to describe computer vision because the problem starts with choosing how to visually "grasp" a situation, so that a quick and effective interpretation can be made (see Chapter 6).

By controlling the image selection and flow, it is possible to begin optimizing the global understanding and local accuracy tradeoff. Pictures can be chosen either close up or at a distance. Strategies can be employed to reduce vibration or to maximize ease of movement. For example, a camera mounted on the end of a robot could be stablized by bracing the robot arm on a nearby table. These optimization strategies must now be captured as a general control system for image acquisition.

Figure 5.3 shows a control system that is suitable for many vision

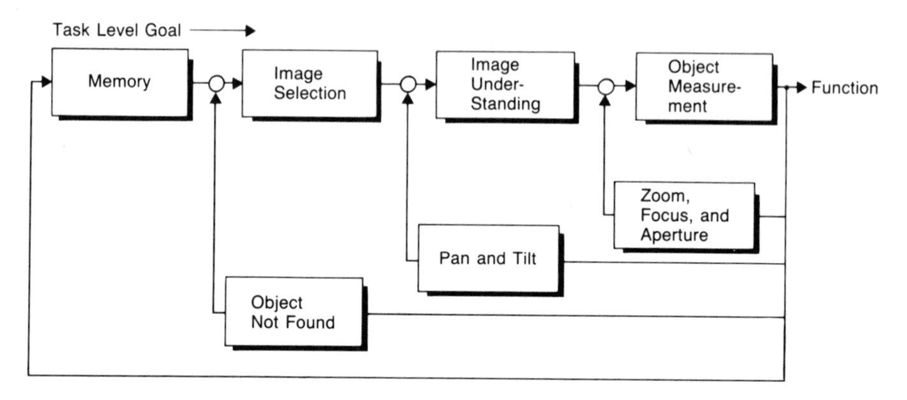

Figure 5.3 Vision prehension and control.

problems. First, the knowledge system provides a goal, which is translated into a visual strategy. This vision strategy is a pre-programmed way to acquire a series of images. The images are interpreted in turn, and if the original goal cannot be satisfied, new images are acquired. If the images within the programmed strategy do not bring results, a negative result is returned to the knowledge system, which responds with a new strategy.Once the goal is roughly satisfied, further adjustments (e.g., focus, aperture, and zoom) are made, and new pictures are gathered. Finally, the goal itself can be obtained by carefully analyzing the goal image.

Note that Figure 5.3 provides a setting for the material in this chapter. Moving across the diagram we cover visual strategies in Section 5.2.1 for solving problems posed by the intelligent machine tool, image understanding in Section 5.2.2, and object measurement in Section 5.2.3. Finally, we also cover the feedback elements on the return side of the control loop in Section 5.2.4.

5.2.1 Selecting Images from a Scene

Selecting the most interesting sequence of images in a scene is the same problem that a movie director faces. Sometimes it is most appropriate to get a closeup of an actor or actress and to dwell on the facial features. Other times it is necessary to track the hero swinging on a rope across the ballroom. In this case, the director's objective is to convey action rather than to dwell on details. Choosing the best sequence for a scene is difficult for both the movie director and the manufacturing eye.

The process of choosing an image sequence is a solution to a vision problem or desire (see Figure 5.3). This desire is a subgoal that must be solved to attain the final system's goal. However, because the vision problem has so many possible solutions their number must be limited by using vision strategies that have already proven to be successful. Figure 5.4 shows a tree of strategies that are task specific at the top and become more general purpose as the tree branches down. The tree's x-axis shows, on the left, the strategies that are useful for achieving a global understanding of a scene and, on the right, the strategies that are more useful for achieving accurate measurements.

On the global side of the taxonomy tree are strategies that are designed to achieve images that capture complete two-dimensional images of an object that are later refined into three-dimensional perceptions. To accomplish this, moving objects must be tracked to keep them in view and to move away from objects that are too close to be completely observed. Image acquisition is not capricious in humans, and it should not be without motive in the intelligent machine tool.

As the tree progresses downward, the visual strategies concentrate on collecting images that can be used to build three-dimensional representations of objects. Apparent motion is a strategy to visualize a three-dimensional representation of an object from moving two-dimensional cues (see "apparent motion" in Figure 5.4). Ullman (1979) has demonstrated that humans perceive a three-dimensional object when they watch a series of frames, each of which is a projection of a set of dots on two concentric, counterrotating cylinders. Only the dots are visible in each frame as their positions change from frame to frame. To see this effect, a viewer does not have to do any work. However, not every object presents its information in this way. Most objects just sit there on the table, and the viewer must take the initiative. For example, the viewer can move around the object to collect different views and then incrementally piece the views together to form a three-dimensional representation. Alternatively, the viewer can pick up and examine the object. In this case, the resulting shape of the manipulator's grasp is a very powerful strategy for determining three-dimensional shape. The shape of the gripper can include features from both the manufacturing hand (e.g., finger positions) as well as the manufacturing eye (e.g., the amount the object overhangs from the gripper, the correlation between the grasp point, and the manipulator feedback).

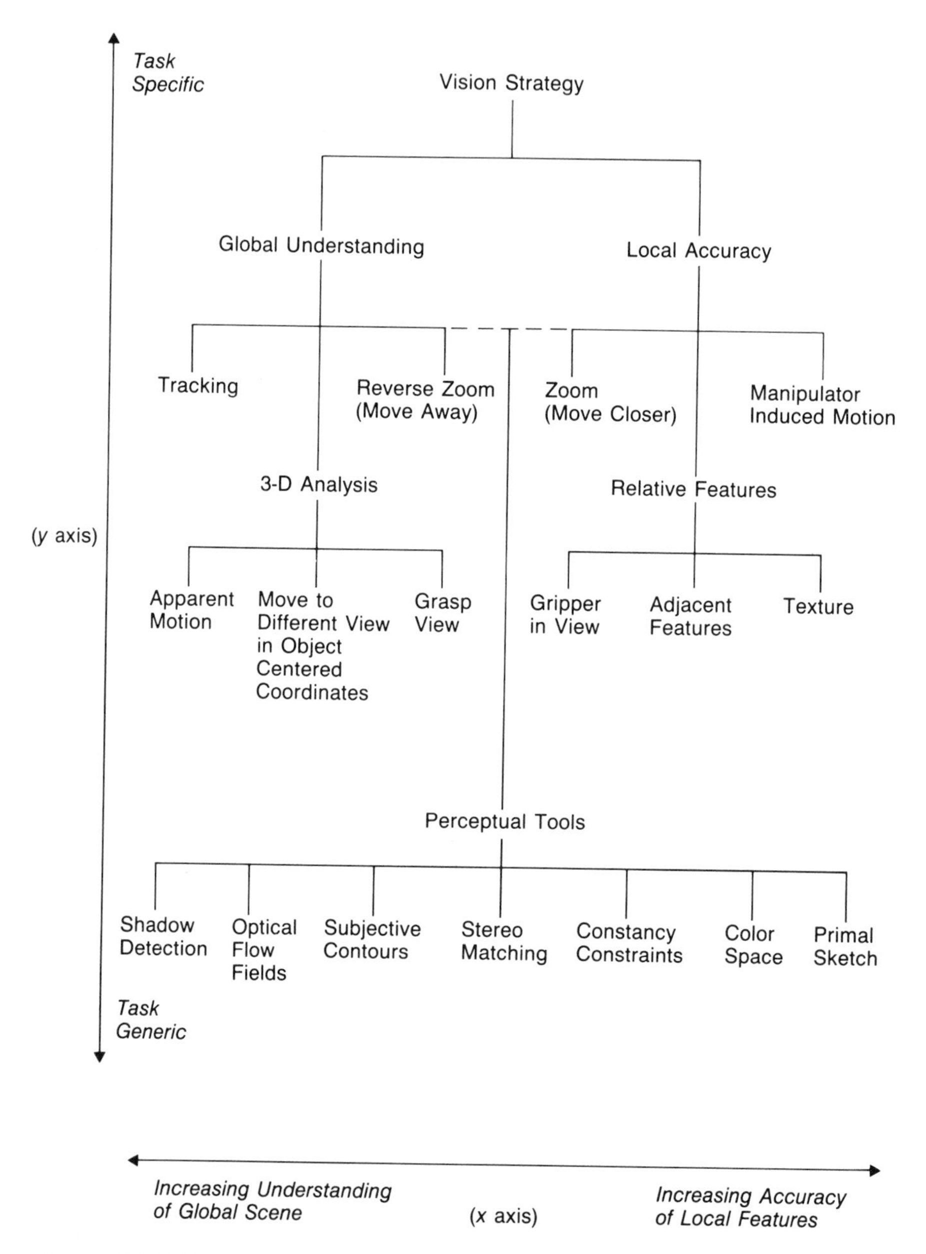

Figure 5.4 Vision strategies available for given tasks.

In the middle of the taxonomy tree are a number of additional vision strategies that are ordered roughly for their suitability in a range of tasks from global understanding to local measurement. For example, shadows are one of the few visual cues that give warning of objects outside the immediate line of sight. On the other extreme, visual systems are filled with simple constraints that make assumptions about the visual world. For example, it is assumed that people remain the same height as they walk off into the distance even though the person's image size will change drastically (see Section 5.2.3 for elaboration of visual measurement).

This tree is not complete, but it shows a range of representative strategies that span a broad spectrum of vision applications. In addition, a tree of strategies such as this could be used and amended by an expert system that couples task types to explicit vision strategies. This same approach could also be taken in the grasping strategies of Chapter 6.

5.2.2 Image Understanding

A single image is not a good place to attempt general understanding because a single image is not a single image at all. It is, instead, a complex composite of many hidden images (Barrow and Tenenbaum 1978). It is possible to ignore the composite nature of images and develop algorithms that work for many applications; Ballard and Brown (1982) offer a good review. However, each algorithm must make very strong assumptions. The problem is that identical scenes can look completely different from one image to another. For example, a shadow can completely divert most edge finders, and texture can complete confound region analysis. And yet, the basic content of the images may not change.

One approach to this problem is to design a system that washes out all the intrinsic images but one. For example, active lighting brings out the orientation image, while almost completely dominating the others. A time-of-flight laser scanner brings out the depth image, while again dominating the other intrinsic images. This approach is very useful in a fixed manufacturing application, but it does not have the flexibility to easily solve several different kinds of vision applications in the same setup.

From the source image in Figure 5.5, five different intrinsic images were extracted, as shown in Figure 5.6. The decoupled images are much easier to analyze than the composites, and the analysis can be accomplished with the same algorithms that are used traditionally in com-

Figure 5.5 The source image.

puter vision. The purpose of this analysis is to extract natural parameters that make up a scene. Figure 5.7 lists some natural parameters and their relationship to intrinsic images. This idea of natural parameters is borrowed from Marr (1982), but we restrict them to features that are strictly related to physical object attributes, so as not to confuse them with the intrinsic images.

Any particular source image may have some particularly good or some particularly bad intrinsic images. Therefore, it is useful to be able to easily characterize the quality of an intrinsic image so that only the strong sources of information are used for any given application. If the algorithm for extracting a particular intrinsic image is good, the image should be empty when there is no information about that domain. On the other hand, the image should be very rich when information is in abundance. Unfortunately it is difficult to perfectly decouple an image into its parts, and thus, single-image algorithms for extracting intrinsic images are probably only part of the required visual process.

Barrow and Tenenbaum recognized the difficulty of the vision problems associated with extracting intrinsic images, so they approached the solution with heuristics and consistency constraints. For example, they observed that object edges are a unique opportunity to measure

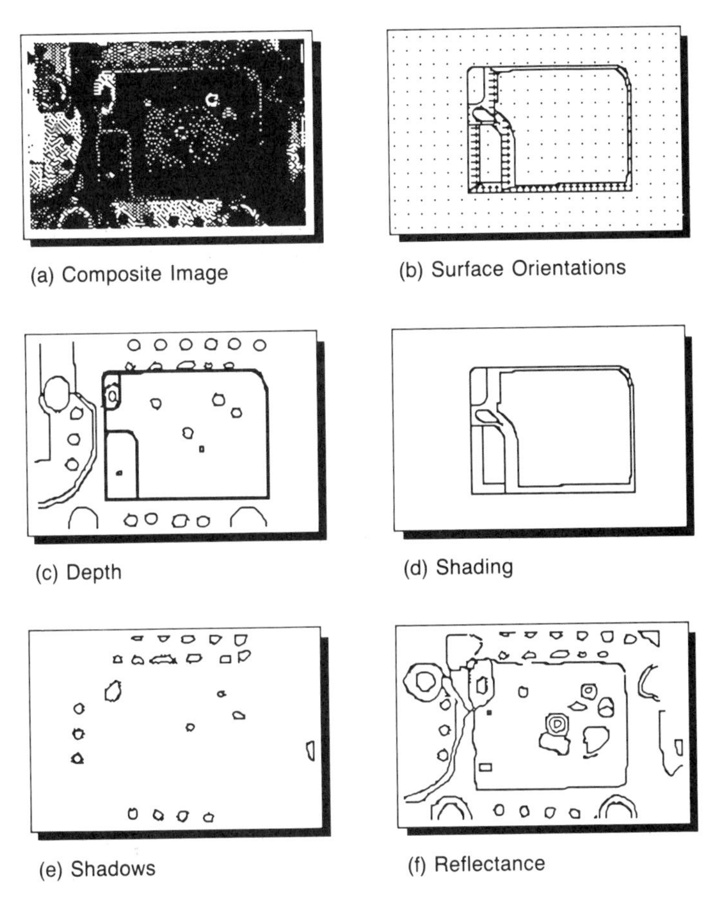

Figure 5.6 Intrinsic images in the machine tool world.

surface orientation; the surface is perpendicular to the camera at the boundary point. Also, they made observations about region and edge qualities to determine whether or not a region should be included in an intrinsic image. Whole arrays of such evaluations must also be checked for consistency. In other words, it would not be consistent to determine that object A occludes object B and that object A is at a greater depth than object B.

To make intrinsic image extraction more feasible, we propose in Section 5.2.4 that a flow of images under robotic control eliminates many of the difficult vision problems. In any event, once intrinsic images are

PARAMETERS	INTRINSIC IMAGES
Light Source	Illumination, Shadows
Object Motion	Optical Flow
Object Coloration	Illumination, Reflectance, Red, Green, Blue
Surface Depth	Stereopsis, Motion, Occlusion, Shadows
Surface Orientation	Perspective Cues, Shading, Texture Gradients
Surrounding Objects	Shadows

Figure 5.7 Natural parameters

gathered, they must be used to determine the natural parameters in Figure 5.7. For example, some of the natural parameters can be determined by making the following kinds of inferences:

- The existence and direction of a light source can be determined from shadows.
- Approximate depths can be determined from optical flow.
- Surface orientation can be approximated from shading.

Once the natural parameters for an image have been established, they can be further related to extract higher order properties: geometry, topology, and object labels. These, in turn, can be used to answer questions that relate to a given task (i.e., task parameters). That is, the task-oriented questions could be looking for answers about position, object clearance, likely danger, likely weight, surface quality, and many other factors.

5.2.3 Measuring Objects in an Image

Once objects have been identified in a scene, they finally can be measured (see Figure 5.3). We are almost never satisfied with knowing how many pixels represent an object in an image, but rather we need to know the correlation between pixels and the external dimensions.

A yardstick or ruler captures an everyday notion of measurement. A measurement is a judgment concerning an object's size relative to other objects, and in the case of a ruler, it is a standardized object. Vision relies heavily on this notion of relative size to make measurements. The famous Ames distorted room (see Ittelson and Kilpatrick 1951) shows us that the human vision system uses some of the same mechanism, even when it is not applicable.

> The floor of this room slopes upward to the right of the viewer, the rear wall recedes from right to left, the windows in the rear wall are different sizes and trapezoidal in shape. When the room is viewed from one vantage point, however, it looks like an ordinary room: the floor appears level, the rear wall is at right angles to the line of sight and the windows are rectangular and the same size. (Wittreich 1959)

When familiar objects such as people are placed in the Ames room, they appear to be greatly distorted in size. So much so that a single person walking in the room actually appears to grow and shrink according to the position in the room. However, this flaw in human perception is caused by the same device that we have learned to rely on from experience. The room is assumed to be rectangular, and all size estimates are made relative to the presumed room size.

Estimates of size by relative measurements can lead to false conclusions, but for the most part it is a very accurate way to judge size. Several of the strategies for taking relative measurements in an image are easy to automate (see right branch of Figure 5.4).

Texture is one valuable way to evaluate measurement. Briefly, this can be understood by imagining a grassy park and concentrating on the grass as it goes off into the horizon. Close up, the grass blades can be seen clearly, but the individual elements completely disappear as the field goes into the distance. On a computer, the size of the texture elements can be computed across a scene. Lieberman showed that approximate depth measurements can be calculated in outdoor scenes by calculating statistical properties of texture in a grass field such as this (Bajcsy and Lieberman 1976). It is assumed that the texture elements would be the same size if they were all viewed under the same conditions. Thus, by looking at the changes in relative size across the image, depth judgments can be made.

It is not necessary to have as many features as are usually found in textured scenes. For example, the distance between two holes in a machined part can be measured. The advantage of this relative measurement is that it can be done much more accurately than calculating the global positions, relative to a machine tool, and then subtracting them. The reason is simply that each device has its own inaccuracies. The vision system is only as accurate as the entire visual task allows.

In the case of vision, the key parameters are: the accuracy of the calibration, the number of pixels in a scanline, the field of view, and the precision with which edges in an image can be marked. If the camera is being moved by a robot or machine tool, the physical manipulator has its own inaccuracies caused by backlash and other mechanical imperfections. This means that global measurements, within the manipulator's coordinate system, compound the errors caused by the inaccuracies of both the vision and the manipulator. This is why the relative measurements within the visual field are more accurate; they do not depend on any of the manipulation parameters. This idea of isolating systematic errors has been carried out in a more general setting by Brooks (1982).

If an image contains no convenient adjacent features from which to make measurement judgments, they can always be added to the scene. It is part of human folklore that particular body parts can be reliably used to measure objects. For example, the last joint in a thumb is said to be one inch, the human span of both arms is said to be equal to a person's height, and a foot is said to be a foot. Of course, there is some statistical basis for these rules of thumb; most people's last joint of the thumb is indeed close to an inch. This same idea can be used in robotic-aided measurement. Dimensional marks can be put on objects so that they are readily visible in an image. But, this is not necessary if a gripper of known dimensions can be included in an image and used as a robotic ruler.

Relating robotic dimensions to visual dimensions is a good way to encourage research that makes the different subsystems of an intelligent machine tool work together as a team. At present, most robotic subsystems work in batch mode: first, one subsystem (e. g., vision) works and then stops, and next, a second subsystem (e. g., move a robot) works and then stops. In these systems, almost all the synergistic effects of multiple subsystems are lost.

5.2.4 Adjusting An Image

The purpose of feedback in a control loop (Figure 5.3) is to find optimal settings for control values. In the case of a robotic camera, these settings include: camera position, aperture, focus, and zoom.

The position of the camera could be in one of a number of coordinate systems: camera-centered, polar, camera-centered Cartesian, object-centered polar, or object-centered Cartesian. The choice of the best coordinate system usually depends on the task. For example, if vision is being used as a feedback element in a robotic manipulation task, an object-centered coordinate system is best because the robot is going to make the corrections by moving the object. If the vision system is designed to calculate stereo disparity, a vision-centered coordinate system is best because the calculation is based on the translational difference of a single object in two images.

By adjusting the camera parameters, different images can be acquired, thus implementing the acquisition strategies discussed in Section 5.2.1. However, a strong assumption is being made in these strategies. The image understanding element in Figure 5.3 must be able to locate the relevant object in the image, assuming that it is there to be located.

We propose in this section that the camera adjustments themselves can often help in the image understanding process. This will involve several demonstrations that the intrinsic images of Section 5.2.2 can be computed almost completely by robotic control. That is, the bulk of the computational effort is spent acquiring the next image, rather than running special-purpose algorithms. To accomplish this, we offer a series of algorithms in the following form:

1. Take a picture.
2. Alter a camera parameter.
3. Take a picture.
4. Subtract the two pictures.

This still presumes that the image understanding element does some work, but that the resulting differenced images have partially unraveled the underlying intrinsic images.

Optical Flow: In Figure 5.8a, we show that by (1) taking a picture, (2) panning the camera, (3) taking a second picture, and (4) subtracting the two pictures element by element, it is possible to produce a rough optical flow field. This can be computed by generating the vector between the left extreme of the gray field and the left extreme of the black field. The right extreme vector can also be computed and compared with the magnitude and direction of left extreme vector. If the vectors are the same, this confirms a translational move; if the vectors are different, the object

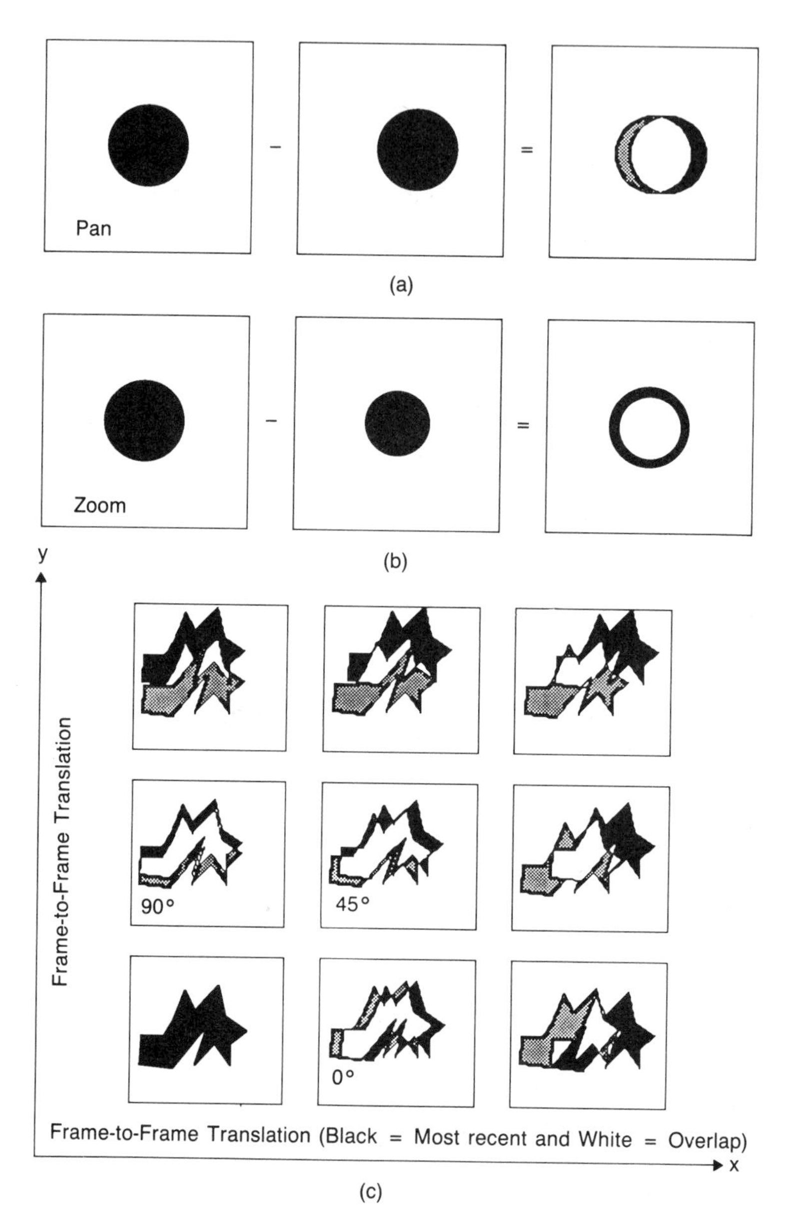

Figure 5.8 Using adjustments to expose intrinsic boundaries.

could have undergone a three-dimensional rotation or the object is moving along the line of sight.

Object Boundaries: There are a number of strategies for computing object boundaries with robotic control. For example, Figure 5.8b illustrates that, when an image is scaled by zooming and subtracted from the original, object boundaries can be detected. This simple approach does introduce additional problems because the objects move in an image, according to their depth, as well as change size. However, if the object is being tracked (i.e., centered and in view), zooming is a good way to detect edges. Another approach that uses the same strategy of edge detection is to focus the camera, take an image, defocus the camera, take another image, and subtract the two images. This approach leaves the images aligned even when the objects are not centered.

To confirm the usefulness of these ideas, we conducted an experiment where images of a machine tool, about to begin cutting, were taken every 35 milliseconds (see Figure 5.9; Figure 5.8e is discussed after this concrete example). These images begin in the upper left corner of Figure 5.9 and proceed left to right top to bottom. In the top center of each image, there is a stationary cutting tool, which is turning at 2500 revolutions/minute. In the bottom left corner of each image, a prismatic part is visible, with the cutting tool posed to cut the top face of the part. As the images proceed, the prismatic part moves to the right into the cutting tool, at which time chips start to fly (image 4). In turn, Figure 5.10 shows each image subtracted from the previous image. In other words, the first image in Figure 5.10 was computed by subtracting the first original image from the second original image (in Figure 5.9). The resulting differences are normalized around a middle-level gray value, so a middle-level gray value means that there is no change in the image values, white means there are brighter values in the newer image, and black means that there are darker values in the newer image. The biggest immediate advantage is that most of the irrelevant background information completely disappears.

Object boundaries can be clearly discerned in these differenced images. Flying chips leave a characteristic wake in the images seen in the upper left corners of the last four differenced images. A simple algorithm, based on thresholding and counting in this subarea, can detect the chips at this point. In addition to the chips, the leading edge of the block shows a clear edge, indicating that it is moving to the right. The block

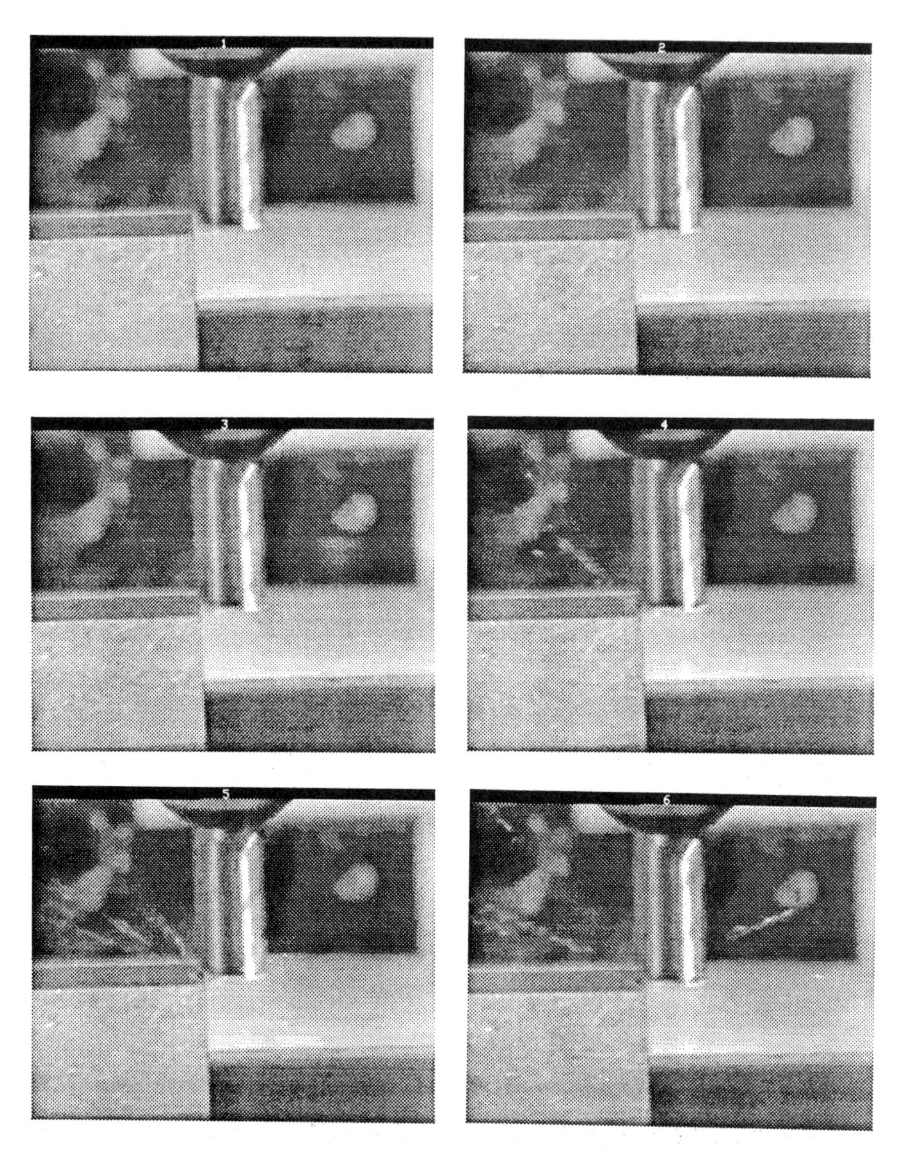

Figure 5.9 Cutting sequence.

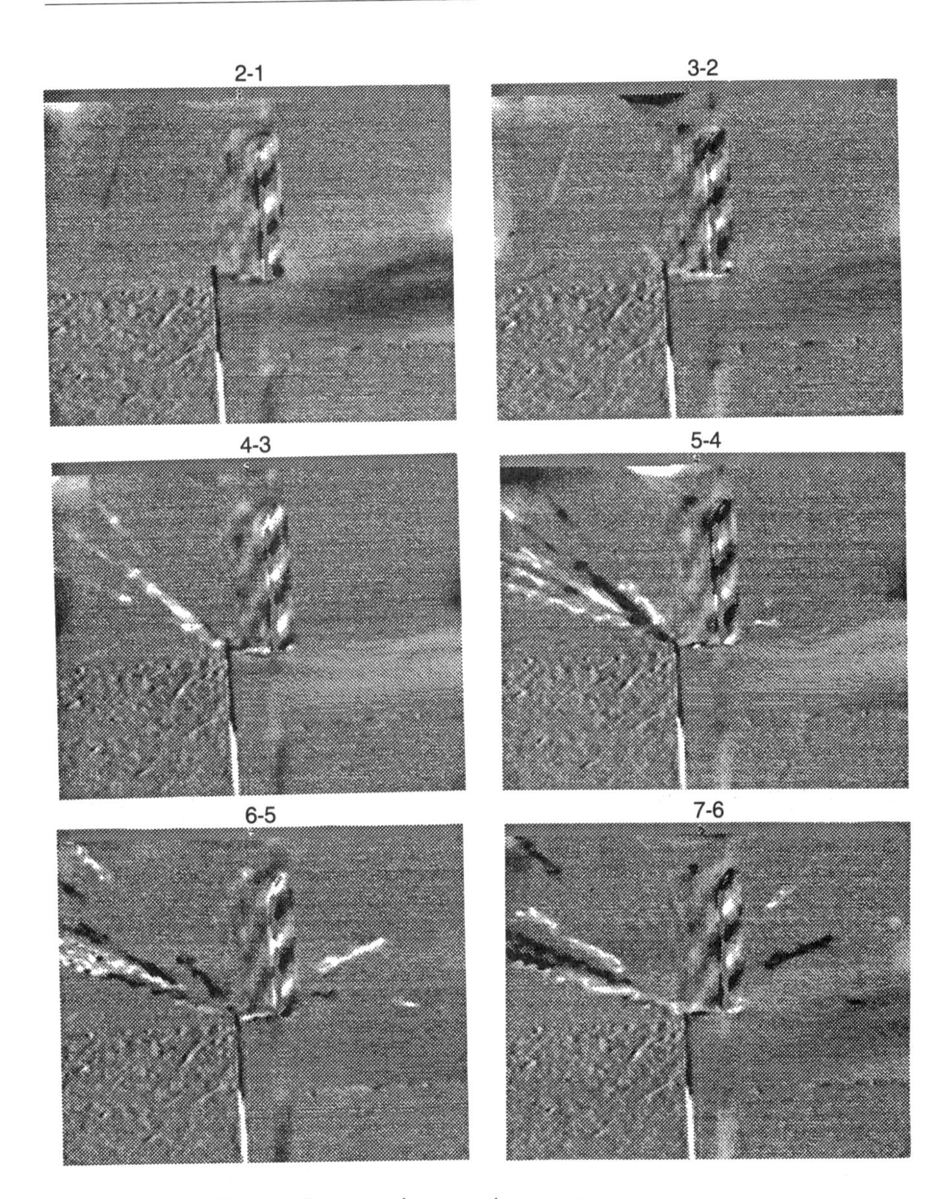

Figure 5.10 Differenced images from cutting sequence.

CONCLUSION	OBSERVATION
1. The machine tool is turned on and a cutting tool is loaded.	• The cutting tool has a barber's pole striped pattern.
2. The prismatic part is moving to the right.	• Leading edge clearly indicated, along with direction and distance. • Texture on part is also moving equally.
3. Cutting begins in frame 4.	• Chip wakes can be easily found. • The relative direction of chip wakes also gives information about part material properties, part direction, and possible chip clogging problems.
4. The tool did not break on part entry and nothing else is abnormal.	• All the activity seen was expected.

Figure 5.11 Task parameters extracted from Figure 5.10.

also exhibits its surface texture, which is moving the same distance to the right as is the leading edge. This simultaneous movement indicates that this subarea of the image is all part of the same rigid object. Finally, the cutting tool shows a "barber's pole" pattern because the tool turns 1.389 times every camera image. If the frame rate and the revolutions of the tool were evenly divisible, the picture would show no change, like a stroboscopic timing light used to tune an automobile. From these images, four practical results can be easily extracted and these are shown and explained in Figure 5.11.

This machine tool environment is an excellent laboratory for work in vision under robotic control. The objects in the machine tool world are confined and well understood. Movement between images is typical, but easy to control. And, the resulting application is extremely useful in the process of mimicing a craftsman. We discuss this application further in Figure 7.8 and Table 7.5, which highlight that a craftsman is especially vigilant as a tool first cuts into a part.

Region Approximations: In many visual applications, the image's details are extraneous, and most of the computational effort is spent removing all but the principal shapes. Elaborate schemes have been developed to ignore everything in an image but the selected pattern. For example, Ballard developed programs that analyzed lung x rays, looking for circular splotches that were indications of black lung disease. To achieve this, he built special algorithms, or filters, to scan an image for circular structures (discussed in Ballard and Brown 1982). In many cases, it may even be easier to search for structures by manipulating the image device, instead of computationally looking in detail at each image. For example, following a road may be a matter of maintaining image values across many images. If the image pattern ever changes, a correction has to be made. For example, a long rectangular block that extends outside of an image can be traced to its end by moving a camera in such a way that the image values are maintained until the far edge of the block is encountered (Potter 1977).

Figure 5.8c shows a complex shape under a range of translations. One idea is that small features can be cut off with small moves, and large features can be cut off with large moves. In other words, controlled camera moves can be used as a directional filter. In this case, the largest intersecting areas between images correspond with the inherent direction of the shape.

Surface Orientation: It has been demonstrated (Bolles, Kremers, and Cain 1981) that surface orientation is relatively easy to determine from active lighting. This method is not necessarily ruled out in the camera adjustment paradigm. For example, the system could take a picture in normal light, project a dominant grid pattern, take a picture, and subtract the two images. In this case, only the distorted grid pattern would remain in the resulting image. In this image, the angle of the grid lines is directly proportional to the surface curvature. In other words, a straight flat line indicates a surface that is parallel to the image plane, whereas a discontinuity in the grid represents a point on the surface that is perpendicular to the image plane. Surface orientation can also be determined from perspective cues. This is simple in a world of rectilinear rigid objects, which is mostly true in the machine tool's world. For example, a square looks like a trapezoid from virtually all nonorthogonal perspectives (Kanade and Kender 1980).

Surface Depth: The depth of the surfaces is one of the most important intrinsic features because many other features (e.g., surface orientation) can be calculated from it. This intrinsic image can be deduced from several sources. For example, depth can be calculated from both stereo disparity and zooming disparity. An object's size is inversely proportional to its distance from the image plane. Therefore, if after zooming an object's size changes only slightly, it is relatively far away with respect to the camera's usual operational range.

Shadows: Shadows are both a curse and a blessing. As was noted earlier, shadows can completely confuse most traditional computer vision algorithms. However, some of the simplest animals—surf clams—have shadow detectors (Kennedy 1963). It is true that these animals do not need to look around and admire the view, but they can take advantage of detecting potential predators before they strike. In the paradigm of robotic control, shadows take on the appearance of a flat sheet laid out on the objects. If the camera moves to different views, the shadow does not reveal any further three-dimensional structure. Beyond this, a shadow does not alter the texture in a surface, whether or not it is in the shadow (Witkin 1982). This is a good clue because most object boundaries mark a change in texture as well.

There are many ways to take advantage of changing images, and these can be used to support the image understanding element. However, there are so many ways to process images that it is important to keep them within the framework of basic strategies. Sometimes it is possible to try two or more strategies at once and then check the results for consistency. At other times, one strategy precludes the others because the strategy changes the environment, which invalidates competing strategies (e.g., panning a camera).

Extracting intrinsic images under robotic control is a research topic that is less than suitable for a typical industrial vision installation. However, some of the ideas can be used immediately, for example, choosing the "right" sequence of images to get three-dimensional inspection information is used in the first industrial case study (see Section 5.4).

5.3 VISION DESIGN CHECKLIST

It can be difficult to choose where vision is practical in the manufacturing environment. This section presents a number of possible vision applications (see Figure 5.12), and a checklist to determine whether vision is an appropriate solution (see Figure 5.13). The problems in this list were chosen to be representative of the manufacturing eye's use spectrum. The purpose of this exercise is not to build new and exciting vision systems that push the technology to the limit, but rather to identify applications that are readily solved with the state-of-the-art methods. The manufacturing environment demands solutions that work virtually 100% of the time.

Problem 1: A batch of cylindrical billets are neatly stored lengthwise in wooden pallets. These billets are metallic and may have been left out in the weather before they are used to make parts. The problem is to use vision to locate the extreme edges, orientation, and center of each billet.

Problem 2: Billets just cooled from being red-hot have been shaped into rough turbine blades, such that there is a lengthwise twist in the blade and each cross-section of the blade is convex throughout the entire length. The problem is to calculate the area of the cross-sections at key locations along the billet.

Problem 3: Two-inch-long ceramic dies for aircraft engine turbine blades can easily form cracks during the casting process. The problem is to find bad dies by finding hairline cracks in the ceramic blank.

Problem 4: Milling cutters become dull and ineffective during use. The problem is to determine whether a tool has enough life to cut a new part with a known machining time.

Problem 5: A robot work area is known to be very dangerous. A recent accident involved a maintenance person who took a shortcut through a manufacturing area at the same time that a robot was making a sudden move. The problem is to watch for intruders within this limited area.

Figure 5.12 Vision problem set.

Check 1: Is the object in question readily identifiable? Determine whether the object of interest has distinct visual features. For example, a person may not even be able to see the hairline cracks in Problem 3 of Figure 5.12. If this is the case, the object may require special treatment (Check 12).

In many cases, the human visual system, combined with our lifetime of experience, is so powerful that it is easy for someone who has not worked with computer vision to think that some problems are easy and others are hard. Quite often, this intuition is misleading. For example, people have no trouble at all going up to a bin of parts, visually picking one out, and then grabbing it. This problem has earned the name "the bin picking problem" because it has proven to be so difficult in application. Some success at finding parts in a bin has been achieved when all the parts are identical (Kelley 1984), but this system does not work well enough to leave unattended in the factory.

Check 2: Is the background pattern easy to control? The bin picking problem has been difficult because the part being searched for is always

1. Is the object in question readily identifiable?
2. Is the background pattern easy to control?
3. Is the object's coloration known and regular?
4. Is the lighting easy to control?
5. Can multiple views of the object(s) be collected?
6. Can the camera be manipulated around the object(s)?
7. How difficult is it to maintain the camera's calibration?
8. Is the required accuracy of the vision results within the capability of the camera system?
9. Are there special optical effects that must be considered in order to get good images?
10. What is the best spectrum of light to get the clearest images?
11. Can the object be readily treated with a foreign substance to improve the image?
12. How difficult is it to interpret the visual information once it is retrieved?
13. Are there reasonable constraints on the task requirements?
14. Are there less expensive sensors that can do the job better?

Figure 5.13 Reference guide to vision checklist.

in a jumbled background scene. In many applications, the background scene is controllable, for example, a conveyer belt carrying mail at a U.S. Post Office. To simplify the situation, the mail can be partially sorted, so that only one piece goes by on the conveyer at a time. This makes an image that is much easier to process because now only the conveyer and a particular piece of mail are in the image at one time.

Unfortunately, controlling the background scene can be quite an expensive proposition. For example, in Problem 1 in Figure 5.12, parts are stored on wooden pallets. In fact, a factory could easily have thousands of these pallets, and to make them inexpensive, they are constructed from wood with no special preparation. The result is broken wood pieces, large knots, and other anomalies that result in "busy" images even before the billets are introduced. In this case, the company considered painting all the pallets black and using higher quality construction materials to make the billets stand out in the image. As it turned out, there were so many pallets that had to be changed, it would have cost about 1/2 million dollars just to solve the vision problem. Needless to say, the economics were the overriding consideration in determining not to use vision.

Check 3: Is the object's coloration known and regular? What made Problem 1 of Figure 5.12 even more challenging was the fact that the company did not have enough storage space in the building for the pallets filled with new billets. After a short time out in the weather, the billets began to discolor and even rust. Then, instead of having to locate billets of one color against a simple background, the billets themselves were heavily textured. Working with textured objects is another difficult visual problem. Again, significant progress is being made in this area (Gool, Dewaele, and Oosterlinck 1985), but current results are not reliable enough to always yield a correct solution, and they are usually computationally prohibitive for real-time processing.

If the coloration is known and regular, the background can almost always be prepared well enough to get good images. This is a dominate factor in determining whether or not vision is practical for an application.

Check 4: Is the lighting easy to control? Vision cannot be used in the dark. Light is a critical part of using computer vision, and the better the control is over the light situation, the easier the computer vision. Active lighting is one way to minimize problems. The idea is to project a known pattern of very bright light on an object, so that the reflection can be seen above the ambient light levels. Once the light is reflected off the object, it is distorted according to the shape and surface properties of the object. A simple light stripe is a favorite choice (Shirai and Suwa 1971) because the resulting shape of the stripe can be used to directly calculate the shape of an object plane; it captures the depth of the front surface in the plane of the projected stripe.

Other systems use a scanning dot or a complete grid pattern to achieve the same effect. All these active lighting approaches are done by lighting the object from the front. Alternatively, an object can be lit from the back, rendering a silhouette. Problem 2 (Figure 5.12) can use backlighting because only one part is being processed at a time, and the cross-sections in the plane are convex. The convexity property of a cross-section means that "hollows" in the part do not exist in at least one viewing plane, and therefore no shape information is being lost that could be collected with a front lighting scheme. However, a simple backlighting scheme can only reveal two dimensions, which is not enough to compute cross-sectional area unless multiple views are taken (see Check 5 and Goldstein, Wright, and Bourne 1985).

Check 5: Can multiple views of the object(s) be collected? Because the processing of a binary picture, a silhouette, is so attractive, we decided to mount the object on a turntable to solve Problem 2 (Figure 5.12) in the industrial setting. Once mounted, the turntable rotated the object around, in front of the camera. Each view of the object readily gave a set of two dimensions, and by the time a full circle had been made, it was easy to reconstruct the vertices of a cross-section and then calculate its area.

In addition to Problem 2, there are many other reasons why using multiple views of an object might be the best way to solve a vision task. For example, the automobile manufacturers would like to inspect a car chassis to find out whether there are the correct number of holes in the right places. A bad chassis that makes it to the assembly line can cause assembly problems. Because a car is so big, it is best to have multiple cameras look at parts of the whole chassis, rather than to try to extract enough resolution from one image of the car.

Finally, stereo vision is a good way to extract depth information from images. If two pictures show the same object from different viewpoints, the distance of the object can be calculated by triangulation. This is proving to be very useful for the navigation of autonomous vehicles, but this should not be surprising. People have been using triangulation for thousands of years to calculate the distance to the stars.

Check 6: Can the camera be manipulated around the object(s)? If the mountain does not move, sometimes the camera can be moved instead. Moving the camera can cause a number of difficult problems that have few solutions that are suitable for industrial applications. The biggest problem is maintaining the camera calibration, the mathematical model that relates real-world dimensions to image dimensions. Unless there is an approach for automatically recalibrating the camera on the fly, it is wise to rule out moving cameras.

Check 7: How difficult is it to maintain the camera's calibration? Camera calibration is one of the hidden problems that makes vision hard. However, calibration becomes much easier when there is an independent source of depth information. This can be in the form of a second camera or some kind of proximity sensor. For example, one application that uses a proximity sensor is a robot that is inspecting the paint finish on a car. The proximity information can be fed directly into the robot control loop, maintaining a fixed distance from the car's surface. In this scenario, the vision system does not have to worry about sudden distance changes and can concentrate on viewing the paint. Actually, this is an advanced system because robot manufacturers are rarely willing to support outside input into their feedback control.

It is not obvious when camera calibration is a problem. For example, in Problem 1 of Figure 5.12, the wooden pallets are stacked up into a pile. As one pallet is moved away, the camera calibration must change to get an accurate view of the next lower pallet in the stack. To maintain calibration, it is useful for the vision system to keep track of the pallet's position on the stack. In this way, the camera calibration parameters can be precalculated for each position. Alternatively, there can be known markings of known dimensions stamped on the pallets. Once one piece of a picture can be easily recognized and related to dimension, it is much easier to compute the dimensions of other objects based on this relative information.

Problem 2 must also maintain calibration. The billet is rotated in front of the camera and, unless the billet is rotated around its center, its distance to the camera is under constant flux. Thus, the relative image dimensions and object dimensions are also changing.

Check 8: Is the required accuracy of the vision results within the capability of the camera system? A vision system has a maximum accuracy when the camera is fixed relative to the object. This accuracy is simply the ratio of the number of pixels in an image to the area covered by the image. However, many factors can further reduce the system's accuracy. The calibration can become inaccurate, and the accuracy of the whole system is reduced as a result (see Check 7). In addition, the camera can be out of focus so that the edge of the object cannot be clearly distinguished; this in essence reduces the number of pixels. The optical (lens) system itself can introduce its own distortions.

There are ways to effectively increase the system's accuracy. For example, a zoom lens can zero in on an object feature. As long as the calibration problem can be overcome, zooming can drastically increase the critical ratio between pixels and area.

Check 9: Should special optical effects be considered in order to get good images? Vision only works well if the quality of the image is good. In Problem 2 of Figure 5.12, the billet was emitting a large amount of infrared light (heat). Since the sensitivity range of most charge coupled device cameras is shifted toward the red part of the spectrum, this rather unusual combination of events caused several image distortions: blooming (whole image columns become saturated) and wash-out (pixels that stop responding). To alleviate these problems, it became necessary to install one of the best infrared filters on the market to mask out the heat.

Problem 1 was also troublesome because the vision-guided pickup station was located next to a large garage door that was used by fork-lifts and other heavy equipment to access the factory floor. Whenever the door opened, sunlight could beam through the door, significantly changing the overall illumination level and casting shadows across the visual targets. Controlling the ambient light becomes part of controlling the image background discussed in Check 2.

Check 10: What is the best spectrum of light to get the clearest images? Removing infrared light from an image was a first step in isolating the part of the spectrum that contained the most information for the task. This idea can be extended to ease visual interpretation. For example, a filter can be used to block out all light but the blue; this can be used to either block out extraneous details (e.g., a fancy pattern in red) or just enhance the blue information so that the intensity levels in the image are more significant. Another application might be to inspect the color mix of paint or the ink on a can's label. Three images can be taken: one with a red band-pass filter, one with a blue band-pass filter, and one with a green band-pass filter. Then an algorithm can be applied to each image, which would add up the image intensities for each picture. The results would be three numbers representing totals for red, blue, and green. These cumulative results can then be matched to goal levels, and the paint or ink mix can be corrected.

This idea can be further extended by intentionally using light outside the visible spectrum. For example, x rays are sometimes the only way to look for defects inside an object. However, special precautions must be taken to protect people, and these often prevent the process from being cost effective.

Check 11: Can the object be readily treated with a foreign substance to improve the image? Quite a bit of folklore is attached to computer vision. For example, one day a research scientist at MIT walked into his lab and discovered half the objects in the room painted with a flat white paint. As it turned out, a graduate student had become very frustrated by the unpredictable effects in images caused by the lighting conditions in the room. Flat white paint reflects light according to a simple reflectance model; the intensity of an object point varies with the cosine of the angle between the surface normal and the line formed between the light source, coincident with viewer, and the object point. This certainly makes it easier to extract the shape of the objects in the image.

A strong argument can be made that this kind of image preparation is simply cheating; that painting every object white in an industrial setting is not practical or reasonable. On the other hand, treating objects in a special way makes inspection possible where before it might not have been possible at all. For example, dipping dies into a florescent particle bath may be the only way to image micro-cracks in the castings (see Problem 3 in Figure 5.12).

Check 12: How difficult is it to interpret the visual information once it is retrieved? Once a picture has been taken and the image is a good quality, it is processed. The output of low-level vision processes (e.g., edge enhancement) can be almost as complex as the image itself. In fact, unless the information can be brought to a level that humans readily understand, it is pointless to use vision.

Perhaps the most successful way to present visual information is in graphics. For example, a circuit board inspection station can paint red over defects on a circuit board, so that the red areas quickly attract attention and can be interpreted by a human operator (Thibadeau 1984). Visual information that is numeric, such as a cross-sectional area, is also at the right level of description for either human or automatic decision making. However, if the information were displayed as a list of 360 width dimensions, the results would probably never be used.

Check 13: Are there reasonable constraints on the task requirements? Problem 5 in Figure 5.12 was included as a red herring, just because it seems "high tech" enough to actually try. It is not a good vision problem, at least, not for industry because it is unbounded; there are no limits of what could be asked of the system. Even worse, it is the last line of defense in personnel safety, which means that if the system fails, somebody could be injured as a result. A conclusion of this section is to strongly suggest that the choice of technology be kept out of the hands of management and left with the engineer who understands its benefits and constraints.

Check 14: Are there less expensive sensors that can do the job better? The design tradeoff presented in Figure 5.2 between global understanding and local measurement offers insight into whether special-purpose sensors are better suited to a task than vision might be. Vision is more generally suited to global understanding tasks requirements, and it must be specifically engineered to achieve useful results in local measurement. However, other special-purpose sensors (e.g., proximity sensors) have

Table 5.1 A List of Typical Manufacturing Applications and Examples of Their Sensor Requirements in the Local-Global Spectrum

GENERAL APPLICATION	LOCAL-GLOBAL TASK REQUIREMENTS
1. Handling during die and investment casting	Local: Load and unload verification Local: Grip and ungrip verification
2. Handling for forging—changing part shape	Local: Load and unload verification Local: Grip and ungrip verification Local: Grip shape after grasp
3. Palletizing and simple parts assembly	Local: Load and unload verification Local: Grip and ungrip verification Global: Locate part Global: Locate vacant destination
4. Spot welding	Global: Identify spot weld location
5. Spray painting and gluing	Local: Identify surface distance Local: Identify suface orientation Local: Measure paint level Local: Measure paint flow
6. Arc welding	Local: Distance from surface Local: Orientation to surface Local: Arc-on verification Local: Filler and wire feed verification Global: Seam tracking
7. Deburring	Local: Force on grinder Global: Geometry edge tracking

just the opposite problem. Typically, they are only designed to measure one task parameter, which makes it possible to finely tune them to achieve the accuracy goals.

In a range of manufacturing applications, there are differing degrees of requirements that span the local-global spectrum (see Table 5.1). The ten manufacturing applications in Table 5.1 were presented as a survey to 40 research and industrial users. The goal was for the respondents to iden-

GENERAL APPLICATION	LOCAL-GLOBAL TASK REQUIREMENTS
8. Inspection of part features	Global: Feature presence
	Global: Approximate feature location (note: this is not an inspection for tight tolerances)
9. Machine cell operations	Local: Load and unload verification
	Local: Grip and ungrip verification
	Local: Machine ready for robot access verification
	Local: Machine integrity checks (e.g., measure cleanliness of hydraulic filter with differential pressure sensors)
	Global: General part acquisition activities
	Global: General inspection activities
10. Automatic assembly	Local: Hole orientation with respect to robot wrist
	Local: Hole orientation with respect to robot fingers
	Local: Grip and ungrip verification
	Global: Locate part
	Global: Locate hole to insert part

tify which of four sensor types were most likely to be important in these applications. These sensors included: (1) simple binary sensors, for example, to indicate a gripped part; (2) force/torque sensing at a robot wrist; (3) tactile sensing at the "fingers" of a gripper; and (4) computer vision. Notice that these sensors range from being extremely local devices to sensors that are intended for applications of global understanding.

The survey was constructed as a set of pairwise questions so that only simple choices had to be made between each step (e.g., comparing vision and force/torque sensors for an assembly application). Further details about this experimental survey (Wright and Englert 1984). The results of the survey are shown in Figure 5.14.

Summarizing, Figure 5.14 shows that for deburring, force/torque is most favored; for inspection, vision is most appreciated; for machine cell operations, the different sensors are more equally ranked; and for automatic assembly, tactile feedback from robotic "fingers" was viewed as increasingly important in addition to force/torque sensing. As these results show, the local and global task requirements in Table 5.1 were clearly given different weights. In general, whenever possible the local task requirements were given priority over the global task requirements.

5.4 INDUSTRIAL VISION EXAMPLES

Vision has become a practical technology in the 1980s. We review two applications that we have developed.

5.4.1 Three-dimensional Inspection of Forgings

In this section, we briefly review our own partial success in applying machine vision to the inspection of open-die forgings. Information from the vision measurements is used to update the hammer positions in the forge. This vision system has replaced a craftsman/overseer who used to use mechanical calipers to measure a few critical locations on the forging. We believe the vision system will eventually give better accuracy, more extensive quality control data by measuring more locations, and ensure that feedback information is available for the operation of the forge. This initial open-die preform geometry is critically connected to the subsequent closed-die step in order to ensure proper die-filling, minimum flash extrusion, minimum die wear, and the prevention of catastrophic die failure.

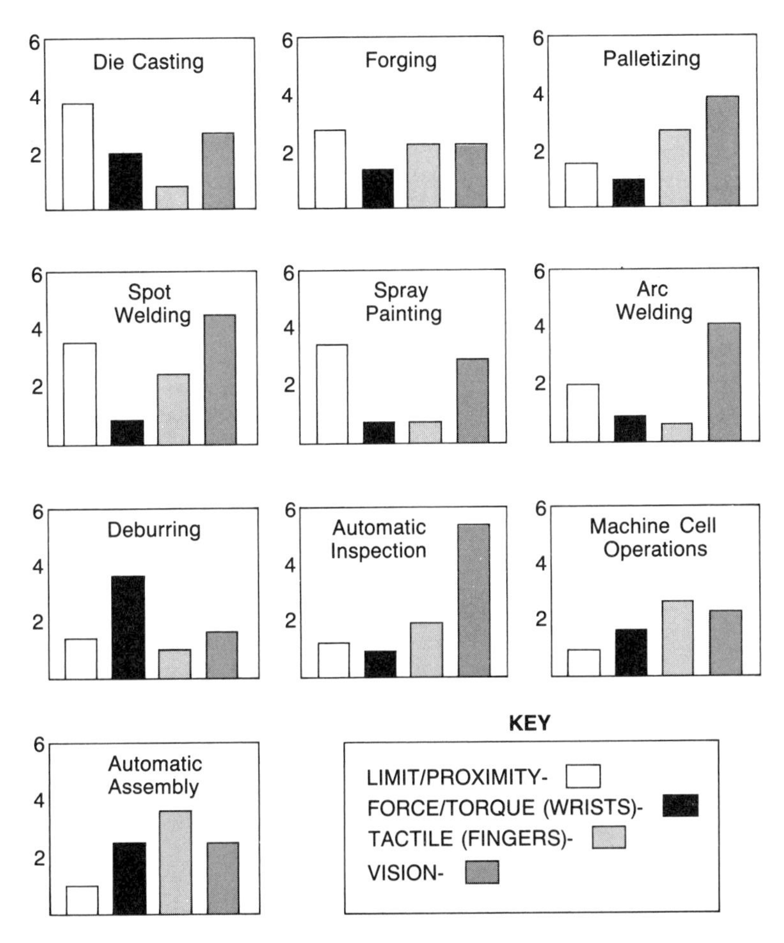

Figure 5.14 The results of a survey to discover what sensors are most suitable for given tasks. (Wright/Englert, " A Review of Sensor-Based Robotic Manipulation" published by ASME, PED Vol. 13, 1984. Reprinted with permission).

The programmer of the NC forge determines the preform dimensions based on his knowledge of the final component geometry. The following open-die preform dimensions are considered for this purpose:

1. **Cross-sectional Area:** The area serves as a measure of the amount of material present at cross-sectional positions. Insufficient material in one area of the open-die preform can lead to gaps and voids in the closed-die forging, whereas too much material can cause excessive die wear and excess flash material.

2. **Thickness and Width:** These are considered jointly because they determine the proportioning (i.e., thickness to width ratio) of the open-die forging. Both the amount of material and its proportions are important. Incorrect proportioning may cause excessive strains in the closed-die forging resulting in cracking. Die wear is also affected by this feature.
3. **Twist:** The twist refers to the angular difference between the orientations of cross-sections. Improperly oriented, open-die preform cross-sections increase closed-die wear and can cause closed-die failure.
4. **Locations of local thickening:** The open-die preform also exhibits areas of local thickening. Such areas of local thickening eventually become the lugs in the closed-die forged airfoil, shown for reference in Figure 5.15. The lugs are the local protrusions in the final

Figure 5.15 Photograph of laboratory setup showing vision system (rear left), backlighting screen (rear right), forging preform mounted on rotary table (center), and cameras on camera stand (front right). Open-die forging preform is being inspected; for reference, the final product after closed-die forging and machining is also shown (on table). The root, lugs, and airfoil are seen. (Reprinted with permission from The Robotics Institute, Carnegie Mellon University © 1983.)

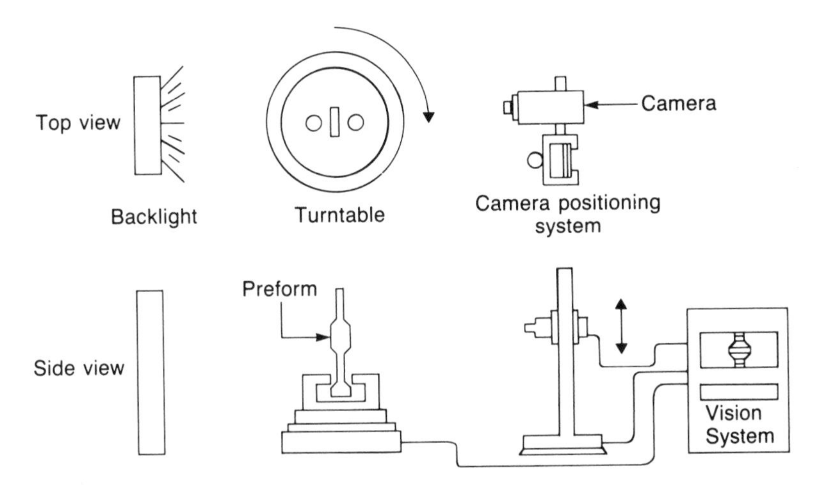

Figure 5.16 Hardware arrangement. (Goldstein, Wright, Bourne, " A Fast Algorithm For High Computer Vision" *Robotics and Computer Integrated Manufacturing*, Vol. 2, No. 2, 1985. Reprinted with permission from Pergamon Journals Ltd).

airfoil that are the sites for interblade stiffening struts in the final steam generator. After the closed-die forge, the lugs are machined to the specified dimensions. Thus, it is important that material intended to become a lug be positioned properly and have the correct cross-sectional features cited in items 1 through 3.

5. **Length:** Knowledge of the overall length of the preform is again important to minimize flash and die wear in the closed-die forge. The bottom of the preform is used as a reference for all other vertical dimensions, including the lugs.

The gauging station makes measurements in all of the preceding five areas. Items 1 through 3 are concerned with cross-sectional features and are measured by performing cross-sectional reconstructions. Items 4 and 5 relate to the vertical dimensions that were measured while analyzing profiles.

The gauging system is built around a commercially available, computer vision system. In the laboratory, the gauging system has been set up to control the position of the camera, orient the forging preform relative to the light source and camera, and to collect and process the data for the analysis of part shape. A backlighting screen is used to form a sharp silhouette of the forging, which is viewed by the camera. The plan and side view of the hardware arrangement are shown in Figure 5.16. A rotary turntable driven by a stepper motor is used to rotate the preform in front of the camera. Since the component must be inspected at different elevations, the camera is mounted on a vertically adjustable lead screw on a stand as seen in Figures 5.15 and 5.16. The orientations of the rotary table and the height of the camera are controlled from the vision system, and closed-loop position feedback is used on both axes to ensure accurate positioning. In an inspection routine, 360 views (one for each 1 degree of the object) are recorded and stored in the vision system.

The first algorithm only used one scan line, AB, from the camera's frame buffer to represent a slice, CD, of the object scene (see Figure 5.17).

When viewed from different angles, the object varied in width, while 360 cone sheets, A'OB', were collected. By superimposing several two-dimensional cones sheets, such as A'OB', it was possible to completely reconstruct the object's cross-section. This scheme begins by choosing two orthogonal views of the object. Then, an algorithm is used to predict which view of the object should be selected in order to optimally approach a "good" polygonal representation of the cross-section. This prediction scheme searches for three adjacent vertices, in the polygonal representation, that form the largest triangle. The object is then viewed in the same direction as the base of that triangle. Finally, the vertical dimension of gauging can be accomplished by "stacking" individual cross-sections from different camera elevations (see Figure 5.18) and then measuring the difference between the resulting cross-sections (e.g., the principal eliptical axes for two cross-sections can be used to measure twist).

Unfortunately, this algorithm is very limited because the resulting accuracy of the gauge is relatively low. Since the entire length of line A'B' was needed for analysis, the complete width of the forging had to always be in view. With the 235-pixel array, this constrained the gauge's resolution. For the system shown in Figure 5.17, the resolution was approximately 0.02 inches (0.5 mm) per pixel. Given a typical size of forging preform and the magnification, the resolution of object thickness and width measurement was ±0.06 inches (1.5 mm). Although the resolution was perfectly adequate for the forging application, the work could not be extended to flexible machining cells where tolerance ranges are 0.002–0.005 inches (0.05–0.1 mm).

In a second algorithm, the cone shown in Figure 5.17 has been represented in another way. The lines A'O and OB' are regarded as linear equations, rather than as sides of a cone sheet. To reconstruct the object's cross-section, it is only necessary to view one side of the cone sheet. However, this approach does require a very careful calibration of the camera relative to the center of the turntable.

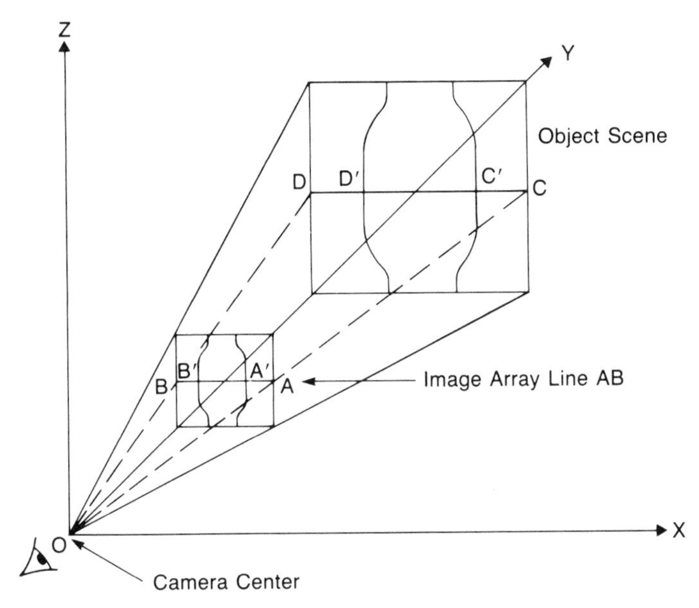

Figure 5.17 Three-dimensional model with image plane inverted. The three-dimensional object scene with a pixel coordinate, say, $C' = X_o Y_o Z_o$, is compressed to a two-dimensional image plane with typical coordinates, say, $A' = X_i Z_i$. (Goldstein, Wright, Bourne, "A Fast Algorithm For High Computer Vision" *Robotics and Computer Integrated Manufacturing*, Vol. 2, No. 2, 1985. Reprinted with permission from Pergamon Journals Ltd).

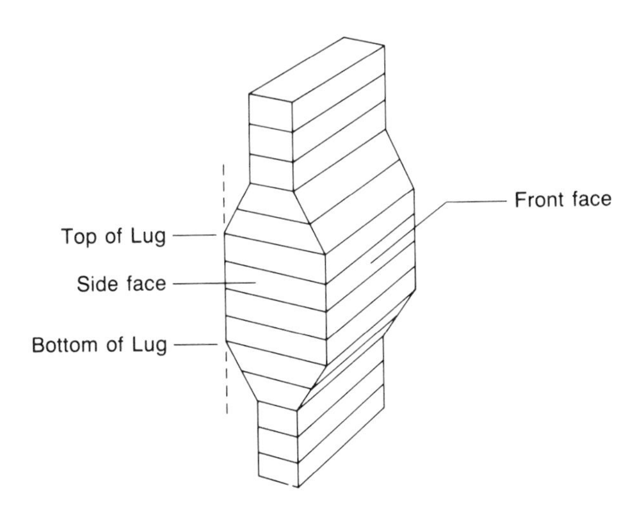

Figure 5.18 Resulting stack of cross-sections with hidden surfaces removed. (Goldstein, Wright, Bourne, " A Fast Algorithm For High Computer Vision" *Robotics and Computer Integrated Manufacturing*, Vol. 2, No. 2, 1985. Reprinted with permission from Pergamon Journals Ltd.)

The resolution of the gauge station is primarily a function of the size of the field of view and the pixel resolution of the vision system. In the first algorithm, the field of view had to be wide enough to keep both sides of a 6-inch object in view, independent of viewing angle. In the second algorithm, only the right edge had to be kept in view, thus increasing the available resolution.

In the forging application, the cross-sections can be eccentric and the parts positioned off the center of rotation; thus, we are only able to halve the viewing field to ensure that the right edge is always visible. In other applications, if the cross-sections were less eccentric and parts were placed more concentric with the center of rotation, the viewing field could be made smaller to ensure that an edge was always in view. Even large objects such as rail wheels could be inspected for roundness using such a setup, provided their eccentricity was small. In a turning operation on a lathe, a small viewing field and hence high resolution would also be possible since the parts are circular. Figure 5.19 shows the relationship between field of view and consequent resolution ability for a system with 235 pixels per scan line. Again, this assumes that one edge of the object being inspected would always lie in view.

5.4.2 The Machine Tool's Sensor Laboratory

To collect and reason about a wide range of sensor information in the machine tool world, the cutting process should be surrounded with a working sensor laboratory. The ideal sensor laboratory will take a number of in-process measurements (i.e., machine vibrations, tool tip temperatures, machine sound's pitch, and clamping force) augmented by quick visual inspections between setup steps. This wide range of tasks

FIELD OF VIEW (IN.)	RESOLUTION (IN.)
3.00	0.0128
1.00	0.0043
0.25	0.0011

Figure 5.19 Resolution = Field of view/235 pixels. (Goldstein, Wright, Bourne, "A Fast Alogrithm For High Computer Vision" *Robotics and Computer Integrated Manufacturing*, Vol. 2, No. 2, 1985. Reprinted with permission from Pergamon Journals Ltd.).

suggests that it is wise to build a modular laboratory so that new sensors can be easily added and integrated into those already developed for the machining process (Chow 1984; Goldstein et al. 1985; Thangaraj and Wright 1987). To create intelligence in the machine tool world, the next step is to construct this more comprehensive sensor laboratory.

The vision problem in the machine tool world is especially challenging because of the range of tasks that must be solved and the accuracy that is required of the results. Vision can be used to accomplish the following goals:

- **Dimensioning:** Dimensioning information can be likened to an electronic caliper determining hole dimensions, slot widths, and edge finding. The information content is primarily numerical and requires little interpretation. In our present system working on small components, we achieve accuracies down to 0.002 inches.
- **Surface Quality:** Machinists view the surface finish of a part to determine a variety of things. At the start of a job, all sides of the initial work piece are examined to see which is the best for initial fixturing. The machinist will determine if sides have been rolled, saw cut, or machined. Then, throughout machining, the viewed surface gives insight into how the machining process is progressing. Vibrations and surface marks give information on the effectiveness of cooling and fixturing. This information is subjective and relies on interviewing machinists to duplicate their judgment.
- **Tool Integrity:** Visual feedback is a primary way in which machinists determine the integrity of their cutting tools. Their decision process is rich in experience concerning how different kinds of cutting tools behave in different stages of wear.
- **Part Verification:** In any machining activity, the final setup must be verified to maintain part tolerances. This requires an understanding of the part specifications and the ability to check part features and make measurements.
- **Machine Monitoring:** Machine vision can be used to monitor the overall machining environment to make sure that basic assumptions are not violated. For example, vision can make sure that chips are not piling up around a part, and vision can make sure that fixtures are in their predesigned position.

Questions about part dimensions must be answered with great accuracy in order to be of any value. This means that the camera must be able to zoom in to get an appropriate field-of-view versus desired-accuracy proportion, while still maintaining calibration. To achieve this, we are building a vision tool that can be mounted in the spindle of the machine tool so that it can move with the same high degree of accuracy as the machine tool (see Figure 5.20).

To achieve very accurate measurements, it is a good plan to measure the distance between part features within a single camera view. These relative distances are then decoupled from the accuracy of the camera positioning (i.e., the machine tool). Our machine tool's accuracy is limited to ±.0005 inches, and this inaccuracy would be further compounded by limitations of the vision system (e.g., pixel size and calibration). Therefore, measurements that produce local distances between part features, like two holes, can be made much more accurately than those taken of each hole position directly in the machine tool's coordinate system.

Other questions do not require the same degree of accuracy, but instead require basic information about the global part machine setup. For example, chips can be randomly scattered over a part, and it may be important to brush them away before continuing the process. This kind

Figure 5.20 A visualization of the machine tool world from tool holder.

of problem does not require any unusual precision, except that the whole part should be in view.

To fully use the sensor lab, the software must be quite sophisticated. In some cases, a single sensor can always do the job, but it may be necessary to automatically switch over to another sensor in extreme conditions. For example, vision may be normally used to verify that the correct number of holes have been drilled, but sometimes the stock part may be so discolored that most vision algorithms do not work at all. In this case, it may be more effective to use a touch probe to be sure that holes in fact exist and that other visual patterns are not holes. Touching each hole is a slower operation than visually looking for holes, but it is definitive.

The ability to use multiple sensors to solve a machine tool's problem extends the idea of intrinsic images beyond the strictly visual domain. Thus, a number of different strategies can be used to detect the underlying natural and task parameters (see Chapter 10 for further elaboration).

5.5 SAMPLE OPEN PROBLEMS IN VISION

In this section, we point the way to a research program that would significantly advance the application of machine vision for industrial applications. Some of the problems are in the center of the "science of vision," and others are strictly engineering tasks; both need to be accomplished.

1. Hand-eye coordination remains an important problem. To date, vision systems have looked at a scene, paused, and then a robot takes an action. Develop an assembly task (e.g., a thread and a needle) where a robot is servoed directly from vision feedback. In an industrial application, this situation would arise while threading wires through fixtures in aircraft and other assemblies.
2. Design and carry out experiments to determine whether vision or touch is the dominant sense in: size determination, shape determination, and feedback for motor control. Rock and Harris (1967) start this project by showing that size determination is dominated by vision.

3. For the process of cutting metal on a machine tool, build a system to perform simple (visual) monitoring tasks. As a starting point, it may be helpful to find out how much visual monitoring a machinist can do when a part is being made by hand.
4. What method of active illumination would be most useful to a mobile robot? Consider a laser spot, a flashlight, and a sonar beam. How does the task (e.g., navigation and object identification) affect the scan strategy? What information can be extracted from the resulting images that are not present in the others?
5. Design a system that determines shape from active shadows. In particular, find the shapes of objects over which a manipulator's shadow is cast. Control the manipulator so that important shape features are significantly enhanced.
6. Choose a set of task-independent vision algorithms that can be pieced together into a process plan to solve either machining or assembly tasks. In doing this, there should be as few algorithms as possible, while still being able to solve the largest number of task problems.
7. Build a planner in an AI language of your choice. The planner should be able to automatically build a process plan that will drive a vision system using primitives from Open Problem 6 to solve a set of diagnostic goals.
8. To achieve a better understanding of image selection in manufacturing applications of vision, interview a movie director and build an expert system to choose image sequences to achieve the best visual effects for a scene. Compare the decisions made by the expert system with scenes not discussed in the initial interviews. What is the role of a director's style in image selection?
9. Based on a commercial vision system, subtract successive images. Build a manipulation system for the camera so that different views can easily be collected. The additional challenge is to control the movement of the camera and to relate the camera's movement to the results in the difference buffer.
10. Develop an algebra for combining different vision operations that are based on robotic control (e.g., zoom, focus, and position), rather than on local image operators (e.g., edge finders, digital filters, and region finders). The final goal of the project would be to automatically develop vision algorithms based on robotic control.

REFERENCES

Bajcsy, R., and Lieberman, L. 1976. Texture gradient as a depth cue. *Computer Graphics and Image Processing*, Vol. 5, No. 1, pp. 52–67.

Ballard, D. H., and Brown, C. M. 1982. *Computer Vision.* Prentice-Hall, Englewood Cliffs, N.J.

Barrow, H. G., and Tenenbaum, J. M. 1978. Recovering intrinsic scene characteristics from images. In *Computer Vision Systems*, edited by A. R. Hanson and E. M. Riseman. Academic Press, New York, pp. 3–26.

Bolles, R. C., Kremers, J. H., and Cain, R. A. 1981. A simple sensor to gather three-dimensional data. Tech. note 249, SRI International, Menlo Park, Calif.

Bourne, D. A. 1980. On automatically generating programs for real time computer vision. *Proceedings of the 5th International Conference on Pattern Recognition*, pp. 759–764.

Brooks, R. A. 1982. Symbolic error analysis and robot planning. *International Journal of Robotics Research*, Vol. 1, No. 4, pp. 29–68.

Bullock, B. L. 1978. The necessity for a theory of specialized vision. In *Computer Vision Systems*, edited by A. R. Hanson and E. M. Riseman. Academic Press, New York, pp. 27–36.

Chow, J. G. 1984. Sensor development for on-line monitoring and the determination of temperature distributions in machining. Ph.D. thesis, Carnegie-Mellon University.

Dickens, C. 1836. *The Posthumous Papers of The Pickwick Club.* Penguin Books, England. Reprinted 1978, p. 573.

Gleason, G. J., and Agin, G. J. 1979. A modular system for sensor-controlled manipulation and inspection. *Proceedings of the Ninth International Symposium of Industrial Robots*, Society of Manufacturing Engineers and Robot Institute of America, Washington, D.C., pp. 57–70.

Goldstein, M. G., Wright, P. K., and Bourne, D. A. 1985. A fast algorithm for high accuracy gauging using computer vision. *Robotics and Computer Integrated Manufacturing*, Vol. 2, No. 2., pp. 105–113.

Gool L. V., Dewaele, P., and Oosterlinck, A. 1985. Texture analysis Anno 1983. *Computer Graphics and Image Processing*, Vol. 29, No. 3, pp. 336–357.

Hildreth, E. C. 1983. *The Measurement of Visual Motion.* MIT Press, Cambridge, Mass.

Horn, B. K. P. 1986. *Robot Vision.* MIT Press, Cambridge, Mass.

Ittelson, W. H., and Kilpatrick, F. P. 1951. Experiments in perception. *Scientific American*, Vol. 185, No. 2 pp. 50–55.

Julesz, B. 1971. *Foundations of Cyclopean Perception*, University of Chicago Press.

Kanade, T., and Kender, J. 1980. *Skewed Symmetry: Mapping Image Regularities into Shape.* CMU-CS-80-133, Computer Science Dept., Carnegie Mellon University.

Kelley, R. B. 1984. Heuristic vision algorithms for bin-picking. In *14th International Symposium on Industrial Robots,* edited by N. Martensson, pp. 599–610.

Kennedy, D. 1963. Inhibition in visual systems. Copublished by IFS (Publ.) Ltd., Bedford, England and *Scientific American,* Vol. 209, No. 1, pp. 122–130.

Marr, D. 1982. *Vision.* W. H. Freeman, San Francisco.

Marr, D., and Poggio, T. 1976. Cooperative computation of stereo disparity. *Science* Vol. 194, No. 4, pp. 283–287.

Marr, D., and Poggio, T. 1979. A computational theory of human stereo vision. *Proceedings of the Royal Society of London B. 204,* pp. 301–328.

Potter, J. L. 1977. Scene segmentation using motion information. *Computer Graphics and Image Processing,* Vol. 6, Dec., pp. 558–581.

Roberts, L. 1965. Machine perception of three-dimensional solids. *In Optical and Electro-Optical Information Processing,* edited by J. Tippett. MIT Press, pp. 159–197.

Rock, I. 1983. *The Logic of Perception.* MIT Press, Cambridge, Mass.

Rock, I., and Harris, C. S. 1967. Vision and touch. *Scientific American,* May. Reprinted in *Readings from Scientific American: Perception: Mechanisms and Models.* W. H. Freeman and Company, pp. 269–277.

Saaty, T. L. 1980. *The Analytic Hierarchy Process.* McGraw-Hill, New York.

Shirai, Y., and Suwa, M. 1971. Recognition of polyhedrons with a rangefinder. *2nd International Joint Conference on Artificial Intelligence,* pp. 80–87.

Thangaraj, A., and Wright, P. K. 1987. Computer assisted prediction of drill failure using in-process measurements of thrust force. *Trans. ASME, Journal of Engineering for Industry,* in press.

Thibadeau, R. 1984. Automatic visual inspection as skilled perception. *Robotics Institute 1984 Annual Research Review,* Carnegie Mellon University, pp. 19–30.

Ullman, S. 1979. *The Interpretation of Visual Motion.* MIT Press, Cambridge, Mass.

Villers, P. 1984. Technologies of robotic and artificial vision systems. *International Journal of Robotics and Integrated Manufacturing,* Vol. 1, No. 1, pp. 125–152.

Witkin, A. P. 1982. Intensity-based edge classification. *Proceedings of the National Conference on Artificial Intelligence,* AAAI-82, Aug., pp. 36–41.

Wittreich, W. J. 1959. Visual perception and personality. *Scientific American*, Vol. 200, No. 4, pp. 56–60.

Wright, P. K., and Englert, P. J. 1984. A review of sensor based robotic manipulation and computer vision in flexible manufacturing cells. *Computer Integrated Manufacturing and Robotics,* American Society of Mechanical Engineers, New York, Vol. PED-13, pp. 53–70.

6 The Manufacturing Hand

Perception of the world requires a sort of tactile flair. . . . Surface, volume, density, and weight are not optical phenomena. It is in the hollow of the palm and between the fingers that man instantly assesses these characteristics. Space is not gauged by sight, but rather by hand and foot which impart an indefinable appreciation and without which recognition remains like a delightful landscape of a dark chamber; inconsistent, flat, elusive, and chimerical.[1]

—Henri Focillon in *The Hand*

IN flexible manufacturing systems, industrial robots can be used for loading machines, inspection activities, and assembly operations. In this chapter, we consider the kinds of robot arms and hands that should be built for such tasks. The human hand has evolved to be a general-purpose manipulator that adapts to the richness of our daily lives, but, in doing so, it is not particularly suited to manufacturing operations. Here, we rely heavily on special tools, fixtures, and protective gloves.

The initial development of robot hands specifically for manufacturing work has involved the construction of an instrumented robotic wrist with several types of grippers. To facilitate the design and fabrication of such industrial robot grippers, some guides have also been proposed that consider a range of manufacturing operations and specify the level of robot control needed.

These design guides emphasize that the actions required of the robot and its gripper depend more on the task being done than on the shape of the object being grasped. The same effect is seen in the way that a human task determines the choice of grip and the actions of the hand. For example, if a person picks up a pencil to hand it to someone, the way the pencil is held is entirely different from the way it is held for writing. The human hand can be adapted to many different tasks: writing with a pencil, groping in the dark, and playing the guitar, to name a few. In addition to being a gripper, the human hand is a sensory organ and an

[1] *The Hand*, edited by R. Tubiana. © 1981, W. B. Saunders Co. Reprinted with permission.

organ of communication. The grippers now used in industry are all very crude in comparison, although recent examples are beginning to show some flexibility. It would be a mistake, however, to assume that the human hand would be the ideal gripper for an industrial robot. The manufacturing world is more demanding and narrower in scope than the everyday world from which the human hand has evolved.

6.1 EVOLUTION OF MANUFACTURING ROBOTS AND HANDS

John Donne's famous statement that "no man is an island" could also be applied to industrial robots and their grippers. The robot gripper, or end-of-arm tooling, is a bridge between the computer controlled arm and the world. The design of the gripper should reflect this role, matching the capabilities of the robot to the requirements of the task. The ideal gripper design should be synthesized from independent solutions to these considerations shown in Figure 6.1.

6.1.1 Robot Arms

The first industrial robot arms, created around 1960, were nearly "islands." They were used primarily as standalone machines for painting, spot welding, or pick-and-place work. In the pick-and-place case, parts were moved from one location to another without giving much

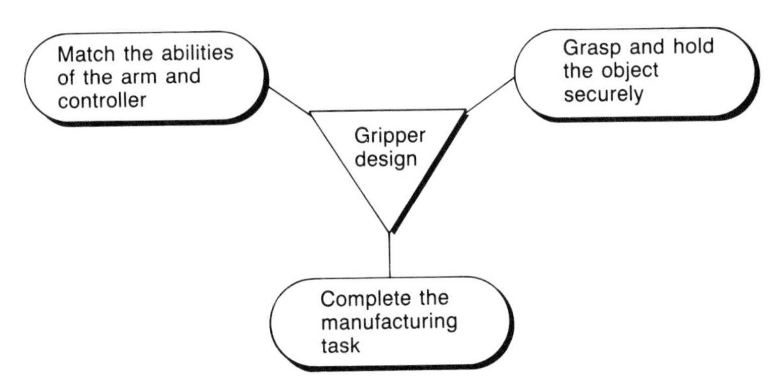

Figure 6.1 Requirements of a gripper. (Wright & Cutkosky, *Handbook of Industrial Robotics*. © 1985 John Wiley & Sons. Reprinted by permission of John Wiley & Sons, Inc.

attention to how the parts were picked up or put down (Engelberger 1980). For pick-and-place work, simple beak-like grippers were used and the ability of the robot arm to grasp and manipulate parts was at best equal to that of a person using fireplace tongs.

Since then, the field of robotics has enjoyed considerable popularity and many advances have been made in mechanical design, in control methods and in the supporting sensory equipment. Many texts are now available that report on the mathematics (Paul 1981), the mechanics and control (Brady, Hollerbach, Johnson, Lozano-Perez, and Mason 1982; Craig 1986) and the applications of robots in manufacturing (Shahinpoor 1987).

Therefore, today's industrial robot arms are put to work in more challenging applications. The objects they manipulate may have complicated shapes and they may be fragile. The tasks the robots perform involve assembling parts or fitting them into clamps and fixtures. These tasks place greater demands on the robot arm's accuracy. Once a part has been picked up, it must be held securely and in such a way that the position and orientation remain accurately known with respect to the arm. While the object is being manipulated during an assembly task, forces arise between the object held by the robot and the mating parts. In such cases, the arm and gripper, as a whole, must be compliant enough to prevent contact forces from doing any harm. Such compliance can be achieved passively by introducing elastic components into the gripper. Alternatively, active compliance can be used to modulate the arm movements according to the force fed back from the task.

6.1.2 Robot Hands

During the last twenty years, industrial robots have evolved in their manipulation ability from simple pick-and-place arms with a pinch-like gripper at the end (Engelberger 1980), to better controlled arms with improved grippers backed up by passive or instrumented wrists (Drake, Watson, and Simunovicsh 1977; Cutkosky and Wright 1986a), to the possibility of developing dexterous hands (discussed and illustrated later in this chapter). The development of dexterous hands is a naturally engaging topic for the research community (Salisbury and Craig 1982; Jacobsen, Wood, Knutti, and Biggers 1984). However, the potential of increased dexterity or flexibility of such hands can only be obtained at an expense: a dexterous hand is more complex, difficult to control, and contains many delicate components.

6.1.3 The Tradeoff Between Dexterity and Strength

Figure 6.2 shows that in nature (McMahon and Bonner 1983) and in robotics there is an inherent tradeoff between dexterity and strength. The four parameters influencing dexterity were introduced in Chapter 3 (see Figure 3.5). A dexterous system has four major components:

1. An experienced knowledge base
2. A graceful control system
3. A large reach potential with many degrees of freedom
4. A high accuracy

Strength can be measured more quantitatively. For this discussion, strength is synonymous with the payload that the complete robotic system can lift, or with the force that the robot system can apply to an external object.

The schematic graph in Figure 6.2 considers how dexterity and strength change as devices are added to a bare arm. On the left, the heavy-duty, bare arm is imagined not to have a force-sensing wrist or hand, but rather just two lobster-like fingers. Such a system is very strong, but lacking in dexterity. On the right, the active finger set in the hand is, by contrast, very dexterous but weak. To be specific, a heavy-duty industrial manipulator (e.g., Figures 2.3 and 2.4) exhibits a payload in excess of 150 pounds. But, this cannot be exploited if the industrial arm is fitted with a dexterous hand of the kind being developed in

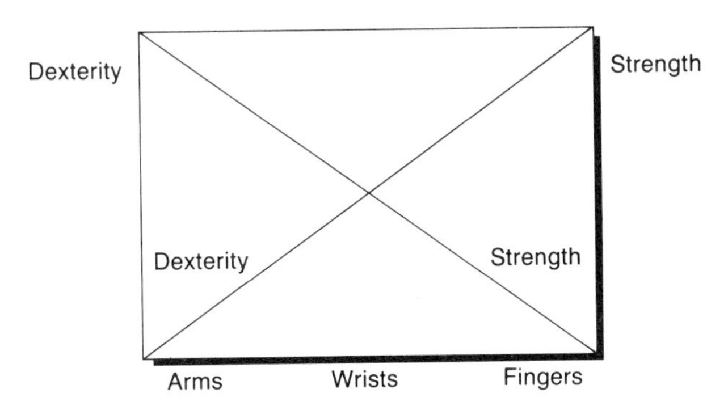

Figure 6.2 The tradeoff between dexterity and strength when an active wrist and then a dexterous hand with active fingers are cumulatively added to a robot arm. (Reprinted with permission from Wright, "A Manufacturing Hand" in *Robotics and Computer Integrated Manufacturing* Vol. 2, No. 1, 1985, Pergamon Journals Ltd.).

laboratories (Jacobsen et al. 1984; Okada 1982; Salisbury and Craig 1982). Hand components would break beyond object payloads of only a few pounds. The arm-wrist-hand system as a whole would fail.

This argument is not meant to be unfair to the designers of robotic dexterous hands: the new hands being developed are specifically for prosthetic needs and light assembly tasks. Rather, the argument is designed to emphasize this dexterity versus strength tradeoff that occurs when the active wrist and active hand devices are cumulatively added to the robot arm.

Human beings face the same tradeoff; we cannot finely manipulate large metal objects in our fingertips. In this context it is relevant to move up the human chain, or arm, and to partition grasping into three levels:

- **Finger tasks**: Fingers are "designed" for light, delicate manipulations. Thus, finger tasks include writing with a pencil, adjusting a wristwatch, and performing fine assembly operations.
- **Wrist tasks**: To work with heavier objects or larger forces, the fingers of the hand tend to adopt a passive clamping position and wrist manipulations take over. In manufacturing, this is seen when using a wrench, hammer, or power tool. The fingers close around the handle in a more or less fixed grip, leaving the wrist to do the work.
- **Arm tasks**: These are used when the object is very heavy. When picking up a chair to carry it across a room, the fingers and wrist are immobilized in favor of manipulations at the elbow and shoulder. As another example, consider picking up a heavy sack of concrete. A person uses both arms, clutched around the sack like two huge fingers. Such "fingers" wrestle the awkward object to the chest, which now functions like a large palm.

This inherent tradeoff between dexterity and strength, as demonstrated by these three task levels, is a recurring theme of this chapter. It guides what can be practically expected from a robotic system and influences other mechanical designs, as will be seen in the chapter on machine tool fixtures.

The primary goal for a manufacturing hand is to solve the dexterity versus strength tradeoff in the context of machine tools, flexible assembly systems, and flexible manufacturing systems. The aim is to build a hand with sufficient manipulation abilities for these environments without unduly sacrificing power. This capability represents a necessary and considerable advance over current industrial grippers.

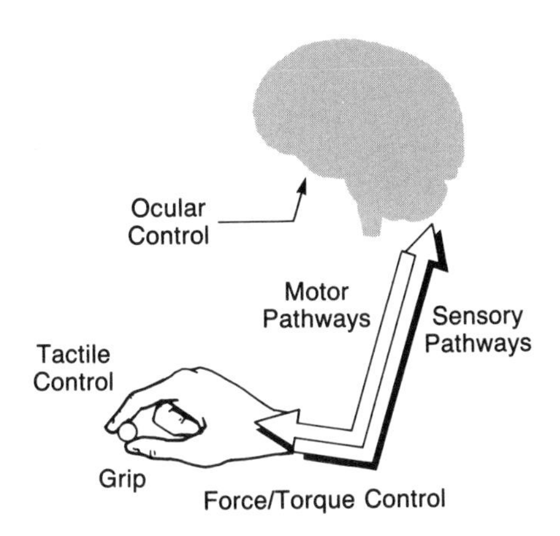

Figure 6.3 The actions involved in prehension: A schematic of the human system. (Reprinted with permission from Wright, "A Manufacturing Hand" in *Robotics and Computer Integrated Manufacturing* Vol. 2, No. 1, 1985. Pergamon Journals Ltd.)

6.2 GRIP PREHENSION AND GRASP CHOICE

Human prehension combines the choice of a grip, the act of grasping or picking up an object, and the control of an object using the hand. A desire, internally or externally generated, triggers responses in the eye and the mind. An instant later, the hand chooses a grip, takes a grasp position over the object, and begins the task. As shown in Figure 6.3, the process is controlled by feedback loops, which include eyesight for approximate positioning of the hand and touch for adjusting the force and position of the fingers.

Prehension is task oriented. As the pencil example in the first section illustrates, choosing a grip is an important step in the process of prehension. The block diagram in Figure 6.4 is greatly simplified, but it illustrates the relationships between the different activities included in prehension.

Note that Figure 6.4 provides a setting for the material in this chapter. Moving across the diagram, we cover prehension and grasp choices in Sections 6.2.1 to 6.2.3, gross position in Section 6.2.4, and fine position in Section 6.2.5.

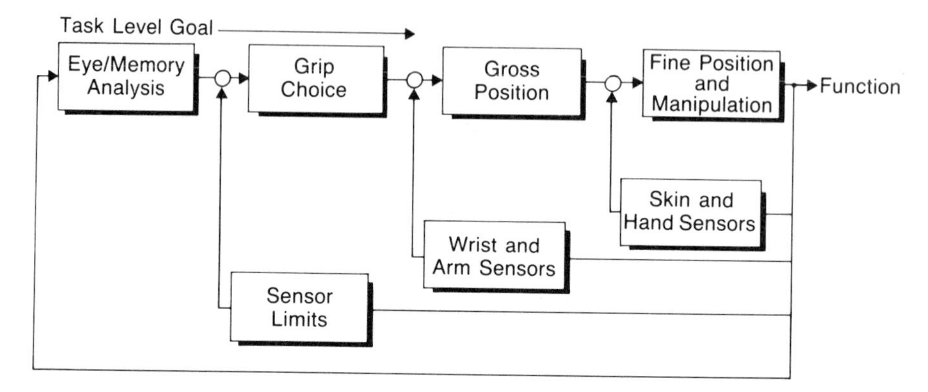

Figure 6.4 The various steps and feedback in gripping and manipulation.

6.2.1 Grasp Choices

A review of pertinent medical literature (Tubiana 1981) reveals as many as eight basic categories of grip but, for manufacturing work, only two are of primary importance: the "three-fingered" grip, shown in Figure 6.5, and the "wrap-around" grip, which is used to hold a large screwdriver or hammer. Tubiana states that the three-fingered grip is used for 90% of light domestic and manufacturing tasks; it is adaptable to many object sizes, e.g., from a pea to a softball. The three-fingered grip emphasizes stability because it uses ligaments in the back of the hand, as illustrated in Figure 6.5. However, it is not as strong as the

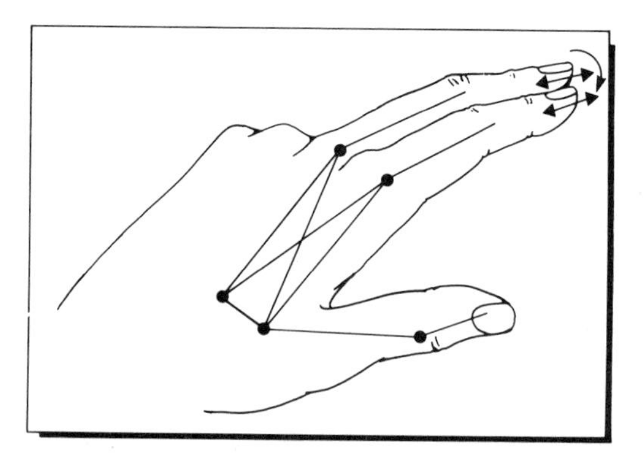

Figure 6.5 The human three-fingered grip. (Reprinted with permission from Wright, "A Manufacturing Hand" in *Robotics and Computer Integrated Manufacturing* Vol. 2, No. 1, 1985, Pergamon Journals Ltd.)

wrap-around grip, which uses the friction between the faces of all the fingers and the tool as well as the power of all five digits, including the very strong muscle running from the small finger along the outside of the hand and into the arm.

The same principles can be applied to industrial grippers, such as those shown later in Figures 6.19 and 6.20. Figure 6.19 is essentially a three-fingered gripper, where the lower face is a very wide thumb. The upper two fingers, by virtue of the ball-joint linkage between them, can conform to the object being gripped. The linkages have the degrees of freedom shown by the solid lines in Figure 6.5. This gripper also has some of the attributes of the wrap-around grip because the spacing between the index and middle fingers is expanded, in comparison to the human hand, and the thumb is widened to resemble a platform or a palm with some scooping ability. As a result, the grip is stronger and more stable. This amalgamation of two primary grip types leads to a design specifically suited to heavy manufacturing.

6.2.2 Stability in Grasping

In recent years, a few papers have dealt with the subcomponents of Figure 6.4. Asada (1979), Asada and Hanafusa (1982), Cutkosky (1985), Salisbury (1982), and Cutkosky and Wright (1986b) consider the force balance for an object held by a gripper with several fingers. This is important for the design of passive robot grippers intended for industrial assembly tasks with objects of arbitrary shape.

Given the three-fingered grip and a particular object with defined three-dimensional shape, what is the best way to immobilize the object given a particular task? Figure 6.6 shows a sample object held by three fingers, which individually have compliance, or elasticity normal and tangential to the contact surface. Consider that this object is part of an assembly operation and that there are interactions both on the object and subsequently on the fingers. For example, Figure 6.6 might represent the head of a short bolt being started in a threaded nut. As a result, a sideways force *(F)* and a twisting moment *(M)* are the reactions on the bolt. This force and moment are, in turn, transferred to the fingers. In the equilibrium state, the grip of the three fingers should resist the externally applied force and moment.

To inquire about the most effective grip position, consider that the externally applied force F is slowly increased. In Figure 6.6, this force acts along a line at the angle θ to the reference frame. This reference frame is also used for the simple, rectangular object in Figure 6.7. In the

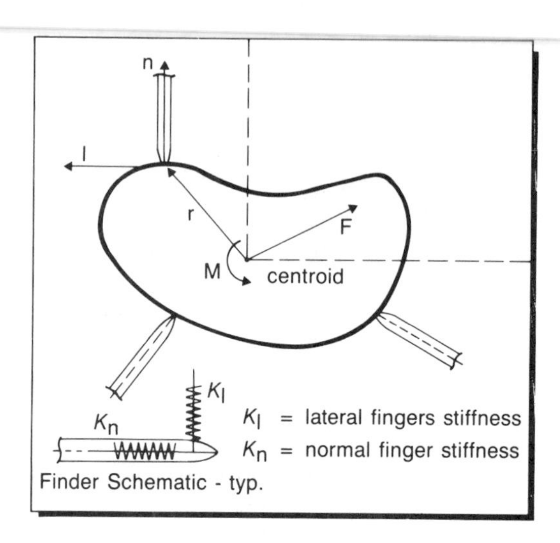

Figure 6.6 A sample object being held by three elastic fingers. (Reprinted with permission from Wright, "A Manufacturing Hand" in *Robotics and Computer Integrated Manufacturing* Vol. 2, No. 1, 1985, Pergamon Journals Ltd.)

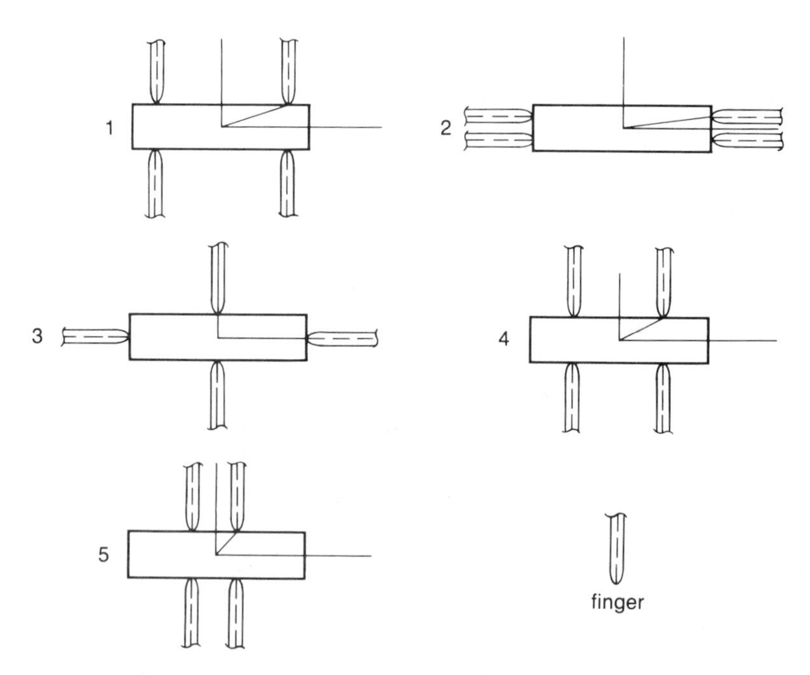

Figure 6.7 Sample grip orientation for a set of four fingers. (Reprinted with permission from Wright, "A Manufacturing Hand" in *Robotics and Computer Integrated Manufacturing* Vol. 2, No. 1, 1985, Pergamon Journals Ltd.)

following example, we consider how the grip stiffness varies with different finger positions and with different application angles θ of the force *F*. The finger-set gripping the object must be stiff enough to resist the applied force, hence a high-grip stiffness is desirable. The example is from Cutkosky (1985), who carried out sample problems on a variety of two-dimensional and simple three-dimensional shapes. To reiterate, the goal is to find the best grip that will enable a passive robot gripper to immobilize an arbitrary object.

Of the five grips on the rectangular block in Figure 6.7, grip 1, grip 4, and grip 5 share the same configuration, but with different finger spacings. As shown in Figure 6.8, four grip types offer the highest stiffness, but at the specific locations of 0° ± 90°. The stiffness of grip 3 is constant, regardless of the direction of the force. Grip 3 is, therefore, the safest choice for arbitrary loads, although other grips offer more stiffness or resistance to slipping when the object is moved in a single direction. This simple example can be extended to more complex situations.

A creative mechanical design allows a single set of such passive fingers to conform to several different object shapes and to be more flexible than

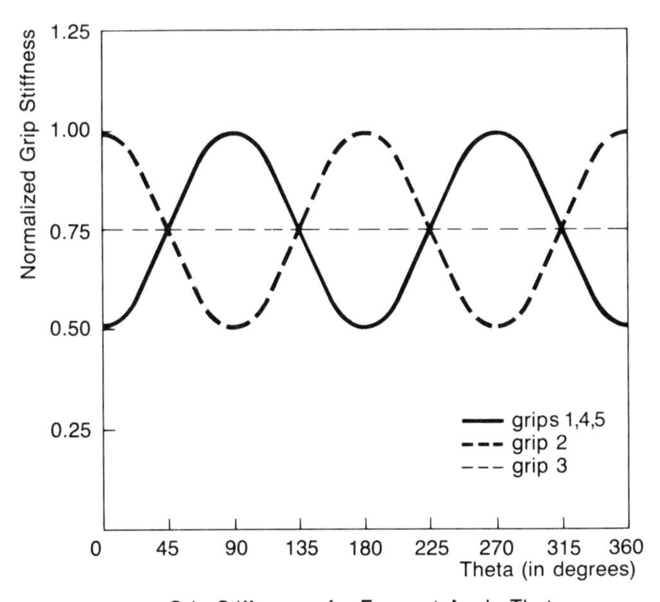

Figure 6.8 Grip stiffness for an applied force at angle θ. (Reprinted with permission from Wright, "A Manufacturing Hand" in *Robotics and Computer Integrated Manufacturing* Vol. 2, No. 1, 1985, Pergamon Journals Ltd.)

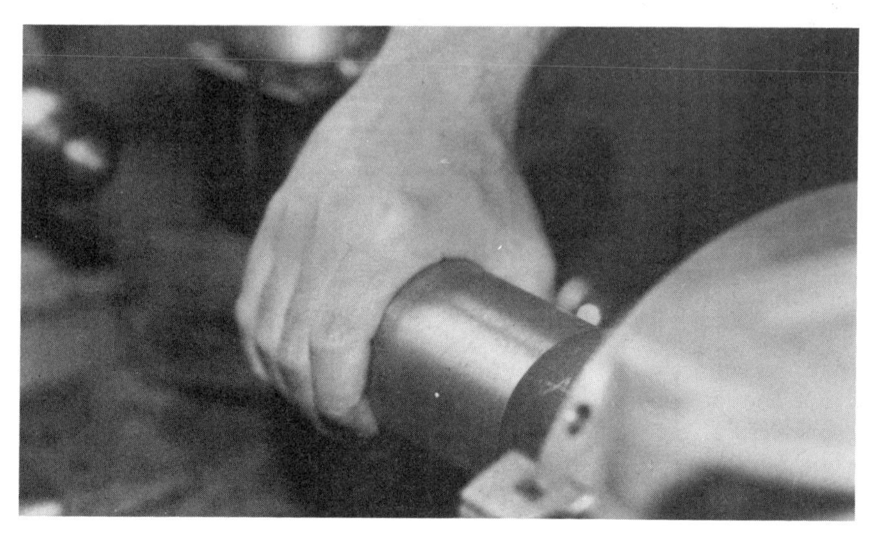

Figure 6.9 Wrap-around grasp for a heavy, convex workpiece. (Reprinted with permission from Cutkosky & Wright, "Modelling Manufacturing Grips and Correlations with the Design of Robotic Hands." Robotics & Automation Conference, San Francisco © 1986 IEEE).

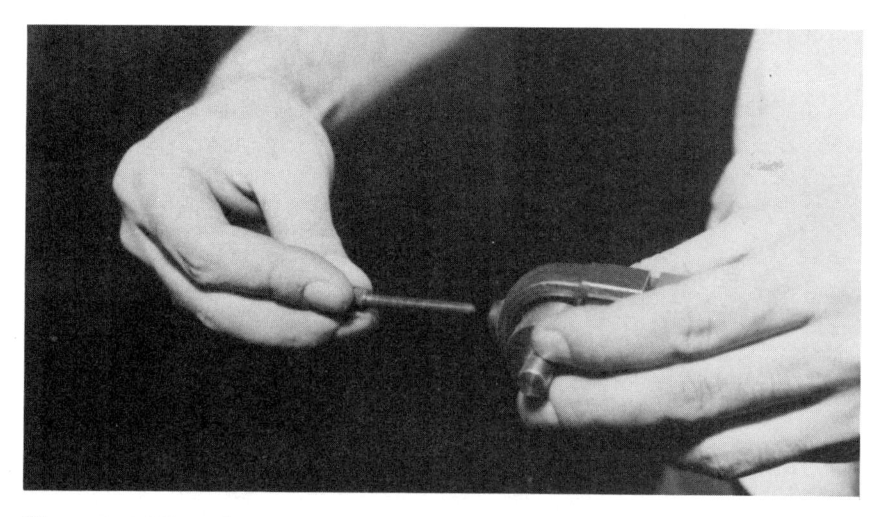

Figure 6.10 Three-finger precision or tripod grip using thumb, and middle fingers. (Reprinted with permission from Cutkosky & Wright, "Modelling Manufacturing Grips and Correlations with the Design of Robotic Hands." Robotics & Automation Conference, San Francisco © 1986 IEEE).

one might expect. Before exploring such passive gripper designs, some of the common grips used by humans in manufacturing are now considered.

6.2.3 Grasps Chosen in a Small-Batch Machining Environment

Figures 6.9 and 6.10 illustrate two grasps used by machinists working in a small-batch machine shop. The accompanying text, explaining when the grip is used, whether motions are made with the fingers or wrist, and how forces are sensed, represents a consensus of several machinists.

Figure 6.9 shows the machinist using a wrap-around grasp to load a heavy billet into the chuck of a lathe. The fingers and palm wrap around the heavy, convex part. Friction between the fingers and the part accounts for much of the grasping force. This grasp is consequently an extreme example of the power grasp. The chuck-loading task is a heavy peg-in-hole assembly and is representative of operations for which robotic assembly aids were originally designed (Drake et al. 1977). The arm brings the billet toward the chuck and provides support as the wrist makes small accommodations that align the billet so that it can be inserted.

The hand in the left of Figure 6.10 illustrates a precision grasp in which the tips of the thumb, index finger, and middle finger hold a small bolt and start to engage it with the thread of a tapped hole. The fingers tilt the bolt, aligning it with the axis of the hole, while rotations about the axis of the bolt are achieved primarily with the wrist. As soon as the thread of the bolt engages the thread of the hole, the machinist switches to a two-fingered grasp so that he can rapidly twist the bolt with his fingers.

Cutkosky and Wright (1986c) carried out a series of protocol analysis experiments with machinists and classified these and other machining grips. Some photographs and more detailed descriptions are given in the reference. These have led to the partial taxonomy of human grasps that is being used in the design of hands serving a CNC machine tool. In the same way, the taxonomy of vision strategies is being used to build inspection and monitoring applications for the machine (see Figure 5.4).

The taxonomy of grasps is shown in Figures 6.11 and 6.12. Moving from left to right, the grasps become less powerful and the grasped objects become smaller. Thus, the wrap-around grips are the most power-

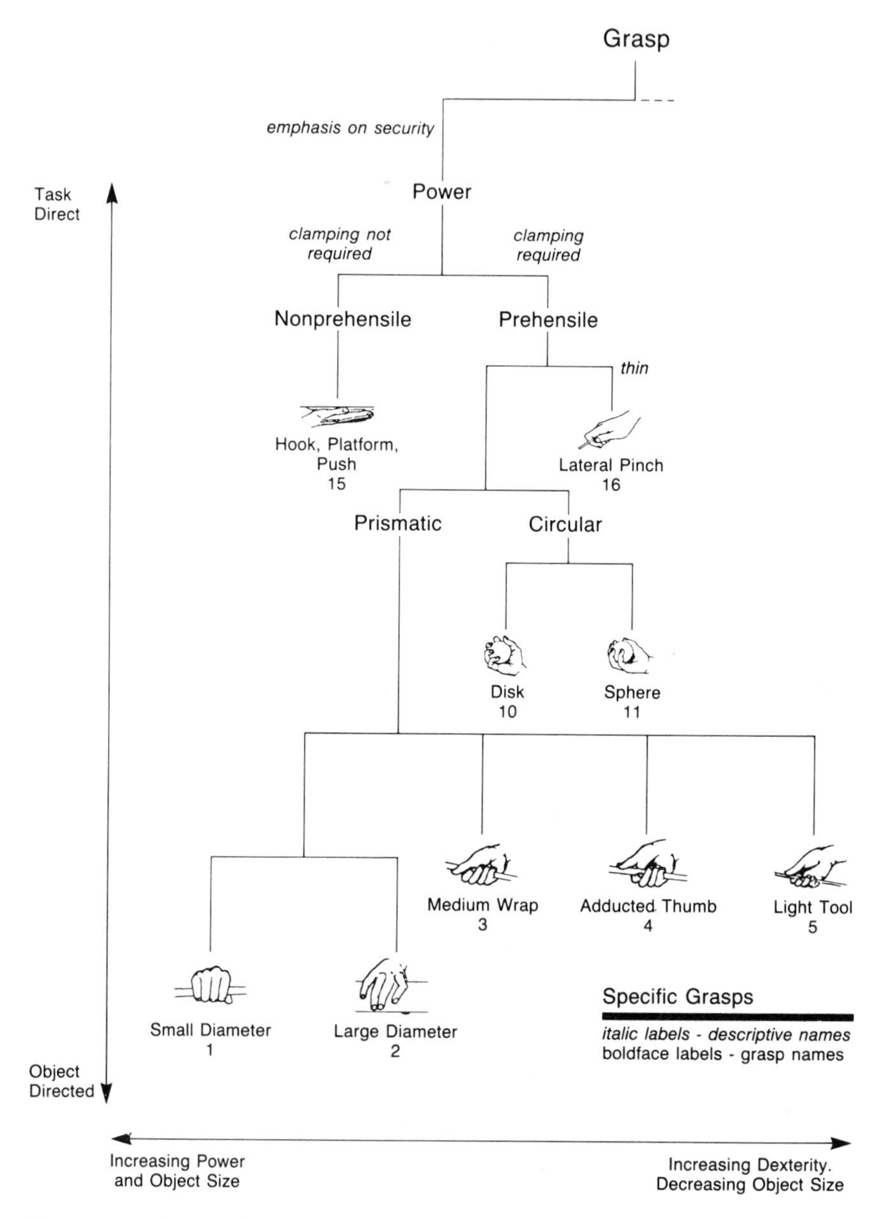

Figure 6.11 A partial taxonomy of human grasps for power. The lower grips are subsets of the four-fingered, wrap-around grasp.

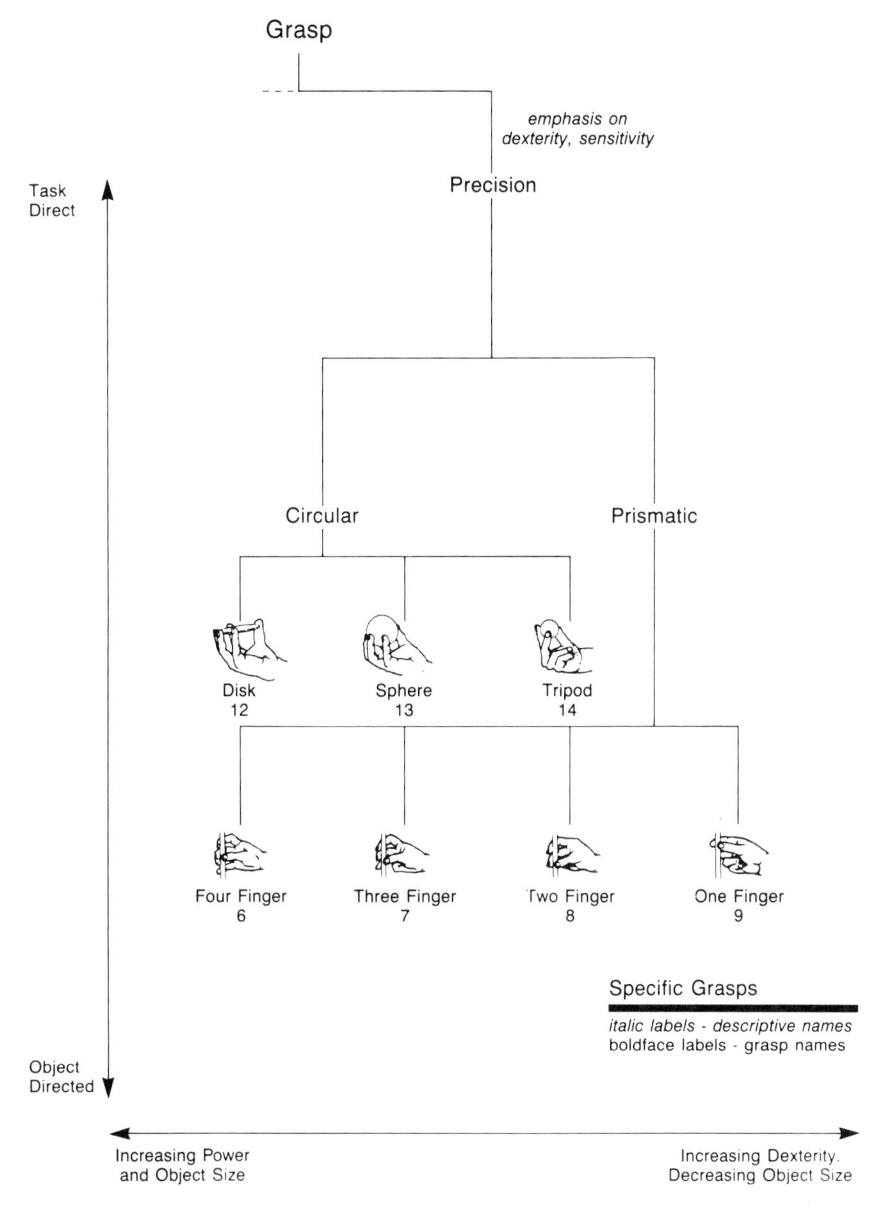

Figure 6.12 A partial taxonomy of human grasps for precision. Note that Figures 6.11 and 6.12 can be combined to form one large taxonomy joined at the top.

ful and least dexterous. In these cases, all manipulation must be done with the wrist, and even the wrist is restricted to a limited range of motions. By contrast, the precision grips, e.g., the tripod grip and the thumb and index finger grips are much more delicate. Unfortunately, any taxonomy that is imposed on nature contains a few exceptions to the rule. For example, a spherical power grasp may be either more or less dexterous than a medium wrap power grasp, depending on the size of the sphere. Moving from top to bottom in Figures 6.11 and 6.12, the choices range from task-directed grasps to object-directed grasps. Again, the taxonomy exhibits some exceptions; for example, a small, flat object may provoke the choice of a lateral pinch grip near the top of the tree.

The influence of forces and torques on grip choice is most apparent when the hand shifts between grips in the course of a task with a single tool. For example, in unscrewing a knob or a jar lid, the hand shifts grips from grip 10 or grip 11, to grip 12 or grip 13. Similarly, when using a screwdriver, if the task-related forces and torques become small, the hand will shift to a light power grip. For example, the grasp may change from grip 1 to grip 5. If the forces become still smaller, the hand may adopt a precision grip. The role of the object size is most apparent when similar tasks are performed with different sizes and shapes of tools. For example, in light assembly work, grip 12 and grip 13 approach the tripod grip 14 as the objects become very small.

6.2.4 Grippers for Gross Motion

One way to accommodate a wide range of manipulation tasks is to provide an array of special-purpose grippers that the robot is programmed to select from a gripper magazine. Using bayonet couplings, the robot can switch its gripper or end-of-arm tooling for different operations (Golden 1980). This coupling method leads to some difficulty in routing sensory information and power between the fingers and the robot arm. Nonetheless, the concept is attractive to many manufacturing engineers because the grippers can be much less complicated than a single universal hand.

The grasp trees in Figures 6.11 and 6.12 suggest, however, that if several grippers are to be used, they should be designed for classes of grasps and tasks—not for different part styles. To design a gripper for part styles is to design a tool, not a hand, like a Phillips-head screwdriver, which can only be used with Phillips-head screws; such a gripper would be a special-purpose device.

A better approach is to start with basic task requirements and let those requirements dictate the basic gripper style. The idea is to design one gripper that can satisfy the requirements in one subregion or in one branchline of the taxonomies. Thus, we can construct one gripper for precision grasps with opposed fingers, another for the power wrap grasp, and a third for pinch grasps. Then, the designs can be modified to fit a wider variety of part shapes, and finally, finger adapters may be used for specific geometric constraints encountered with any unusual parts. Specific examples of such industrial grippers are shown later in Section 6.3.

When designing a gripper, it is also unnecessary to achieve all of the different grasps in Figures 6.11 and 6.12. While it suits the machinist with his human hand to bring out a full repertoire of grips, using all five fingers in Figure 6.9 and three fingers in Figure 6.10, the grip shown in Figure 6.9 would not necessarily be needed for a robot. A three-finger wrap-around with strong fingers could well achieve the task.

6.2.5 Robotic Wrists for Fine Motion Control

The next evolutionary step in improving robots comes from compliant wrist units that can be mounted between the gripper and the robot arm. Compliance can be defined as the ability of a mechanism or structure to elastically deform in response to forces and torques. When such deflecting forces are removed, the structure returns to an equilibrium position. In particular, compliance allows the end effector of a robot to absorb minor impacts when elastic deformations occur within the wrist and/or gripper structure. For example, compliance keeps contact forces from becoming excessive if the gripper interacts with guides or fixtures during assembly.

A specific type of compliant wrist unit is the Remote Center of Compliance (RCC) device. It consists of two metal discs separated by angled springs (Figure 6.13). One disc is fitted to the arm and the other to the gripper so that the gripper can float with respect to the arm. It is especially suitable for assembly tasks where a peg that is held by a gripper and backed up by the RCC wrist is to be inserted into a hole e.g., a spindle into a pulley.

By choosing an appropriate angle and stiffness for the springs, the "center" of compliance can be projected out to the tip of the peg. This is done in such a way that initial contact between the peg and a chamfered hole produces no tilting and consequent jamming. Instead, compliant deflections caused by the contact forces orient the axis of the peg in the

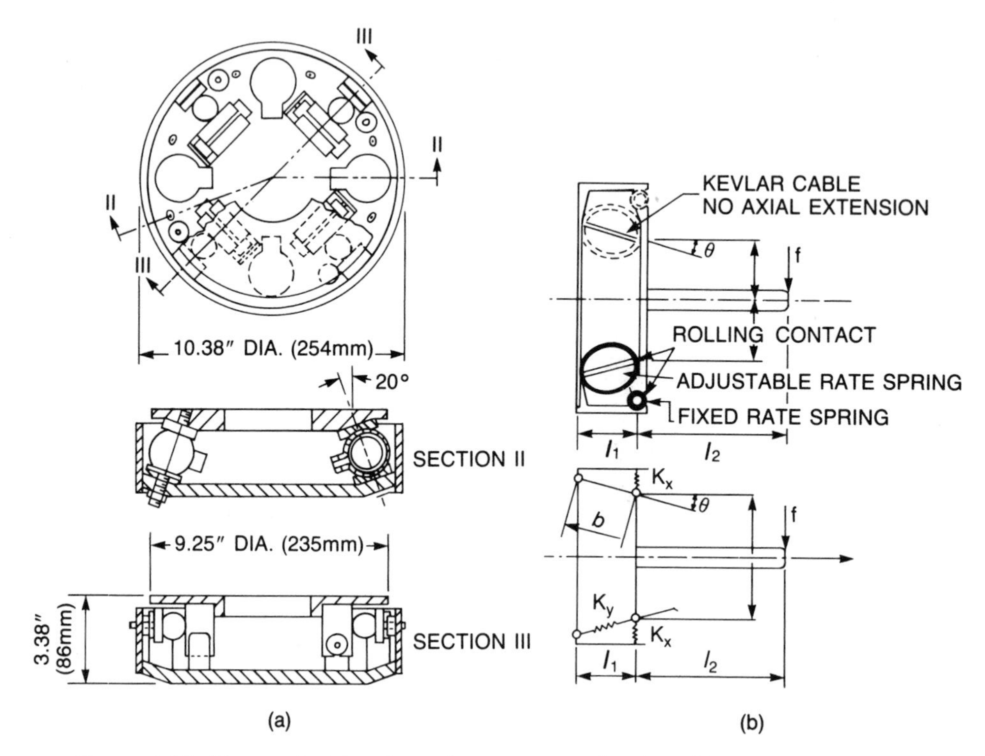

Figure 6.13 Plan and side view of instrumented adjustable remote center of compliance (IARCC) wrist. (Reprinted with permission from Wright, "A Manufacturing Hand," in *Robotics and Computer Integrated Manufacturing*, Vol. 2, No. 1, 1985, Pergamon Journals Ltd.)

right direction for smooth assembly (Drake et al. 1977; Whitney, Gustavson, and Hennessey 1983).

An additional advantage of compliance is increased safety. A compliant structure can absorb impact forces. Errors resulting from programming mistakes, robot inaccuracies, and misaligned fixtures may arise and, without compliance, damage may result. In fact, it is usually desirable to go a step beyond compliance and make the gripper capable of breaking away or collapsing in the event that a major collision occurs.

In the construction of the wrist, shown in Figures 6.13 and 6.14, (Cutkosky and Wright 1986a), the objective was to expand the capability of the existing passive remote center of compliance devices by adding two features:

1. Modifying the design so that the center of compliance length is adjustable. The center of compliance is the point in space at l_2 in Figure 6.13, also shown corresponding to the peg length.

Figure 6.14 Wrist "finding" a contour. In the lower figure, a grinding tool follows the edge of a turbine blade. (Reprinted with permission from Wright, "A Manufacturing Hand," in *Robotics and Computer Integrated Manufacturing*, Vol. 2, No. 1, 1985, Pergamon Journals Ltd.)

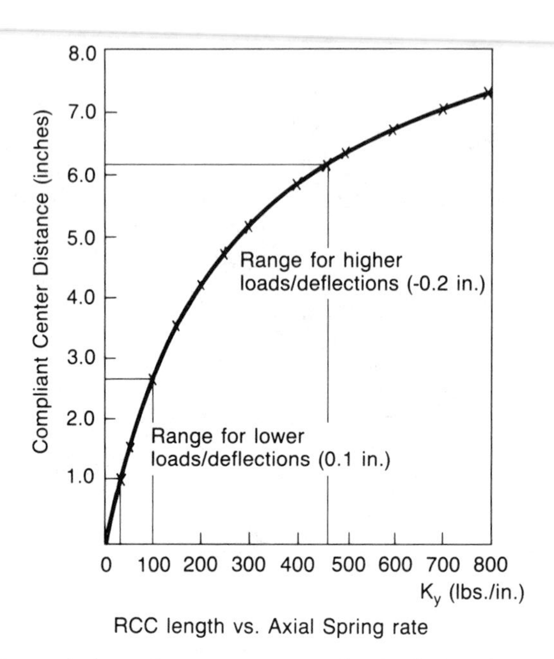

Figure 6.15 The variation of remote center of compliance length with sphere stiffness. (Reprinted with permission from Wright, "A Manufacturing Hand," in *Robotics and Computer Integrated Manufacturing*, Vol. 2, No. 1, 1985, Pergamon Journals Ltd.)

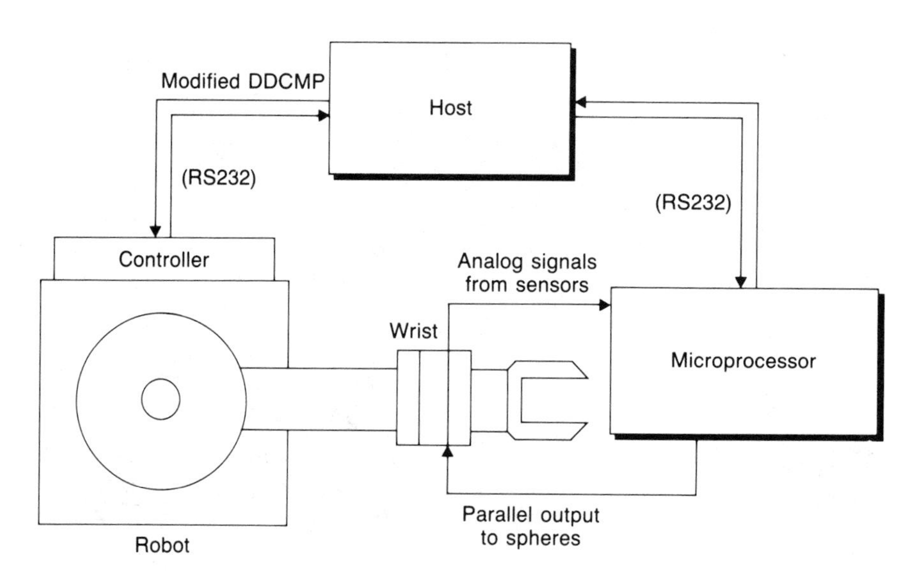

Figure 6.16 A schematic outline of the wrist control. (Reprinted with permission from Wright, "A Manufacturing Hand" in *Robotics and Computer Integrated Manufacturing* Vol. 2, No. 1, © 1985, Pergamon Journals Ltd.)

2. Providing instrumentation to sense deflections and give feedback to the robot controller.

In an RCC device, the projection of the remote center length is given by

$$l_2 = \frac{dk_y \tan\theta}{k_x + 2k_y \tan^2\theta}$$

where d is the diameter of the wrist, k_x is the stiffness of the side springs, k_y is the stiffness of the main springs between the face and the base, l_2 is the RCC length, and θ is the angle of inclination of the main springs (see Figure 6.13).

The position of the remote center depends on the angle θ, the stiffnesses, and the diameter. The design goal was to be able to vary l_2 in order to handle a variety of peg lengths and weights. It was decided that variations in the spring stiffness k_y, were the easiest to alter dynamically. The unit thus contains four hollow spheres that can be pressurized. Therefore, the force-deflection curves are deliberately nonlinear, though monotonic, and different pressures in the spheres lead to varying stiffnesses.

The net result, shown in Figure 6.15, is that the RCC length can be varied dynamically from 1 to 6 inches to respond to different requirements. Additionally, the wrist's sideways motions can be stiffened to suit the task. These automatically adjustable, reinforced rubber spheres used for the springs mimic limited aspects of our human wrists. We can tense our muscles to varying degrees depending on the strength or precision needed for different tasks in everyday life. The "mechanical programmability" of the robotic wrist allows it to be used in more manufacturing applications.

The primary function of the eight linear variable differential transformers (LVDTs) is to produce feedback information for hybrid position/force control (see Section 6.3.2). Deflections of 0.18 inches in the radial plane and 0.2 inches in compression are possible. The accuracy of the sensors in the given range is within 0.001 inches for translational motions and 0.001 radians for rotations. A simplified system layout is shown in Figure 6.16 (Cutkosky and Wright 1986a) .

6.2.6 Dexterous Hands with Active Fingers

The next evolutionary step comes from the addition of a dexterous hand with active fingers. Control laws for multijointed active hands have been developed by Asada (1979), Salisbury and Craig (1982), Okada

(1982), Jacobsen et al. (1984), and Hollerbach, Narasimhan, and Wood (1986). Dexterous hands have been built and controlled on these bases. They have been used to demonstrate gesticulations, everyday tasks like holding a cigarette, and simple assembly operations like screwing a large nut onto a bolt. For such movements, a kinematic analysis of the rolling between the fingertips and the object is needed (Cutkosky and Wright 1986b). Delicate motors and/or tendons must be installed to provide active, or closed-loop motions, of the finger joints.

The Utah/MIT hand is shown in Figure 6.17. For laboratory work concerned with developing control algorithms, it offers considerable flexibility. The four-fingered hand has sixteen joints, individually powered by tendons. Air valves in a remote power source control the tendon tension, which can be modulated to adjust both the position and stiffness of an individual joint. Hall effect sensors in the joints give feedback information on angular position. See Jacobsen et al. (1984) and Hollerbach et al. (1986) for further details of the control methods. While this particular hand is not intended for industrial applications, it is a precursor of slightly simpler, more robust models that will find industrial usage in the next ten years.

6.3 DESIGN GUIDES AND CASE STUDIES

6.3.1 The Evolution of Industrial Task Levels

A robot arm capable of repetitive point-to-point operations is now a well-established industrial entity. The active wrist unit (Figure 6.14) has been used for grinding and deburring, and similar wrist units are commercially available. Active fingers are a subject of advanced research (Figure 6.17). With this perspective, it is appropriate to examine industrial tasks and classify them in a way that suits the capability of robots and sensors.

Intersections of abilities with tasks provide design guides for both investors and production engineers. To illustrate this point, consider Figure 6.2 again. We have seen that dexterous tasks require an active finger arrangement but at a loss of power, as might be measured by gripping force, tool/work force, or payload. At the other extreme, heavy-duty, point-to-point tasks are likely to lack dexterity. Application of this principle at any time in the chronological development of robotics can answer the question: What can robots do in industry? Four tasks are now considered:

1. The pick-and-place of a heavy object

Figure 6.17 The Utah/MIT, four-fingered dexterous hand. (Photograph courtesy of Barry Hanover.)

2. A large peg-in-hole assembly
3. Grinding along a metal panel
4. Finer assembly, say, of a nut onto a bolt

These are somewhat arbitrary, but they typify industrial tasks and the classifications. Can an industrial robot do these tasks? The first task merely requires an electrically or a hydraulically operated arm with a passive gripper. If accuracy is not a concern, the ± 0.05 inches repeatability, typical of the heavy-duty industrial robot arms, is more than sufficient. As the second task, the robot is upgraded for an insertion. A peg of any length between 1 and 6 inches may now be inserted accurately into a hole.

In the third task, the grinding operation, the instrumented wrist can now be used to locate a randomly oriented plate and trace along its curved edge while applying the grinder with a prescribed force against the surface (Figure 6.14). Such a task requires hybrid position/force control of the robot (Craig 1986). If the center of compliance is adjusted to be near the tip of the grinder, the problem can be described by lateral deflections in the plane of the wrist and a rotation about the central axis of the wrist, i.e., translations x and y and rotation, θ_z.

To begin the task, assume that the approximate position, orientation, and curvature of the plate's edge are known and that an initial estimate for the robot path has been computed offline. As the robot begins to move along the estimated path, differences arise between the desired wrist deflections in the x, y, and θ directions and the actual values obtained by periodically reading the wrist sensors. The differences result from two kinds of errors (Cutkosky 1985):

- **Long-term errors:** These produce deflections in the wrist that show a statistically significant trend over several measurements. They are the result of misalignment of the work piece's edge in Figure 6.14.
- **Short-term errors:** These arise as the robot follows the surface and is subjected to fluctuating forces and torques resulting from surface roughness. These random fluctuations are superimposed on the long-term errors.

The wrist is unable to distinguish between the different sources of deflections and, consequently, the errors and disturbances combine to produce wrist-deflection data with systematic and random components. The task of the control algorithm is to extract the long-term trends from sensor data. This is a filtering problem in which any random fluctuations are thought of as "noise" corrupting a systematic signal. Whitney and Junkel (1982) describe similar problems involving noisy sensor data and discuss the advantages of using a Kalman filter to extract the systematic trends.

In the fourth task, finer assembly of a nut onto a bolt, it might still be possible, depending on the size, to rely on active wrist motions and a passive finger grip to get the nut started on the bolt's thread. Indeed, the protocol analysis of the task in Figure 6.10 indicated that this was the case. However, as previously described, the machinists did switch to fine finger manipulations once the thread was engaged. Thus, the protocol analyses of lighter assembly tasks do point to a potential need for dexterous hands with active fingers.

The automation of subcomponent assembly in the electronics industry is an area where sensors in fingers are often used. But, this brings out challenging control issues, beyond the capabilities of today's machines operating solely under position control (see Engelberger 1980; Hogan 1984; Yen and Nagurka 1987). We now review these and refer to Craig (1986) and, Whitney (1987) for a deeper discussion.

6.3.2 The Evolution of Control Methods

In the grinding operation of Figure 6.14, the robot must follow the uncertain position of the plate's edge, while still maintaining an acceptable grinding force against the edge. Simply put, the robot is asked to do two things at once: operate under position control and operate under force control. This is referred to as hybrid position/force control (Craig 1986).

As humans, we spend a lifetime developing and refining our own control systems to master such dual demands. Writing with a pencil requires position control in the plane of the paper and force control normal to the paper with a slight frictional drag in the paper's plane. Sporting skills and musical skills particularly demand the mastery of hybrid position/force control. We suggest that, while the dexterous hand (Figure 6.17) could in the near future be programmed to play a piano, it will take longer to play music that is aurally sophisticated. The execution of piano scales relies on subtle force-controlled movements, coupled with the positioning and timing of the fingers.

Returning to the grinding task, the position/force control duality can first be responded to in an approximate way. As the initial step for this method, Cutkosky (1985) and Jourdain and Wright (1986) carried out an ad hoc "calibration" of edge grinding. This involved trial-and-error experiments in which the rotating grinding tool was pushed against the plate's edge in order to determine an acceptable depth of cut that lead to a good surface finish. This empirically obtained depth of cut was then defined as a pre-position for the wrist. As a result, the initial pre-position, corresponding to an initial preload for acceptable grinding, was then set with the instrumented wrist in Figure 6.14. During the edge tracking and grinding process, the wrist deflections thus consisted of the long-term and short-term errors (discussed in the previous subsection) superimposed on the as-calibrated pre-position.

Although this method functioned well in the laboratory, it has a practical weakness. It assumes that grinding has a constant stiffness: that it obeys Hooke's law so that the initial pre-position always corresponds to an acceptable applied grinding force. Referring to Craig (1986, pp. 271-274), it assumes that the force control law for edge grinding is a linear and nonchanging proportionality between applied force and position, i.e., depth of workmaterial removed. The control problem is still predominantly related to tracking the contour under position control.

Whitney (1986) describes the grinding process model more realistically. The relationship between applied force and depth of material removed depends on many practical considerations. These include the natural variations in the hardness of the workmaterial, the amount of "clogging" that occurs on the grinding tool's surface as the process continues, and the particular vibrational characteristics of the combined robot/wrist/grinder structure. The force control law in the hybrid position/force controller must account for these physical variations before a reliable, industrial system is developed.

The design of a hybrid controller first involves the general specification of the complete position trajectory in all the degrees of freedom of the manipulator, and also the force related trajectory in these same degrees of freedom. Second, the control system must be built to operate under position control in some of the robot's degrees of freedom and under force control in others. In some of the degrees of freedom, it may be necessary to switch back and forth from one control modality to the other to suit the task demands. These task demands vary; for example, washing a window with a sponge is relatively easy because the robot can operate exclusively in position control in the plane of the window. When approaching the glass, along the normal to its plane, the robot can operate under position control in this degree of freedom. Upon contact, it must switch to force control and apply an appropriate load against the fragile glass. At the end of the stroke, it can switch back to position control in this axis.

The grinding of a plate's edge to a specified dimension is much more difficult. It requires the control of both force and position in the direction normal to the plate's edge. Unfortunately, in any physical system, we cannot independently control force and position in an independent way along the same coordinate axis. In this situation, force and position are intrinsically coupled via the mechanical stiffness of the robot/workpiece system. For this more challenging situation, the hybrid controller will need to rapidly, yet smoothly, alternate between the control modalities and do this in the same axis of the manipulator.

We now summarize this discussion of the control of active wrists and dexterous hands. The grinding tool in Figure 6.14 can be viewed as a single finger with a grinding tool at its tip. We have described how difficult it is to cope with the hybrid control issues involved when this finger tries to track a variable edge contour, while also responding to the grinding forces. Cutkosky (1985) and Jourdain and Wright (1986) simplified the situation by pre-positioning, i.e., preloading, the grinding finger by a fixed amount and then moving along the contour under position control. Analogically, they created the finger for a mechanical piano player: position control was graceful, but force control was crude. As already stated, realistic emulation of industrial grinding will require the implementation of Whitney's (1986) force control laws in a hybrid position/force controller. Likewise, realistic emulation of piano playing by a dexterous hand requires an experimentally obtained position/force relationship for piano key actuation and its implementation in a hybrid controller. And, note that this must be done for the many degrees of freedom of the hand, not just the single "grinding digit."

This discussion leads to the observation that simple, everyday tasks like drinking coffee from a styrofoam cup will require hybrid control schemes of great complexity if we expect to mimic every movement with a robotic dexterous hand (also see Mason 1985). When humans first grasp and lift the cup with a thumb and two or three fingers, they also dynamically "weigh" the contents. As the cup is placed against the lower lip and the contents tilted into the mouth, complex position/force adjustments are made. Then, the cup is repositioned on the table with a soft, force-controlled landing. It may be that for medical prosthetic situations, the dexterous hand designers will always be required to address these complex hybrid control issues. But for manufacturing, it will be preferable to redesign parts and manufacturing processes to avoid these complexities. (This is introduced again in the last two open problems of this chapter.)

6.3.3 Design Guides

At this point, it is possible to summarize the important considerations in designing a gripper. The design guides 1 through 6 are for current industrial grippers with passive fingers; guides 7 through 9 are given for the next generation of industrial grippers with active feedback.

Guide 1. Study the task to be performed and the geometry and characteristics of parts to be grasped. Focus on the task because it influences grip choice the most.

Try to express the gripping requirements as abstractly as possible. This will require a force analysis of the kind discussed by Wright and Cutkosky (1985). It is also possible at this stage to decide whether contact forces must be distributed over large areas or not. Are the parts fragile? Compliant? Slippery? Irregularly shaped? The flexibility and compliance required of the gripper will depend on these characteristics and on task-related criteria. Finally, what sort of sensory information is required from the gripper? Simple on/off devices will be adequate for most current applications.

Guide 2. Study the local environment, working space, conveyor positions, and human factors that influence the robot and its hand.

These aspects are not necessarily related to the act of grasping, but they also influence the design. In particular, if the part must be picked up from a conveyor belt or vee blocks, the gripper fingers must gain access to enough surface area of the part for the retrieval.

It may be necessary for the gripper(s) to be automatically disengaged from the wrist of the robot. Environmental conditions including high temperatures or abrasive dirt should also be introduced at this time. Other factors may include stringent weight allowances or a cramped working space for the gripper. Basic design decisions are made in response to these requirements. For example, it may be necessary to locate the actuators remotely from the fingers of the gripper, or to make some parts of the gripper from compliant materials, or to design the gripper so that it will break away in the event of a crash.

Guide 3. Determine specific solutions to the requirements in Guides 1 and 2.

Use the following subcategories in modular form:

- **Articulation and number of joints:** Review the part geometry and task movements to determine the number of joint links and their configuration geometry.
- **Materials for construction:** For example, does the work require aluminum for lightness, stainless steel for corrosion resistance or nickel for heat-resistance.
- **Actuation:** For high strength, hydraulic actuators are best; for control, electric motors are best; for ease and low cost, pneumatics are best. Also, for the task being considered, how should the actuating power be transmitted to the finger?
- **Compliance and sensors:** Compliance and safety sensors should be considered for assembly work and for good part holding.

The idea here is to develop independent solutions to the individual design requirements. At this stage, no single combination of sensors, mechanisms, or actuators should be considered. This modular approach keeps the design flexible and open to innovation and makes it easier, at a later stage, to evaluate how well competing designs satisfy each requirement. For example, if power must be transmitted from remote actuators, any sort of flexible transmission device, including cables, gear trains, chains, hydraulic lines, or rotary shafts, may do the job. Similarly, if the task requires force information, foil strain gauges, piezo-electric load cells, or piezo-resistive devices will work. As Lundstrom, Glemme, and Brooks (1977) show, there are many solutions to gripping requirements. Another common source of inspiration is nature. The human hand is the most obvious example, but, as discussed, it is unnecessarily complex for most manufacturing tasks. Simpler designs from nature include a bird's beak, a dog's mouth, a lobster's claw, an elephant's trunk, and an octopus's tentacles.

Guide 4. Begin to develop designs combining the preceding modular solutions so that the final gripper design as a whole is taking shape.

Experiment with different combinations of the solutions determined in Guide 3. Additional concerns invoke the kinds of questions habitually asked by designers: Is the design serviceable? Robust? Economical? How could it be made with even fewer moving parts?

Guide 5. Consider designs with two or three grippers mounted together at the end of the arm.

Doing this makes it unnecessary to build a single gripper to perform all the robot tasks. Another advantage is that the robot becomes more productive because it does not have to move back and forth as often. For example, a robot can use one gripper to carry a rough part over to a finishing station, use a second gripper to pick up the finished part, load the rough part with the first gripper, and return with the finished part.

Guide 6. Redesign the part and/or the task.

At this stage, it should be possible to see how changes in the part or in the robot task could simplify the design of the gripper and improve the robot's ability to accomplish the task. As pointed out in Section 6.1, the gripper is a bridge between the robot and its environment. Therefore, the part design should not be isolated from the design of the fixtures and robot gripper with which it interacts, but should be part of the total design. For example, if a gripper is required to pick up castings, it is often useful to design the castings with a tab or other identifying marks to make it easier to establish their orientation. The tabs or marks can be removed later, when the casting is machined into a finished part. As another example, it has been shown that parts become easier for robots to assemble when particular chamfers are chosen for the mating surfaces (Whitney et al. 1983).

This concludes the design guides for passive wrists and grippers mounted on a bare arm. These are summarized in Table 6.1 for convenient reference. As shown in the tradeoff Figure 6.2, the dexterity of the robot system increases as a wrist and then fingers are added. However, a key feature is that going from an active wrist with passive fingers to an active wrist with active fingers does not, per se, make the

Table 6.1 Summary of Design Guides

A. For passive wrists with passive fingers:
   1. Focus first on the task, second on the part geometry.
   2. Study the local environment and accessibility of the part.
   3. Determine number of joints, construction materials, actuation methods, and compliance.
   4. Integrate these subcomponents into a preliminary design.
   5. Consider two or three grippers for the end of the arm.
   6. Consider redesigning the part and/or the task.

B. For active wrists with passive fingers:
   7. Ensure that a person would do the task primarily with wrist manipulations.
   8. Ensure that the payload or force interactions make this a wrist task.
   9. View other factory tasks within the tradeoff among arms, wrists, and fingers.

robot "better." Appropriateness for the task is the most important design consideration from an industrial viewpoint.

It is first reemphasized that many industrial tasks, not yet automated, can be performed with a current industrial arm, a passive wrist, and a passive set of fingers. Second, for heavy manufacturing tasks, an active wrist with passive fingers will be able to do the job better than a gripper with active fingers.

These ideas are recapitulated as a set of design guides for future work:

Guide 7. Heavier manufacturing tasks that are currently done by humans using a wrap-around grip holding hand tools are the first candidates for sensor-based robots because such tasks are best done by an active wrist and passive fingers.

There is a gray line between these wrist tasks and lighter operations that would best be done with active fingers using the three-fingered grip. However, this guide can be coupled with Guide 8.

Guide 8. Manufacturing tasks that involve forces or moments greater than 2 to 5 pounds-force suggest an active wrist with passive fingers, rather than active fingers.

The 2 to 5 pounds-force quantity is derived from preliminary protocol analysis experiments that identify the force level at which the three-fingered grip slips during a task, suggesting reorienting the fingers for a wrap-around grip. In general, factory assembly tasks should be viewed in terms of Guide 9.

Guide 9. Manufacturing tasks can be categorized according to the tradeoff among arms, wrists, and fingers, and practical applications can be selected in the factory.

The last three design guides show which manufacturing tasks are the first candidates for active wrist designs. The final guide evokes the Bauhaus "form-follows-function" motif.

6.3.4 Industrial Gripper Examples

Figure 6.18 shows a simple, compact design of the most common gripper style used today. Grippers resembling this one are sold in many sizes and are made from steel, plastic, or aluminum. The actuator for closing the gripper may be a hydraulic or pneumatic cylinder or an electric motor. Only a short motion of the actuator is required to completely open and close the jaws and, because the linkage works as a toggle mechanism, the gripping force is high at the end of the actuator stroke.

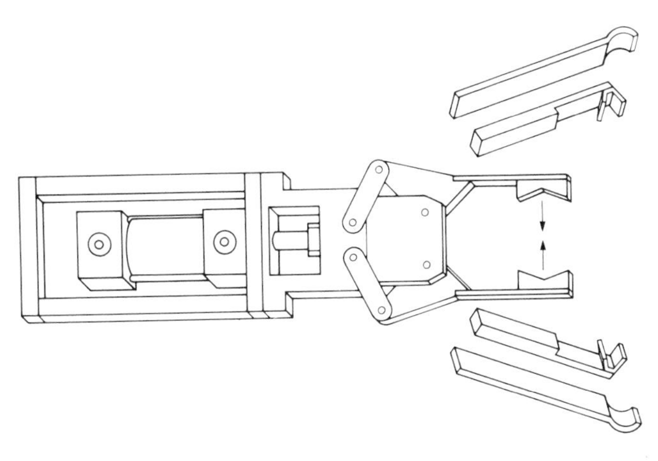

Figure 6.18 Two-fingered gripper with changeable fingertips adapted from designs in Lundstrom et al. (1977). (After Lundstrom, *Industrial Robots: Gripper Review*, © 1977, International Fluidics Services, Ltd.).

There are several variations on this basic two-fingered gripper. The actuation can also be achieved by a rack and pinion mechanism, or by a conical punch driven by a cam. A force analysis is available in Wright and Cutkosky (1985) to approximately decide on hardware selection, e.g., cylinder size in Figure 6.18. Such two-fingered grippers act on the external diameter of an object. By contrast, for gripping tubes, jars, and other items with holes in them, an internal gripper can be used, for example, with an expanding membrane.

For objects with complex or irregular surfaces, a gripper with more flexibility is required. The gripper shown in Figure 6.19 is designed to hold turbine blades. The blades come in a variety of shapes and sizes, but they all taper and twist between the root and the tip of the blade. As discussed in Section 6.2, this gripper combines some of the attributes of the human three-fingered grip and wrap-around grip. The upper passive fingers, driven by a single pneumatic cylinder, pull the blade backward until the rear edge of the blade rests against two teeth mounted in the lower fingers. This ensures that the part is correctly aligned within the gripper. Microswitches are used to detect whether the upper fingers are fully closed and whether the rear edge of the blade is pushed against the teeth in the lower fingers. The gripper shown in Figure 6.20 is designed to handle rough forgings. Like the gripper in Figure 6.19, it uses two upper fingers that are connected by a ball-joint linkage.

These grippers have a greater flexibility and compliance than the gripper in Figure 6.18. In fact, many task-based grippers have been built to increase flexibility. They include the designs shown in Figures 6.21 and

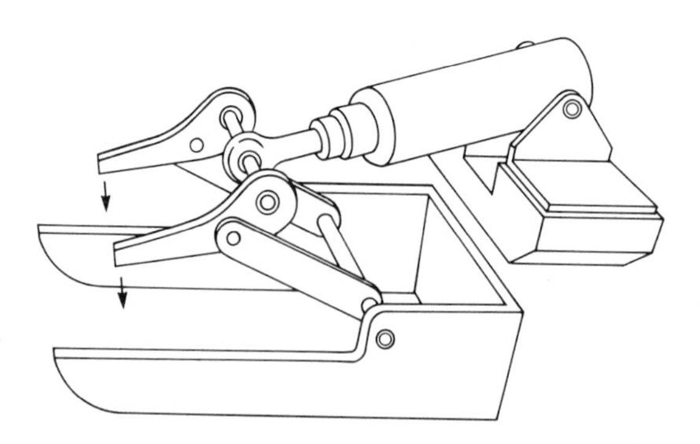

Figure 6.19 Flexible gripper for turbine blades. (Wright & Cutkosky, *Handbook of Industrial Robotics*. © 1985 John Wiley & Sons. Reprinted with permission of John Wiley & Sons, Inc.).

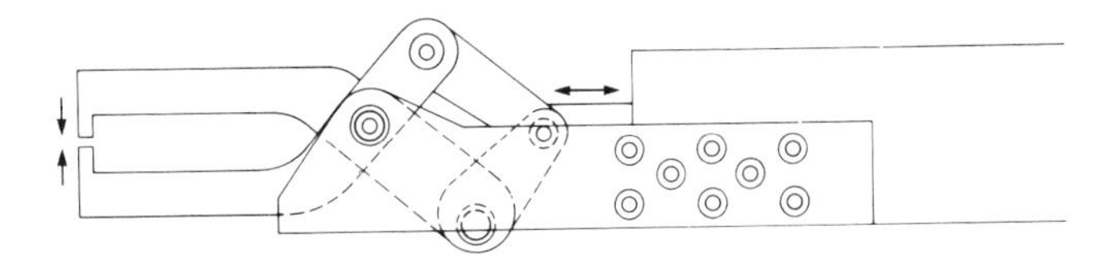

Figure 6.20 Gripper for irregular forgings. (Wright & Cutkosky, *Handbook of Industrial Robotics*. © 1985 John Wiley & Sons. Reprinted by permission of John Wiley & Sons, Inc.).

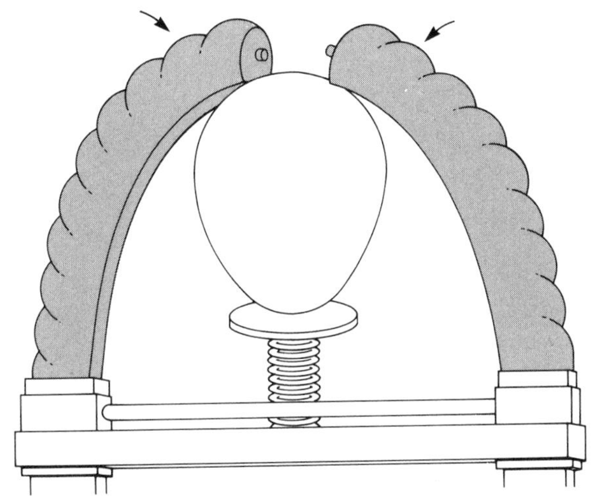

Figure 6.21 Soft pneumatic gripper and palm. (After Lundstrom, *Industrial Robots: Gripper Review*, © 1977, International Fluidics Services, Ltd.).

6.22 where a soft pneumatic gripper and adjustable "palm" can pick up a variety of shapes, and the three-fingered hand by Skinner (1975), which can alter its finger positions to suit different object shapes for passive gripping. The review by Lundstrom et al. (1977) shows the range of flexible, but passive, grippers including vacuum grippers, magnetic grippers, fabric grippers, and conformable chain grippers.

The flexibility in these designs allows the fingers or actuators to respond to the rough shapes of objects. For example, in Figure 6.19, the upper fingers are connected by a ball-joint linkage, allowing the upper fingers to settle independently against curved or irregular surfaces. The compliance that can be obtained by mounting the gripper on springs between itself and the robot arm allows small elastic deflections to compensate for accidental bumps between the gripper, its object, and assembly fixtures.

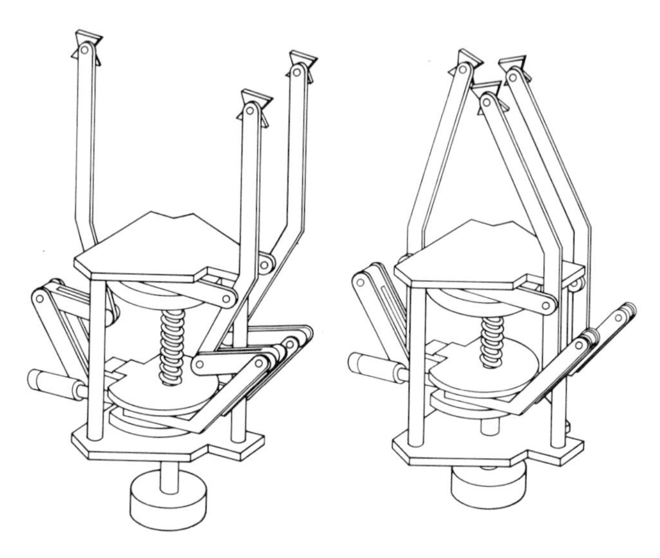

Figure 6.22 Skinner's three-fingered hand. (After Lundstrom, *Industrial Robots: Gripper Review*, © 1977, International Fluidics Services, Ltd.).

Sensors are being used increasingly in industrial grippers and can be divided into three categories of increasing cost and complexity:

1. **Binary Sensors:** These include microswitches, optical and magnetic switches, and bimetalic thermal switches. These are generally inexpensive, rugged, and easy to interface with robot controllers. They are used to sense the presence or absence of a part, or to check whether or not a variable such as pressure or temperature is within a permissible range.

2. **Analog sensors:** These include thermocouples, linear variable differential transformers, strain gauges and piezo-electric sensors. These are more expensive than the binary sensors and, generally, require some signal processing and analog-to-digital conversion. They are used when the magnitude of a quantity is required.

3. **Sensor arrays and sensors requiring low-level signal processing:** These include pressure sensitive arrays used on the fingers and palm of a gripper (Hackwood, Benni, Hornak, Wolfe, and Nelson 1983), optical arrays, and actively stimulated piezo-electric devices. For the most part, sensors of this type are confined to research laboratories, although arrays of pressure-sensitive rubber are beginning to be marketed for industrial use. All these devices

require considerable signal processing. This processing is usually accomplished with a dedicated microprocessor that communicates over serial or parallel lines with a higher level computer such as the robot controller.

6.4 SUMMARY

In Figure 6.2, an inherent tradeoff between strength and dexterity was emphasized. This tradeoff directly influences the design of industrial grippers (Wright and Cutkosky 1985). Figures 6.18 through 6.20 show strong grippers that are not particularly dexterous. The use of active fingers, of the type shown in Figure 6.17, as a substitute for the upper fingers in the turbine blade gripper of Figure 6.19 would increase dexterity, but sacrifice gripping strength. The soft pneumatic gripper in Figure 6.21 conforms to many objects and is thus fairly dexterous but weak. Skinner's three-fingered hand (Figure 6.22) has a certain measure of dexterity with its movable fingers, and, although its strength could be increased with more powerful actuators, the links and moving parts could break. It is thus not likely to offer as much gripping strength as the simple hydraulic design in Figure 6.18.

The availability of the instrumented, adjustable remote center of compliance (IARCC) wrist has increased the capabilities of industrial robots for assembly and manufacturing tasks. The wrist completes the first phase of the construction of a manufacturing hand. Thus far, this active wrist has only passive fingers. Although work is in progress to construct active fingers for the manufacturing hand, it has been emphasized that this will not make the hand "better," but only make it appropriate to another subset of manufacturing tasks. A differential between arms, wrists, and fingers has been constructed and compared with four levels of tasks ranging from simple pick-and-place work to fine assembly with dexterous hands. The cost-effective application of robots in manufacturing depends on matching the proposed task to the appropriate robotic end-effector through an awareness of the strength versus dexterity tradeoff. Design guides have been proposed that, while still being developed, provide some framework for designers of industrial robots and end-effectors.

The protocol analysis of machinists at work has identified some basic manufacturing grips. However, robots do not have to mimic the idiosyncrasies of all grip types. The taxonomies in Figures 6.11 and 6.12 categorize grips and their hierarchies, but the aim is to focus on "tasks" rather the "grips." By designing hands that satisfy all the grips in one subregion or in one branch line of the taxonomy, it is hoped that simple hands will accommodate many tasks.

6.5 SAMPLE OPEN PROBLEMS

As in the previous chapters, the sample open problems are believed to be useful starting points for research projects in either industries or universities. This field of hand research is wide open for advances in sensors, control theory, and innovative mechanism design. The list obviously has our own bias toward heavier manufacturing work and certainly does not presume to be a "top ten" set of issues for all robotics research.

1. Using the taxonomy in Figure 6.11, or a modified version to suit your experience, design a hand for another domain of life other than manufacturing. Repair work in nuclear power stations, hands for mining, and hands for lunar exploration are some ideas. The aim is to focus on a task-based, not necessarily a grip-based, design.
2. The fingers of the Utah/MIT hand are tendon operated. This design provides for a streamlined operation, but tendon drives suffer from friction and backlash. Rare earth motors are an alternative drive source, but they are rather bulky. The open problem remains of developing compact direct-drive actuators that can fit into industrial hands.
3. Design a fingertip sensor that will provide both normal and shear force feedback. Although a number of tactile array sensors are now being sold commercially, they largely give information only on the force normal to the fingertip. In fact, the human hand relies heavily on shear forces to give information on surface interactions, texture, and most importantly on whether an object is slipping. Miniaturized sensors for normal and shear force will, therefore, be very useful.
4. Simulate the control algorithms and then construct the appropriate hardware and feedback elements for the force control of a finger, based on signals from the sensor in Problem 3. The implementation of force control for hands and robots in general remains a major area for investigation. Whitney (1987) has reviewed the

various control methods for robots and hands. The list includes impedance, force, position, stiffness, compliance, and hybrid control. Research is needed on the feasibility and practical implementation of the various schemes, especially in combination.

5. Develop a variety of artificial skin materials that will enhance the capability of robotic fingers. If a skin material has a high coefficient of friction, it will be easier for lightweight, medium-force finger actuators to pick up objects. Ideally, the skin materials should exhibit whorls like our own fingertips so that moisture or oils in the manufacturing environment do not drastically reduce the coefficient of friction.
6. Develop design guides for robot hands that can be "locked up" during one phase of a multiphase manufacturing task. For example, when humans carry out tasks like carrying a suitcase, giving a blow in karate, or holding a piece of wood while drilling a hole in it, a limb or a hand locks into a desired position. In particular, fully extended limbs carry more weight than partially extended limbs because the weight is directly carried by the member.
7. Using these design guides, construct a gripper that is both dexterous and able to be put into the locked-up configuration to increase its power potential. This problem can be extended to the more general case of dynamically varying the stiffness of a robot hand to match all the different phases of a manufacturing operation. This is a logical development of the adjustable spheres in the IARCC wrist (Figure 6.15). For this work, control algorithms are needed for the various phases of the task—some tending toward locking, others tending toward a high degree of compliance. A task like manipulating a folding lawn chair requires such variability.
8. Extend this concept of variable stiffness to variable shape, i.e., develop compliant hand structures that conform to the shape of the environment in a task. For example, a plumbing snake conforms to the shape of pipes. Apply this idea to a manufacturing domain with a limited work volume.
9. Extend the "design for manufacturability" motif to "design for handling" and consider a variety of common domestic and office objects. Imagine that these will be manufactured in a fully autonomous factory. How can geometrical features and mating

subcomponents be redesigned for manipulation by simple robot hands? Is it possible to redesign parts so that no manipulations by a hand are necessary? Could gravity feeders or bowl feeders be substituted for humans and robots (see Boothroyd and Dewhurst 1983)?

10. Extending the "design for handling" motif, consider whether robots will be needed 50 years from now for manufacturing and assembly work. Think about the following simile: if one looks at a modern combine harvester it is virtually impossible to see any connections, in a mechanical engineering design sense, with a hand-held plough or scythe from 200 years ago. However, some large, mechanized agricultural devices made 75 years ago do resemble the 200-year-old human tools. As a technology moves from exclusive human operation to exclusive automation, there is an in-between phase where the developing hardware mimics the human tools. This is where we are now in the evolution of robotics. This is not a lack of design imagination, but is related to the fact that a newly installed "one-off" robot must blend into the surroundings just recently occupied by the human worker. Will robots and clever hands in fact be needed for fully autonomous factories where part designs, process designs, and production designs will be radically different from today's methods?

REFERENCES

Asada, H. 1979. Studies on prehension and handling by robot hands with elastic fingers. Ph.D. thesis, Kyoto University.

Asada, H., and Hanafusa, H. 1982. Stable prehension by a robot hand with elastic fingers. *In Robot Motion: Planning and Control*, edited by M. Brady et al., MIT Press, Cambridge, Mass. pp. 323–336.

Boothroyd, G., and Dewhurst, P. 1983. *Design for Assembly Handbook.* University of Massachusetts, Amherst, Mass.

Brady, M., Hollerbach, J. M., Johnson, T. L., Lozano-Perez, T., and Mason, M. T. 1982. *Robot Motion: Planning and Control.* MIT Press, Cambridge, Mass.

Craig, J. J. 1986. *Introduction to Robotics, Mechanics and Control.* Addison-Wesley, Reading, Mass.

Cutkosky, M. R. 1985. *Robotic Grasping and Fine Manipulation.* Kluwer Academic Publishers, Boston, Mass.

Cutkosky, M. R., and Wright, P. K. 1986a. Active control of a compliant wrist in manufacturing tasks. *Trans. ASME, Journal of Engineering for Industry*, Vol. 108, pp. 36–43.

Cutkosky, M. R., and Wright, P. K. 1986b. Friction, stability and the design of robotic fingers. *International Journal of Robotics Research*, Vol. 5, No. 4, pp. 20–37.

Cutkosky, M. R., and Wright, P. K. 1986c. Modelling manufacturing grips and correlations with design of robotic hands. *IEEE Robotics and Automation Conference, San Francisco*, pp. 1533–1539.

Drake, S. H., Watson, P. C., and Simunovicsh, S. H. 1977. High speed assembly using compliance instead of sensory feedback. *Proceedings of the 7th International Symposium on Industrial Robots*, Tokyo, Japan, pp. 87–97.

Engelberger, J. 1980. *Robotics in Practice*. Amacom Press, New York.

Focillon, H. 1981. *The Hand*, edited by R. Tubiana. W. B. Saunders, Philadelphia, Pa.

Golden, H. D. 1980. Robotic system for aerospace batch manufacturing—Task A. *Air Force Systems Command Report*, No. AFWAL-TC-80-4042, U.S. Air Force Systems Command, Washington, D.C.

Hackwood, S., Benni G., Hornak, L. A., Wolfe, R., and Nelson, T. J. 1983. A torque sensitive tactile array for robotics. *International Journal of Robotics Research*, Vol. 2, No. 2, pp. 46–50.

Hogan, N. 1984. Impedance control of industrial robots. *Journal of Robotics and Computer Integrated Manufacturing*, Vol. 1, No. 1, pp. 97–113.

Hollerbach, J. M., Narasimhan, S., and Wood, J. E. 1986. Finger force computation without the grip jacobian. *IEEE Robotics and Automation Conference*, San Francisco, pp. 871–875.

Jacobsen, S. C., Wood, J. E., Knutti, D. F., and Biggers, K. B. 1984. The Utah/MIT dexterous hand: Work in progress. In *1st International Conference on Robotics Research*, edited by M. Brady and R. P. Paul. MIT Press, Cambridge, Mass., pp. 601–653.

Jourdain, J. M., and Wright, P. K. 1986. A predictor method for robotic contour following, using position feedback information. *Manufacturing Simulation and Processes*, American Society of Mechanical Engineers, Vol. PED-20, pp. 237–248.

Lundstrom, G., Glemme, B., and Brooks, B. W. 1977. *Industrial Robots—Gripper Review*. International Fluidics Services Limited, Kempston, Bedford, England.

Mason, M. T. 1985. The mechanics of manipulation, *IEEE Robotics and Automation Conference*, St. Louis, pp. 544–548.

McMahon, T. A, and Bonner, J. T. 1983. *On Size and Life*. W. H. Freeman, New York, pp. 56, 124–145.

Okada, T. 1982. Computer control of multijointed finger systems for precise handling. *IEEE Transactions of Systems, Man and Cybernetics*, SMC-12(3).

Paul, R. P. 1981. *Robot Manipulators: Mathematics Programming and Control.* MIT Press, Cambridge, Mass.

Salisbury, J. K. 1982. Kinematic and force analysis of articulated hands. Ph.D. thesis, Stanford University.

Salisbury, J. K., and Craig, J. J. 1982. Articulated hands: Force control and kinematic issues. *International Journal of Robotics Research*, Vol. 1, No. 1, pp. 4–17.

Shahinpoor, M. 1987. *A Robot Engineering Textbook*, Harper and Row, New York.

Skinner, F. 1975. Designing a multiple prehension manipulator. *Journal of Mechanical Engineering*, Vol. 97, No. 9, pp. 30–37.

Tubiana, R. 1981. *The Hand*, W. B. Saunders, Philadelphia, Pa.

Whitney, D. E., and Junkel, E. F. 1982. Applying stochastic control theory to robot sensing, teaching and long term control. *Proceedings of the 12th International Symposium in Industrial Robots*, Paris, pp. 445–457.

Whitney, D. E., Gustavson, R. E., and Hennessey, M. P. 1983. Designing chamfers. *International Journal of Robotics Research*, Vol. 2, No. 4, pp. 3–18.

Whitney, D. E. 1986. Intelligent robot grinding system. *Proceedings of the 13th NSF Conference on Production Research and Technology: Manufacturing Processes, Machines and Systems.* Society of Manufacturing Engineers, Dearborn, Mich., pp. 189–193.

Whitney, D. E. 1987. Historical perspective and state-of-the-art in robot force control. *The International Journal of Robotics Research*, Vol. 6, No. 1, pp. 3–14.

Wright, P. K., and Cutkosky, M. R. 1985. Design of grippers. In *Handbook of Industrial Robots*, edited by S. Y. Nof. John Wiley and Sons, pp. 96–111.

Yen, V., and Nagurka, M. L. 1987. A sub-optimal trajectory planning algorithm for robotic manipulators. *Proceedings of the ROBEXS-87 Conference*, Pittsburgh, Pa, June.

Part Three

THE SKILLS OF THE CRAFTSMAN IN MACHINING

Things men have made with wakened hands, and put soft life into
are awake through years with transferred touch, and go on glowing
for long years.
And for this reason, some old things are lovely
warm still with the life of forgotten men who made them. . .[1]

—D.H. Lawrence in *Things Men Have Made*

Part III reconsiders the skills of the craftsman specifically within the context of machine tool work. The skilled machinist, familiar with CNC milling machines and lathes, possesses years of experience about process planning and the day-to-day operation of a machine as described in Chapters 7 and 8. The machinist also knows how to best operate the fixtures, what signals to be alert to during real-time chip formation and what decisions to make if things "feel" wrong. These issues are discussed in Chapters 9 and 10.

[1]From *The Complete Poems of D. H. Lawrence*, collected and edited by Vivian de Sola Pinto and F. Warren Roberts. Copyright © 1964, 1971 by Angelo Ravagi and C. M. Weekley, Executors of the estate of Frieda Lawrence Ravagi. Reprinted by permission of Viking Penguin, Inc. and Laurence Pollinger Ltd.

7 Extracting Skills from The Craftsman

One man was operating what I knew to be a vertical mill of giant proportions. It towered above him, and its slotted table stretched out to left and to right. . . . In front of the table his hands danced on the controls, so that the roaring cutter rose and fell and swerved around contours of the component while the iron turned to water before it. The howling cutter ran over the casting like a mouse, it was impossible to believe that it had not eyes and ears and a quick mind of its own. Then in a instant it had lept off into the air and the jig was open, and the cycle of the man's dance had begun again. Where did the man end and the machine begin, I wondered? Is not a musical instrument in the hands of a great player as much a part of him as his vocal cords, tongue and mouth? Ask a man how he operates his vocal cords, and he will say, "I speak, what else can I tell you?" Similarly, another man might say, "I play," and this one would shrug and say, "I mill." [1]

—Peter Currell Brown in *Smallcreep's Day*

CRAFTSMANSHIP is more than simple physical labor backed up by a few sensor-based decisions. Peter Currell Brown worked for many years in a factory and, as captured in his novel, became fascinated by the intimate, literally personal, relationship that people develop with machine tools. The opening quotation, Lawrence's poem from the previous page, and Robert Pirsig's novel (see Chapter 1) touch the "soul" of craftsmanship. How do machinists acquire this? At the beginning of their apprenticeship, they are "taught-by-showing" them many skills. They develop an ability to plan ahead before arriving at the machine tool; they create a setup sheet that minimizes setups and maximizes accuracy; they are shown how to seat parts in fixtures and how much to tighten vises and clamps; they are taught to listen for vibrations and watch the surface finish of the part to ensure good machining practice; and they are instructed on the use of micrometers and gauges to achieve good quality control. But it is the deep instinctual acquisition of

[1] Peter Currel Brown, *Smallcreep's Day*, © 1973. Reprinted with permission of Victor Gollancz Ltd.

these skills that ultimately creates the master craftsmen. These craftsmen can stand beside master chefs and master musicians knowing that they are in total control of their environments.

Is it really possible to build an autonomous machine tool that will do all this instinctual work for itself? Not in our lifetime. But we can aim for some middle ground where, as time goes on, two trends will meet. The first trend in manufacturing is an increasing predictability in processes due to the effects of "design for manufacturability," better NC programming aids, more consistent heat treatments for workmaterials, and more homogeneous tool materials. The second trend is that machine tools are becoming smarter and more sensor intensive to the extent that, while they may not be a match for the master craftsmen, they will be able to cope in the machining world with the aid of more predictable manufacturing plans.

7.1 AN INTRODUCTORY DEFINITION: THE MACHINIST

Imagine a blueprint or CAD drawing of a simple prismatic component in view of the machinist. As a working definition, the machinist generates the plan for making this part, writes the NC program, machines the part on a milling machine, and then inspects it by making comparisons with the original drawing. In a small tool and die factory, all these functions are likely to be performed by one person. Although the functions are usually performed by many different people in a large corporation, all the master craftsman's skills must be covered by this manufacturing team. In other words, these individual skills, whether held by one person or by many, make up a whole that must be considered in an integrated way in order to obtain a good component.

7.1.1 The "Brain" Anatomy of a Craftsman

Scientists interested in both human and mechanical intelligence are divided between those who advocate the strategy of copying the mechanisms of the brain and those who believe it is easier to invent intelligence from scratch. The argument for invention states that it is often more difficult to understand a mechanism and duplicate it than it is to invent a new mechanism from scratch and then try to understand the original mechanism using the invention as a model. Despite this controversy, it is engaging and thought provoking to look more deeply into the human brain's solution to intelligence. With this intention, we take a brief look at the human brain and compare it with the present and future "brain" of the intelligent machine tool.

Figure 7.1 shows a schematic of the human cerebral cortex taken from Guyton (1976) and from Penfield and Rasmussen (1950). In the following overview, we describe those areas of the brain responsible for motor coordination, sensor integration, and decision making. In view of the current goals, this overview cannot be done with rigor or depth. However, the discussion serves as a useful introduction to what is needed in a machine tool controller that purports to be intelligent.

The top surface of the brain contains a strip known as the primary motor cortex. Electrical stimulation within this area can cause contractions in the muscles responsible for the actuation and control of individual parts of the body. Penfield and Rasmussen (1950) have studied the subregions of this motor cortex in detail. They have identified the points from which each muscle group is controlled, as shown in the central-left schematic of Figure 7.1. The top of Figure 7.1 also shows the degree of representation of the different muscle groups within the motor cortex. For example, the human hand requires a great deal of motor support because most of its muscles can be controlled individually, whereas the arm's muscles act together in teams. Similarly, a manufacturing hand will require a great deal of individualized control to support the required degrees of flexibility necessary for mixed control strategies (e.g., hybrid position/force control, see Chapter 6).

The somatic sensory areas are adjacent to the motor cortex and are expanded on the right of Figure 7.1. These areas support the tactile and kinematic sensations in the main muscle groups, the sense of changing temperatures, and the sense of pain. Again, the hand is a dominant feature with a very fine-grained tactile appreciation. Craftsmanship is obviously dependent on this sense of feel in the hand. All evidence and intuition points to the need for a high degree of sensor intensiveness in the intelligent machine tool's hands and fixtures.

The lower diagram in Figure 7.1 indicates the supreme importance of vision among the human senses. At the rear of the brain, a huge area known as the primary visual cortex deciphers visual information arriving from the optic nerve fibers of each retina. Guyton (1976) presents a detailed review of the subregions in the visual cortex, and, as expected, the academic research in computer vision (see Figure 5.1) attempts to mimic the functionality of some of these subregions.

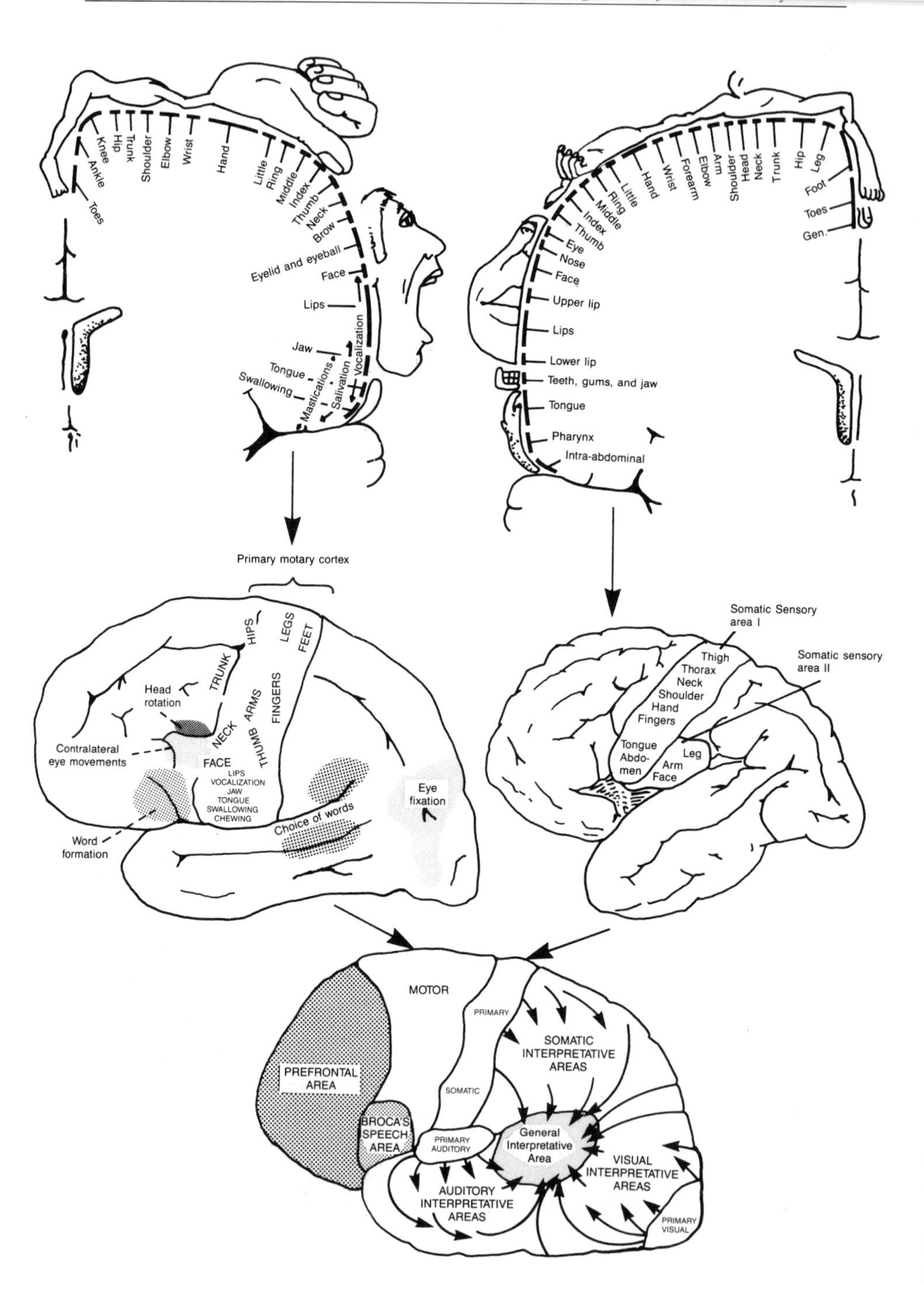
Toes
Ankle
Knee
Hip
Trunk
Shoulder
Elbow
Wrist
Hand
Little
Ring
Middle
Index
Thumb
Neck
Brow
Eyelid and eyeball
Face
Lips
Jaw
Tongue
Swallowing
Mastications
Salivation
Vocalization
Leg
Hip
Trunk
Neck
Head
Shoulder
Arm
Elbow
Forearm
Wrist
Hand
Little
Ring
Middle
Index
Thumb
Eye
Nose
Face
Upper lip
Lips
Lower lip
Teeth, gums, and jaw
Tongue
Pharynx
Intra-abdominal
Foot
Toes
Gen.
Primary motary cortex
HIPS
LEGS
FEET
TRUNK
Head rotation
ARMS
FINGERS
NECK
THUMB
Contralateral eye movements
FACE
LIPS
VOCALIZATION
JAW
TONGUE
SWALLOWING
CHEWING
Word formation
Choice of words
Eye fixation
Somatic Sensory area I
Somatic sensory area II
Thigh
Thorax
Neck
Shoulder
Hand
Fingers
Tongue
Abdo-men
Leg
Arm
Face
MOTOR
PRIMARY
SOMATIC
SOMATIC INTERPRETATIVE AREAS
PREFRONTAL AREA
BROCA'S SPEECH AREA
PRIMARY AUDITORY
General Interpretative Area
VISUAL INTERPRETATIVE AREAS
AUDITORY INTERPRETATIVE AREAS
PRIMARY VISUAL

The visual, somatic, and auditory interpretative areas all meet one another in a general interpretative area, which is shown in the center of the brain at the bottom of Figure 7.1. Guyton states that this region of confluence brings together several interpretative areas and is responsible for many high-level intellectual functions that make up the process of thinking. Memories of complicated visual scenes are also associated with this region. In Chapters 3 and 4, work that had the goal of bringing together different domains of experience was described. This can be regarded as a general interpretative function.

In addition to the general interpretative area, the prefrontal area also plays a vital intellectual role. First, it seems to be associated with shorter term immediate memory, with arithmetic tasks, and with keeping mental functions focused toward a goal. This is reminiscent of the control system presented in the manufacturing brain (see Figure 4.5). Second, the prefrontal area is associated with the elaboration of thoughts, controlling their depth and abstractness. Guyton postulates that the prefrontal area's ability to store and recall many types of information simultaneously could well explain the many functions of the brain that we associate with higher intelligence. Examples of tasks carried out in the prefrontal area include planning for the future; delaying action in response to one incoming sensory signal, so that a variety of sensory inputs can be weighed before deciding on the appropriate course of action; considering the consequences of motor actions even before they are enacted; and correlating all avenues of information in diagnosing errors and problems. The intelligent machine tool must master all these abilities before it can truly fill its role in the factory.

We now match these physiological descriptions to internal descriptions of the present and future machine tool control. Table 7.1 describes a top-of-the-line commercial machine control (supplied by Taylor 1987). Given the current controller's configuration, he found it convenient to subdivide the software system into eight categories. Three columns in

◀**Figure 7.1** Regions of the cerebral cortex from Guyton (1976) and Penfield and Rasmussen (1950). (Top: reprinted with permission of Macmillan Publishing Company from *The Cerebral Cortex of Man* by Wilden Penfield & Theodore Rasmussen. © 1950 by Macmillan Publishing Company, renewed 1978 by Theodore Rasmussen. Bottom: from A. C. Guyton, *Organ Physiology, Structure and Function of the Nervous System*, W. B. Saunders Co. Philadelphia: 1976. Reprinted with permission).

Table 7.1 Today's Machine Tool Controller in Eight Categories.[1]

CATE-GORY #	DESCRIPTION	CPU POWER	MAIN MEMORY	SECON-DARY MEMORY	AMALGA-MATED
1.	Program storage and management	15%	25%	50%	24%
2.	Motor control; including closing servo-loops, path control interpolation, and table drives	25%	15%	1%	17%
3.	Sensor-based mechanism control; including tool-changer, touch probe, and torque controlled machining	25%	20%	1%	20%
4.	NC data processing; including geometric transformations and cutting tool diameter offsets	10%	25%	1%	14%
5.	Operator interface	15%	15%	10%	14%
6.	Process-related tables; including tool offsets, fixture offsets, and tool life data	0	0	15%	3%
7.	Communications interface to the factory network	10%	0	1%	5%
8.	Other	0	0	20%	3%

[1]The columns with % values show different kinds of computer support within the controller (Taylor 1987). The last column amalgamates these using the intuitive formula; 3CPU ≡ 2 main memory ≡ 1 secondary storage. This has been done only to gain a general perspective and to aid the description of Figure 7.2.

Table 7.1 show the percentages for each category: Central Processing Unit (CPU) computer usage, main memory usage, and secondary storage (disk) usage. The fourth column of Table 7.1 combines the rows into one representative value to allow gross comparisons between the different categories.

At present, the machine tool controller has none of the intellectual functions concerned with planning or decision making. Also, computer vision has not yet been exploited. Thus, in the central diagram of Figure 7.2, the majority of the "machine tool's cerebral cortex" is hollow, except for the five areas shaded in as being partially represented. The areas that have been partially "transplanted" into the machine's control are:

1. Part of the prefrontal area
2. Some of the motor cortex
3. Some of the somatic cortex
4. A small part of the general interpretive area
5. A small area spanning both the speech and auditory regions

The storage of the active CNC program seems, to us, to approximate short-term, immediate memory and, in the spirit of naive analogy, we have allocated this to the prefrontal area. This accounts for a small part of category 1 in Table 7.1. Unfortunately, the precise location of long-term memories in the human is still an unresolved problem in science. Actually, the usual notion of location may actually be misleading when it comes to human memory because it is unlikely that single facts are stored in single places. However, there is growing evidence that at least some kinds of memories are based on long-lasting chemical changes to a neuron's post-synaptic membrane (Lynch and Baudry 1984). Thus, in Figure 7.2, there is currently no way to determine where to put the long-term storage and management of NC programs that will be used for subsequent cuts and other part programs. Category 6 has not been positioned for the same reason.

At the top of Figure 7.2, category 2 is concerned with motor tasks such as closing servo-loops, interpolation, and machine tool table drives. The regions of the machine tool controller that take care of these motor tasks correspond to the human motor cortex. But, referring to Penfield and Rasmussen's diagram at the top of Figure 7.1, since manufacturing hands and flexible fixtures do not yet exist on commercial machines, today's machine tool motor "cortex" is exclusively occupied with the torso and major limbs.

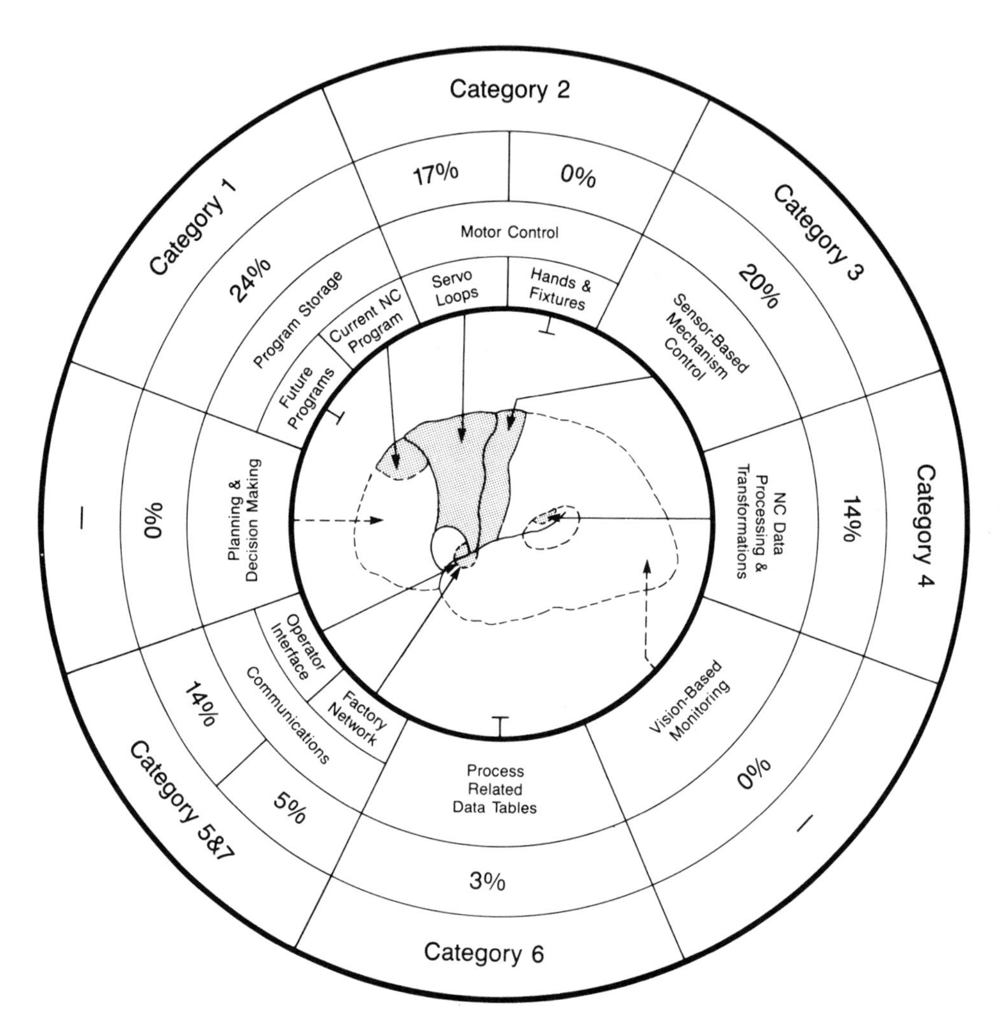

Figure 7.2 The "cortex" of the present-day NC machine tool controller. The diagram includes data from Table 7.1 showing the percentages of computer support for individual parts of the machine.

Category 3 corresponds to the somatic area of the human cortex. On today's machines, there is an increasing use of sensors for tool change mechanisms, touch probes, and torque-controlled machining. These add up to a significant computational need.

The general interpretative area is assigned to category 4 in order to account for the ability to make real-time decisions based on incoming sensor data. For example, the execution of the current CNC block (i.e., statement) brings together all the information and sends out a wave of triggering outputs. In the spirit of bringing together information and getting it ready for motor control, the geometric transformations of a CNC controller translate from one coordinate system to another. Another similar example is the ability to make offsets to compensate for cutting tools that have different lengths and diameters than were originally specified.

The operator interface, category 5, can be viewed as the communication medium with the outside world. Since humans use speech and then aural reception for communication, we have allotted this category to a region spanning the speech and auditory areas in Figure 7.2. Since category 7 is also concerned with communication, in this case to the factory network, we have also allocated it to this part of the machine's "cortex."

Chapter 1 concluded with two growth areas that will eventually lead to autonomous systems: knowledge mechanisms and motor/sensor mechanisms. As these areas mature, the intelligent machine tool's controller will more closely mimic the various subregions of the human cerebral cortex. Figure 7.3, estimates how much relative CPU usage each subsystem will need in order to build the future intelligent machine tool. These are rough estimates, but they can be compared with today's values shown in Figure 7.2.

7.2 THE MACHINIST'S KNOWLEDGE

The previous section compares and contrasts the craftsman's brain with machine tool's "brain." The goal was to highlight the many and varied functions that are needed in intelligent machines. Since the human brain already exists as a near perfect device, one can sympathize with Mary Shelley's Dr. Frankenstein and other futurists (Fjermedal 1986) who play with the idea of literally transplanting parts of a human's brain, possibly brain "software," into a robotic machine. However, this section takes the more practical approach of systematically copying the knowledge held by the machinist. The task of exposing this knowledge is difficult, partly because the knowledge falls into several categories (i.e., factual knowledge, motor skills, and sensory skills), and partly because machinists are not used to verbally expressing their thoughts about machining. Each area of knowledge is elaborated in a subsequent chapter (see Figure 7.4):

- **Chapter 8:** The intelligent machine tool will have to make its own setup plans.

- **Chapter 9:** It will have to perform its own setup with programmable fixtures.
- **Chapter 10:** It will need to have a variety of sensors that that collect data on real-time machining.

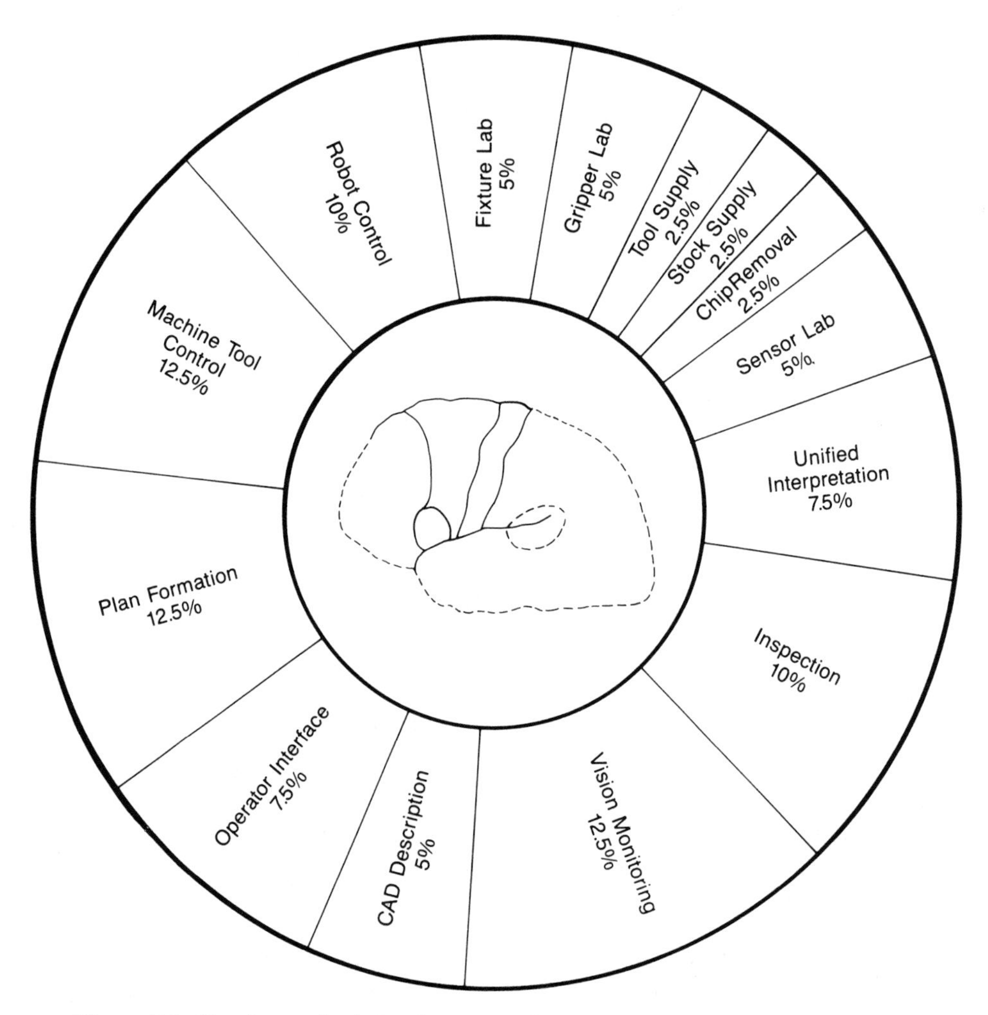

Figure 7.3 The "cortex" of the future intelligent machine tool controller, surrounded by segments that represent its major subsystems. The percentages estimate the relative CPU usage of each subsystem in the proposed intelligent machine tool controller.

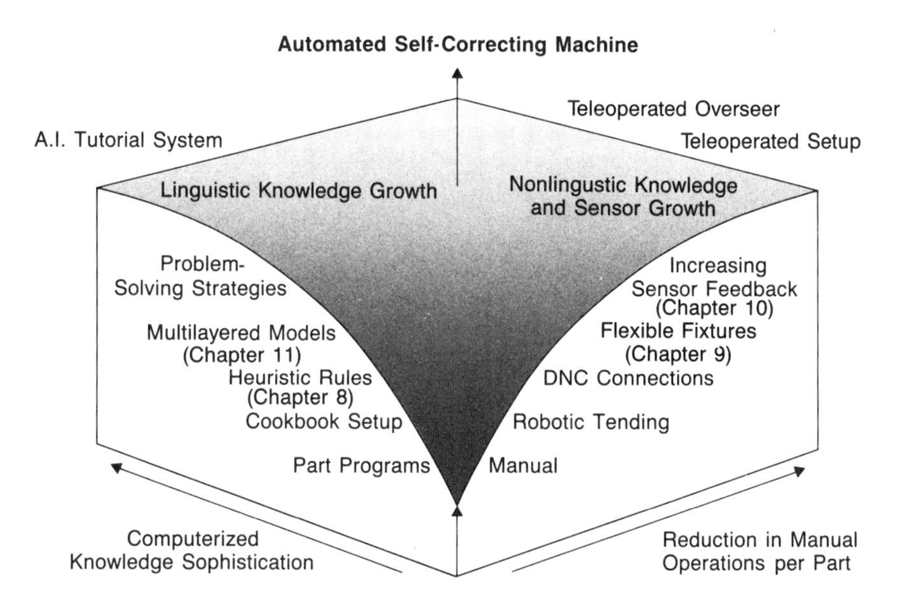

Figure 7.4 The autonomy of the intelligent machine tool. Developed from Figure 1.7 to show corresponding chapters in Part Three.

- **Chapter 11:** A team of software experts will need to manage the overall coordination of machining activities.

Just as a laboratory physics experiment is concerned with setting up a hypothesis, running tests, collecting data, and drawing conclusions, we have run experiments to extract the machinist's knowledge in the same way. It is our view that this knowledge is essential to the creation of the intelligent machine tool. This view is stated as a general hypothesis, in both its positive and its negative forms (in Figure 7.5). The positive hypothesis states that craftsmen possess unique skills that are the missing elements of full automation. Like the secret ingredients in food products and perfumes, these missing elements must be duplicated to fully automate the manufacturing process. Conversely, the negative hypothesis states that machinists are needed only to cover for the mistakes of others. For example, if the other team members of the manufacturing plant are all doing a perfect job, the master craftsman would never be needed. These hypotheses are tested in the following experiments.

Positive Hypothesis: When skilled machinists set up and operate machining centers they depend on craft knowledge and rules-of-thumb to get "a good part right the first time." To create autonomous factories for small-batch production of high-accuracy components, it is necessary master and automate the craftsman's skills.

Negative Hypothesis: The craftsman's skills do not contribute in a meaningful way to manufacturing and are only required to account for the mistakes of others. Thus, once people are removed from the factory, the craftsman's skills will no longer be required.

Figure 7.5 Positive and negative formulations of a general hypothesis.

7.2.1 Linguistic and Nonlinguistic Knowledge

Before the experiments can be started, it is necessary to recognize that one part of a machinist's knowledge is a list of facts that can be readily written down and another part of the knowledge is embodied in the motor and sensory skills that come into play during live machining operation. For example, when a violinist performs a piece of music, part of his or her knowledge is associated with reading the notation on the musical score. The violinist can manipulate this knowledge by using thought processes. It is also possible to verbally convey to someone else what the notes and the timings mean. For this reason, we define this kind of knowledge as linguistic.

To make the piece of music sound colorful and pleasant to the ear, however, the violinist has other knowledge associated with the arm movements and finger positions on the violin, when carrying out such techniques such as vibrato. Generally speaking, the best way for a musician to acquire this knowledge is to watch an expert and then repeat the actions. It is not so easy to capture or convey this kind of process knowledge because it includes both continuous actions coupled with intensive sensor interpretations. When a musician becomes an expert, these actions become so intuitive that they are difficult to describe to someone else. In other words, it is almost impossible to write down in words what to do and expect somebody else to do it. Riding a bicycle and swimming are other tasks that rely on well-developed motor and sensory skills. They are tasks that are especially influenced by the somatic interpretation and knowledge shown on the right of Figure 7.1. We define this kind of knowledge as nonlinguistic to deliberately contrast it with the factually oriented linguistic knowledge.

Eventually, these two kinds of knowledge must be combined. If we made a robot hand that could play a violin, the linguistic and the nonlinguistic knowledge would have to finally control it. Thus, Figure 7.4 has three axes: the linguistic knowledge, weighted toward the left of the figure; the nonlinguistic knowledge and sensor developments, weighted toward the right; and their combination, which builds an autonomous system shown on the z-axis. Figure 6.3 had a similar set of main components. The linguistic knowledge is combined with the sensor-based nonlinguistic knowledge to create an autonomous hand at the bottom. In Figure 7.1, linguistic knowledge is associated with the speech and general interpretative areas, nonlinguistic knowledge is associated with the somatic area, and the pragmatics of hand control correspond with the motor cortex.

In machine tool work, the values of speed and feed can be regarded as linguistic knowledge; they can be processed and can be conveyed by word of mouth. Speeds and feeds are fixed quantitative values that can be set a priori on the machine's controller. By contrast, carefully handling parts during fixturing requires a great deal of nonlinguistic skill. This skill originates from tactile, kinematic, and intuitive learning. Finally, if these tasks are going to be automated, mechanical hands and fixtures that can achieve this skill level must be constructed.

To acquire linguistic knowledge, machinists were interviewed "across a mahogany table." We captured as much as possible about their planning knowledge while they were thinking about the way in which they were going to make a particular part, create the CNC subroutines, and the setup sheet. Table 7.2, Table 7.3, and Chapter 8 summarize such work.

After these interviews, the part was made on the machine tool to expose nonlinguistic skills. To accomplish this, video tapes recorded the physical actions during setup (see Table 7.5) and the intuitive hand/eye-oriented tasks that were carried out for getting the part right the first time (see Tables 7.6 through 7.8). It was these intuitive motions, especially those that occurred during part refixturing between setups, that were of particular interest for building manufacturing hands and programmable generic fixtures (see Chapter 9). From studying the machinist in the video tapes, it is possible to gain some insight into what sensor measurements are needed in the fixturing process.

There are times when a knowledge engineering task such as this is an experiment within an experiment. For example, the machinist might make a claim at an interview table: "I never allow the dimension of a part outside of the vise to be greater than the dimension of the part being clamped by the vise." This claim is a hypothesis that the machinist is making because there is a good chance that the idea has never before been tested or even verbalized. In this case, the machinist's hypothesis can be later tested in actual setup circumstances by comparing the verbal claim against video-taped actions. From this, the machinist may find that his actions say something different from his original words.

7.2.2 Experimental Work Using Protocol Analysis

The first experiments were conducted with two machinists who knew nothing about computers, computer scientists who knew nothing about machining, and mechanical engineers who knew a little about both. A controlled experiment using protocol analysis was constructed so that parameters of the experiment could be well understood (Bourne 1986; Hayes and Wright 1986).

Simple parts that could be easily manufactured on a three-axis, vertical machining center were initially considered. Figure 7.6 shows some examples. However, the machinists and knowledge engineers did not know about these parts before the interview sessions. At the beginning of a typical session, one of the machinists was presented with a sealed copy of one of the components, represented as a conventional engineering drawing (the plan view and the front and side elevations). After

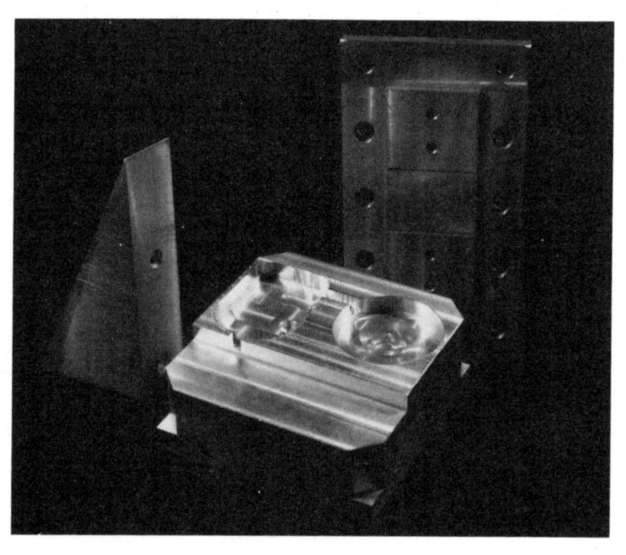

Figure 7.6 Sample aluminum parts used during the protocol analysis experiments.

1. I'm looking at the shape of the part. (*This is his way of saying that he is getting an idea of three-dimensional shape.*)
2. I'm looking for major material (*removal*) areas. (*He is looking for any unusually large features.*)
3. I'm looking at (*the*) overall size: 2 x 3 x 4 (*of the part*). (*He is looking at the envelope of the piece.*)
4. I'm checking the stock to see where the stock is with respect to the part. (*He is making sure that the stock is larger than the part.*)
5. (*I'm looking at the*) perpendicularity (*of the part*). (*This is a type of tolerance.*)
6. (*I'm checking the tolerances of the feature*) angles: plus or minus 30 seconds. (*This is another type of tolerance check.*)
7. Since the angles are plus or minus 30 seconds, I wouldn't machine them. I would send them to the grinding department. (*He is joking a little. We have already made the restriction that everything has to be milled; he knows that he can't send it to the grinding department. He is merely expressing the opinion that, since the tolerances are high and difficult to produce by milling, he would rather make it someone else's problem.*)
8. I first noticed (*the*) large material removal, but (*I am*) now noticing all of the strange angles. (*Both of these areas are of interest because both make it difficult to clamp the part.*)

Figure 7.7 An experienced machinist's thoughts during protocol analysis (Hayes 1987). The text shown in regular font is the experimental data, that is, the machinist's spoken words. The text in italics has been added for grammatical clarification or for explanation. (Courtesy of Caroline Hayes, Robotics Institute, Carnegie Mellon University, 1987).

scanning the drawing for only a few seconds, the machinist was asked about his thought processes and some of the key features that he was looking for in order to get started with the planning process (see Figure 7.7). In these first few seconds, the machininst noted the tolerances to which the part had to be made, the type of work material, and the general geometry of the part. In the subsequent 10 to 20 seconds, he concentrated on machining problems that might come from the interacting features of the part. More details on interacting features are given in Chapter 8. Next, he considered whether a simple parallel-sided vise could be used for clamping, or whether toe clamps or some special-purpose fixturing device would be needed. The knowledge engineer continued to ask questions and elicit comments during all the phases of process planning and machine action planning at the machine tool level. The machinist was observed to "picture" the part being made. He first visualized his machine tool, the available cutting tools, and the geometry of the fixtures, and then built a step-by-step machining plan in his mind. More information about how to capture the mental models used by people is given in an excellent series of papers (Gentner and Stevens 1983).

Video tapes and taped conversations represented the experimental data that was then studied with protocol analysis (see Figure 7.7). In addition, this data has become a library resource that we return to for studying other areas (e.g., hand and eye design). The vocal transcripts were best obtained by a court stenographer who, by professional training, was adept at capturing three or four simultaneous conversations and intertwining them into the chronological order of events. We found that a conventional tape recording did not work well in the main sessions. If the conversations became unfocused or overlapping it was virtually impossible to transcribe the tape later. However, if a few days later the knowledge engineer had to get the machinist to clarify a few details, a tape recording was an acceptable medium in this simplified context.

A dolphin swimming through water is a colorful metaphor for protocol analysis (Hayes 1981). Occasionally, the dolphin rises above the surface and can be seen. His journey under the water is less evident. Our verbal communications during the interview sessions were like that. When words were used, we could get a good picture of what the machinist was thinking about. Then the machinist lapsed into silence, colloquialisms, or "ahs." It was during this time that the knowledge engineers had to try to reconstruct the thought processes by analyzing the video tapes and returning to the machinist for follow-up questions. As a side benefit and goal, the knowledge engineers were being taught how to machine so that they could eventually repeat some of the steps, rather than just record facts.

After an initial planning session, generally taking one to two hours, the machinist prepared the individual CNC subroutines in the language of the machine tool. Based on the setup plan, these subroutines were then ordered, numbered, and combined into a setup sheet that contained all the step-by-step instructions for the work on the machine tool.

In subsequent sessions at the machine tool, the part was machined. Video taping continued as the knowledge engineer questioned the machinist about craft activities, monitoring procedures, measurement equipment, and general machining practices. As might be expected, there were some mistakes (i.e., errors in judgment) in the machinist's original plan. Thus it was particularly important to get good video-tape footage and transcripts during these times when the plan had to be altered. The machinist knew that there were some cases where machining would be extremely difficult and other cases where success was virtually guaranteed.

Figure 7.8 shows a range of sensitivities that the machinist used in different situations. For example, drilling a 1/4-inch hole in a block of aluminum could be done without any feedback. In this case, the operation was so safe that no contingencies had to be considered. The machinist made sure that as many operations as possible would fit within this safe or relaxing area during the part setup. The extra care taken in the setup phase of machining demonstrated the level of respect that the machinist had for managing "on alert" and "on error" situations.

Machining is plagued by errors that "might" happen, and this puts pressure on the machinist to be always vigilant (see row 2 of Figure 7.8). For example, some materials are very soft and can jam a tool, thus stalling it. Tall skinny parts can easily start vibrating in the middle of a cut, thus impairing the surface finish. A skilled machinist takes a number of precautions in this state of alert and watches the situation very carefully as it develops. The machinist reacts to very subtle and differing cues that are received from the senses. These cues include a change in chip

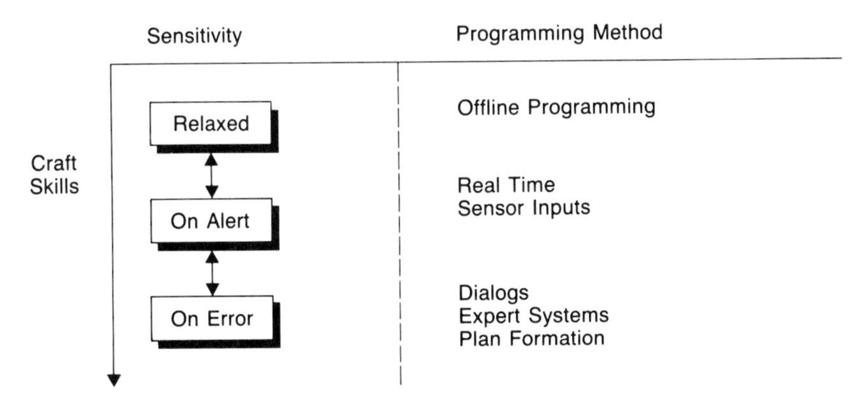

Figure 7.8 Levels of craft skills and methods.

color, a change in pitch of the machine tool, or a change in the surface finish of the part. The intelligent machine tool will need many different kinds of sensors that are preconditioned to notice such "on alert" situations. If a problem arises, intelligent sensors will have to interrupt the machining process and the controller will have to plan and execute actions to rectify the problems. For example, the controller may have to change feeds and speeds, or even initiate an emergency stop.

The machinist's skill faces the greatest demand when an error situation actually occurs (see row 3 of Figure 7.8), for example, a broken tool stuck in a part. The video tapes obviously reveal these details, often to our embarrassment. It is certainly worthwhile to rescue the part from the error and recover completely, but it is more important to discover the cause of the error situation and rectify it before another part is ruined. This discovery process must be automated, and so the system must guess the cause of the error and develop a plan to verify it. Sometimes the correction is very simple, like slowing the feed rate. But sometimes a basic misconception of the manufacturing process is uncovered that may require a completely different part setup.

7.2.3 Experimental Results

The results from the protocol analysis fall into several domains that are discussed individually. One domain is concerned with planning setups, and Chapter 8 describes this skill in more detail. Other knowledge domains are concerned with NC programming techniques, machine and part setup (Chapter 6 and Chapter 9), real-time monitoring of machining (Chapter 10), and part inspection (Chapter 5). Figure 7.4 pictorially covers this range of skills. Tables 7.2 through 7.8 list the chronology of events that occurred on the video tapes.

From these experiments, seven domains have emerged as useful categorizations of the various stages of machining:

1. Planning activities (Table 7.2)
2. NC programming (Table 7.3)
3. On-machine setup (Table 7.4)
4. Real-time machining monitoring (Table 7.5)
5. Between-pass inspections (Table 7.6)
6. Refixturing and resetting (Table 7.7)
7. Final inspection and gauging (Table 7.8)

Table 7.2 Planning Activities

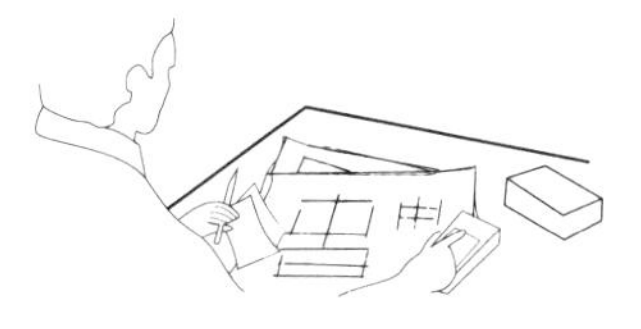

- The machinist studies part feature interactions and carries out a thoughtful ordering of the setup procedures on the machine tool. This includes refixturing from one part side to another.
- Tooling selections are made from the available database of tools to achieve the desired component shape, engineering accuracies, specified surface finishes, desirable tool life, and ease of chip disposal.
- The machinist may back up and replan so that the number of intermediate setups is minimized and so that accuracy is maximized.
- Emphasis is placed on the fixturing of parts to get an accurate part the first time. This includes the correct choice of fixture type; its correct positioning on the machine tool table; checking that the fixture will not interfere with the tool positions both during cutting and as tools come in and out of the tool changer; ensuring that all the parts can be machined given the fixture arrangement; ensuring that fixtures do not obscure areas of the part during machining; and final checking that the fixtures are reasonably flexible so that change over from one part style to another does not require a great deal of human intervention.
- The initial decision is made on the process plan. This includes how much roughing versus finish-machining, and how to order the use of the tools in the tool changer to get the maximum flexibility out of the system.

7.2.4 Summary of Results

The machinist begins by carefully planning a part's setup (see Table 7.2), attempting to always follow safe machining practices (see again Figure 7.8). The next step is to encode the process plan into a CNC program (see Table 7.3), once again rethinking the details of the setup plan. The machinist then prepares the machining environment to ensure that everything is ready for cutting (see Table 7.4). If any part of the

Table 7.3 NC Programming and Computer-Aided Process Planning Skills

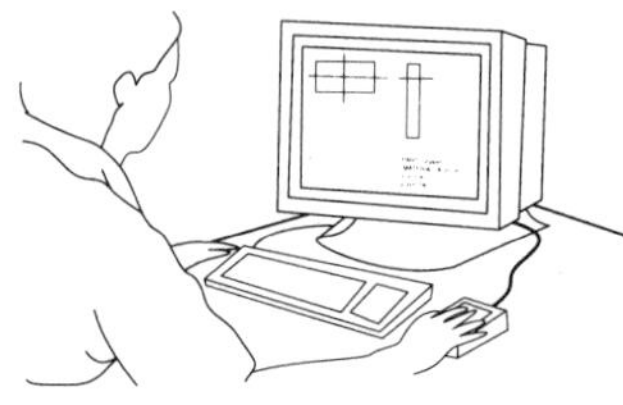

- In mental pictures, the machinist matches the volumes swept out by cutting tools to the stock material. Various combinations of roughing cuts and finishing cuts are required. Features frequently interact, and thus careful planning and programming are required to ensure that good machining practices occur.
- The tool paths and reference points are summarized, and then the various program statements are written. This involves specifications on tool movements, speeds and feeds, special commands such as interpolation routines, and commands that describe the needs for coolant or drilling cycles.
- The machinist plans a setup sheet that ties together the ordering of the CNC subroutines and explains additional data for fixturing and tooling selections. The setup sheet can be viewed as the instructional information to the operator at the machine tool. The machinist follows this setup sheet in much the same way that we follow an automobile repair guide.

environment is neglected, the first attempt at cutting may fail for a small technical reason that has nothing to do with making a complex part. The machinist then starts the cutting process, mentally checking off each operation as it begins and completes (see Table 7.5). A small misjudgment in machining technology can usually be detected before the part is spoiled. At the end of each cutting pass, rough measurements are taken to verify that everything is working as planned (see Table 7.6). If the machinist has planned a rough cut followed by a finishing cut, small dimensional errors can usually be detected and corrected in between these cutting passes. After a part has been machined on one side and checked, it is carefully refixtured, avoiding situations that could reduce part accuracy (see Table 7.7): chips on fixture plates could cock a part. And finally, after all of the sides have been machined, the finished part is carefully inspected.

Table 7.4 On-Machine Setup Skills

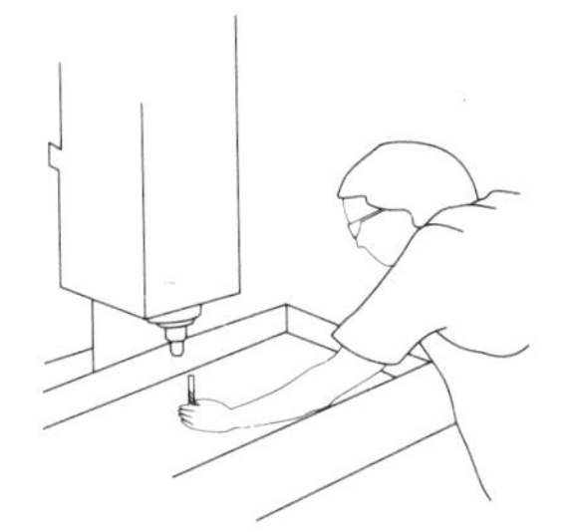

- At the machine tool, the machinist loads the cutting tools into the tool changer. In some manufacturing operations, the lengths of the cutting tools are preset in the tool room. The tool room comprises a group of highly skilled operators who understand tooling characteristics and have many years of experience in this field. The tool room acts as the central resource for tool maintenance, care, and setting. If the characterization of the tools has not been done in an offline tool room, the machinist does this with depth gauges and micrometers. The tool setting information is typed into the controller so that part programs are adjusted according to the tool dimensions.
- After tools have been selected and set, the fixtures are arranged on the machine tool. If necessary, the height of step jaws or other features of the fixtures are measured again to ensure compatibility between tool lengths, fixture positions, and desired features from the CNC subroutines.
- The machinist checks the overall workstation for coolant, safety features, and correct settings of probes or measurement devices.
- The machinist acquires the rough stock and puts it into the fixtures for the first cut. This involves the correct seating of parts and fixtures to maintain accuracy (even how much to tighten vises or clamps to ensure proper seating).
- To operate the first CNC subroutine, the machinist initializes the stock to the tool position. This involves using edge-finding tools or touch probes to find the location of part corners.
- Once the tools, fixtures, and CNC subroutines have been initialized, the machinist begins the first set of cutting operations and keeps a very watchful eye on the chip formation to ensure that tools and fixtures are behaving as expected.

Table 7.5 Real-Time Machining Monitoring Skills

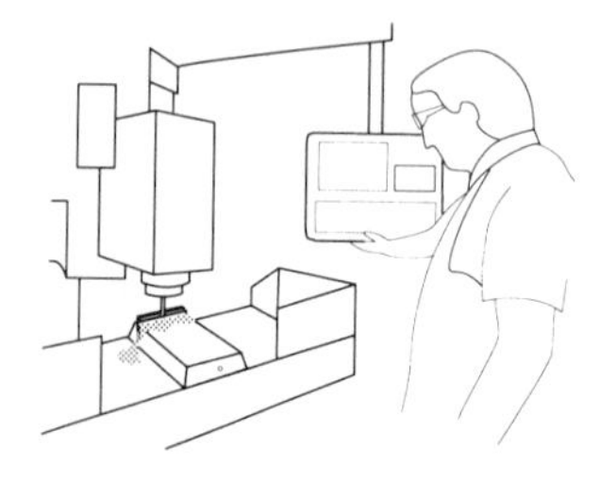

- During the cutting process, the machinist monitors chip-tool interactions. For example, the color of the formed chip indicates, in a very general way, how efficiently the cut is going. Generally speaking, a chip color that is gray or silver indicates a very conservative choice of cutting speeds and feed rates. By contrast a chip color that is purple indicates that the cutting rates may be too high and that, consequently, the tool life may be lower than desirable.
- The machinist judges the surface finish of the machined component and also listens to the vibrations from the chip-tool interactions. Readings of these situations may suggest different cutting speed, tooling selections, or minor changes in the fixturing setup. Broad judgments of the cutting forces can be made by making sure that the fixtures are not moving under the influence of these forces.
- Checks are made for adequate coolant flows and good chip formation. Short, curled chips are much more preferable than long, stringy chips, even for manned operations.
- The machinist also looks at the general shape of the features that are being formed by the tools. Even at this stage it may be possible to detect minor errors in the NC program.

7.2.5 Discussion

For the machining of quality components in small batches, the knowledge engineering experiments have verified the positive hypothesis in Figure 7.5. It has been found that the machining steps and the entire machining cycle are filled with judgment calls followed by careful inspection activities. The need for critical judgments increases when the part being machined has a complex shape that is being specified with tight tolerances (less than 0.001 inch).

For example, aircraft parts always exhibit complex profiles that are originally stipulated by the designer for aerodynamic efficiency. These

Table 7.6 Between-Pass Inspection Activities

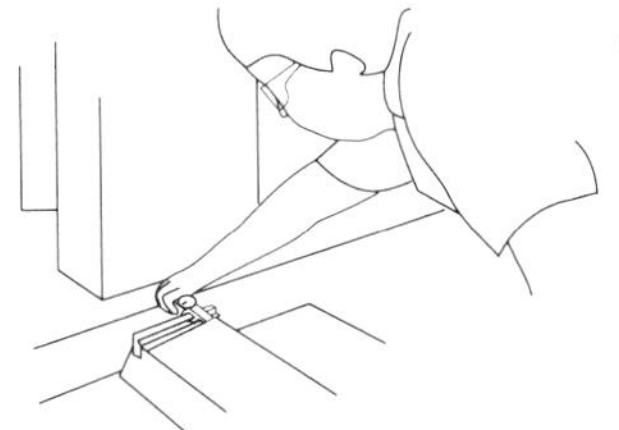

- After each cutting tool is used, the machinist may pause in the production of the part and check general features of tools and fixtures. For example, from time to time the cutting tools are checked for the amount of flank wear and crater wear, monitoring, in a very qualitative way, the amount of tool life that has been used up thus far.
- Intermediate measurements of the part are made with a micrometer to ensure that various features and accuracies have been obtained from the NC program. The machinist watches for inaccuracies that may come from backlashes in the machine or the fixtures. Other inaccuracies may come from the warm-up period of the machine tool, if it has not been used for an extended time.
- Between passes, the machinist inspects the fixtures and other aspects of the machine tool to ensure that everything is functioning according to plan.

shapes in turn require specialized NC programming techniques, unusual tooling and unique fixture designs. At the same time, to reduce weight in flight, internal pockets are designed into such parts, deliberately creating thin outer walls. As a result, the machinist needs to carefully monitor cutting speeds and fixturing techniques to ensure that such thin-walled parts will not buckle or distort both during machining (due to tool pressures) and after machining (as residual stresses in the metal are released).

The major frustration is that the majority of aircraft parts are made from titanium and nickel alloys that are very "difficult-to-machine." First, the materials work-harden during machining (Trent 1984). If the machinist chooses an inappropriate feed rate during the roughing cuts, a hard skin will be formed on the part's surface that will make the finishing cut almost impossible. Second, such alloys impose very high temperatures and stresses on the tool. This means that tool wear is very noticeable and continuous throughout cutting. After each cut on such

Table 7.7 Refixturing and Resetting Skills

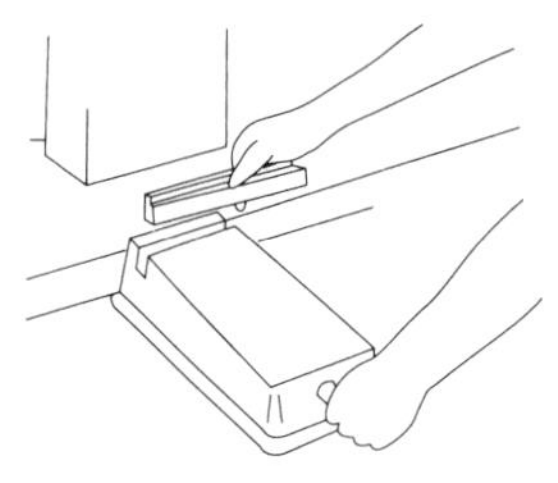

- Prior to removing the part from the fixture, the machinist removes burrs on the edges of the part since these may affect subsequent refixturing.
- The machinist then releases the fixtures and repositions the stock for subsequent cuts. In the process of doing this, the machinist makes a number of critical judgments. The device is cleared of coolants and chips so that subsequent locations are properly seated.
- The operator takes pains to settle the block into the vise, listening for slight impact interactions between the step jaws of the vise and the part. The machinist may rattle the part a little in the fixture to make sure that it is well seated. Perhaps, the stock may be hit several times with a rubber mallet to make sure that it has settled onto the step jaws of the vice. If the machinist is not safisfied, the part may be released, readjusted, and reclamped, while still using the rubber mallet and looking and listening for alignment. These are obviously very subtle movements and need to be accounted for in any automation of such a process.
- The machinist then reindexes the next tool, acquires the next CNC subroutine, indexes it to the stock with the probe, and proceeds with the next cut. This process is repeated enough times to obtain the final part.

alloys, a wise machinist checks the tool length and diameter. If this is not done, a tool that is significantly worn will produce faulty parts. Conversely, when machining soft aluminum alloys and common steels, such wear problems are barely an issue. The tools will last for long periods without changing their dimensions through wear.

Thus, the hypotheses in Figure 7.5 are strongly influenced by batch size, by the complexity of the part, and especially by the workmaterial being used. Almost anybody can set up a manufacturing facility to

Table 7.8 Final Inspection and Gauging Activities

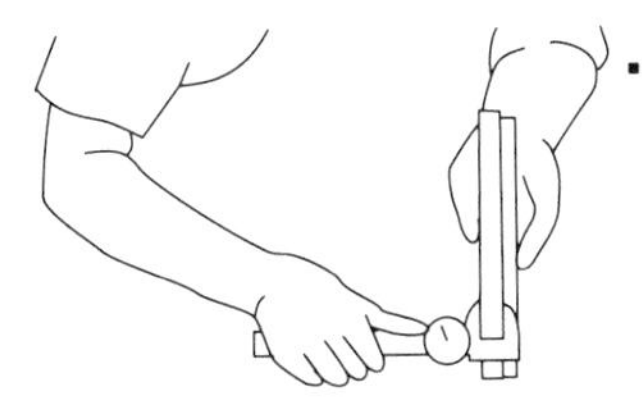

- When all features have been obtained, the machinist removes the part from the fixture and begins the final inspection. This includes dimensional measurement and checking surface finish. During this process, the NC program is reviewed and part features checked against the initial part description.
- Between components, tool quality, fixtures and other features of the machine tool are checked.
- If necessary, some quality control information is recorded for the part, with potential feedback to the process planners in the factory, the tool room, and the designer. But, without good computer networks in the factory, it has been traditionally difficult for manufacturing personnel to communicate with the designer.

produce large batches of aluminum ashtrays with undemanding tolerances. These machine actions are so predictable that an NC program can be constructed, downloaded, and executed without any checking: the craftsman's skills are not needed.

To conclude, once difficult-to-machine alloys are used, while attempting to achieve high accuracies, the cutting process itself is very unpredictable. This open-loop system is nondeterministic and unstable. The skilled machinist's linguistic and nonlinguistic knowledge become indispensable feedback elements that focus and stabilize the system's output. The machinist's judgments and abilities to plan ahead constitute the irreplaceable secret ingredients of expert machining. For this environment, the positive hypothesis in Figure 7.5 has been verified.

7.3 OPEN ISSUES

The knowledge engineering experiments captured information about linguistic knowledge with verbal transcriptions and information about nonlinguistic knowledge with video tapes. This information demonstrates that it is not adequate to build a system that exclusively uses well defined analytical formulae in an open-loop control system. Rather, it is necessary to apply a craftsman's heuristics in a feedback control system. This automation of the craftsman makes it possible to cope with the unpredictable events that occur in small-batch, high-accuracy manufacturing (see positive hypothesis in Figure 7.5). After having tested the original hypothesis, there remain open research issues that extend well outside the machining domain.

1. It is not well understood how process-oriented knowledge (nonlinguistic) can be readily extracted from an expert. However, our analyses of the video tapes have led to some insights. Some of the first science-like categorizations are shown in Figures 5.4, 6.11, and 6.12. These figures show hierarchies of visual and manual strategies that are the subcomponents of machine tool operations and of craftsmanship in general. The craftsman uses these strategies in a variety of combinations to complete various phases of the task. In the case of vision, the craftsman may first select some global understanding strategies to scan the scene and decide on a local task plan. As the craftsman concentrates on the task, the local accuracy strategies come into play. Later on, for general monitoring, the craftsman may revert back to the global understanding strategies.

2. To account for uncertainties in the environment, visual strategies and literal hand movements must be modified in real-time by sensor feedback (see Figure 6.3). A challenging area for research is the quantification of this sensor feedback. This will be necessary to understand how humans use their sensors to measure and modulate their environment.

3. Nonlinguistic and linguistic knowledge must be employed in unison. However, the parameters of their interactions have never been explored. Research that studies these interactions may have to resolve the same dilemmas reviewed in Chapter 4. For example, Figure 4.5 shows that a linguistically formulated plan may have to be altered by nonlinguistic feedback.

Because of these open issues, it has been easiest to extract strictly linguistic knowledge from machinists. Planning setup steps (see Table 7.2) uses mostly linguistic knowledge as do other tasks that can be put simply into words. The next chapter provides more details on automatically planning part setups.

REFERENCES

Bourne, D. A. 1986. Manufacturing: Acquiring craft skills through dialogues. In *Intelligent Robots and Computer Vision: Fifth in a series,* Proceedings of SPIE. Edited by D. Casesent. October, Vol. 726, pp. 481–489.

Brown, P. C. 1965. *Smallcreep's Day.* Pan Books Ltd. (Picador Edition), London, England, p. 30.

Fjermedal, G. 1986. *The Tomorrow Makers.* Macmillan, New York.

Gentner, D., and Stevens, A. L., editors. *Mental Models,* Hillsdale, N.J.: Lawrence Ellbaum Association, 1983.

Guyton, A. C. 1976. *Organ Physiology—Structure and Function of the Nervous System.* 2nd Ed. W. B. Saunders, Philadelphia.

Hayes, J. R. 1981. *Complete Problem Solver.* Franklin Institute Press, Philadelphia, Chapter 3.

Hayes, C. 1987. Planning in the machining domain: Using goal interactions to guide search. M.Sc. thesis, Carnegie Mellon University.

Hayes, C., and Wright P. K. 1986. Automated planning in the machining domain, *Knowledge based expert systems for manufacturing,* American Society of Mechanical Engineers, New York, Vol. PED-24, pp. 221–232.

Lawrence, D. H. 1968. *Selected Poems.* Penguin Books, Middlesex, England, p. 144.

Lynch, G., and Baudry, M. 1984. The biochemistry of memory: A new and specific hypothesis. *Science,* June 8, pp. 1057–1063.

Penfield, W., and Rasmussen, T. 1950. *The Cerebral Cortex of Man.* Macmillan, New York.

Taylor, R. C. 1987. Private communications on data from Cincinnati Milacron Inc.

Trent, E. M. 1984. *Metal Cutting.* Butterworths, London and Boston.

8 The Machine Tool's Setup Plans

When the ten thousand things are viewed in their oneness, we return to the origin and remain where we have always been.

—Sen T'sen

A skilled craftsman possesses a natural instinct for creating a setup plan for a random part almost without thinking. It is ironic that, for us to understand this plan, we must first break it down into "ten thousand things" only to again reassemble it into one. However, during the reassembly process, the science of setup crystallizes and evolves. The first discoveries are chains of events that lead to the accomplishment of subgoals within the overall, grand setup plan. Then, correlations and interdependencies between these subgoals are discovered. As the subgoals are linked, the complete architecture of the planning process solidifies. When this is accomplished, it is possible to see how the expert pauses, ponders on conflicting constraints, and either backs up in the plan or moves forward.

Knowledge engineering has identified chains of events that lead to a setup plan. This classification of knowledge is the first step in developing the science of setup. The setup plans are an exercise in chronological, three-dimensional geometry that are governed by practical constraints. Since they are derived from research in its infancy, they do not possess the elegance or rigor of an Euclidean geometrical proof, but these proof methods have been refined for over two thousand years. Thus, as time goes on, the science of setup will be refined and extended to more complicated parts. Already, the new awareness of the science of setup has provided the following advantages:

- Formalized setup plans can be used to educate novice machinists and designers.
- Formalized plans provide a benchmark for the further development of the science as shown along the left side of Figure 7.4.
- The plans create a language for communication. Already, in our knowledge engineering sessions, the machinists converse using such labels as major and minor reference surfaces.

- The setup plans provide chains of events that are needed to build a working expert system. In essence, the templates developed later in this chapter are one of the knowledge representations that are useful for this particular application.

8.1 AUTOMATING SETUP

In order to be completely autonomous, the intelligent machine tool must plan many kinds of actions. However, this chapter presents only the setup process as a working example, and it is based on an experimental expert system.

For today's machining on the factory floor, this software aims to provide better instructions to the machinist as he or she carries out actions on the machine tool. Current machine tool controllers only contain CNC subroutines that govern tool paths, cutting speeds, feed rates, and simple error functions. There is no information in control programs to describe how a particular part should be set up. The machinist is provided with English instructions for choosing tools, fixtures, and setup steps. These instructions are usually written or typed in a folder of papers known as the setup sheet.

For future machining systems, the software system discussed in this chapter replaces this setup sheet and is called EXSUS (EXpert SetUp Sheet). Besides being a reference source for the expert, EXSUS provides information that will make it possible for the novice machinist to set up a part. EXSUS will eventually be used by the machine tool itself to perform its own setup (see Figures 7.3 and 7.4).

Once this software has been created for the machinist, the knowledge behind it will also be used by a designer working in a completely different part of the corporation. During the design phase, the engineer will be able to interact with the system to see whether it is possible and/or easy to machine the part. And, acting as a craftsman, the system might then suggest minor design changes to simplify the part's fabrication.

Many mechanically inclined professionals have had the experience of taking a "back of an envelope" design into a model shop and discussing the design and the desired part with a skilled machinist. In such a conversation, a "spirit of compromise" occurs between the designer and the machinist. At first, only the designer has a clear idea of the part that is needed. The designer usually does not have knowledge of machining techniques and also cannot fully appreciate the shop's capabilities. On

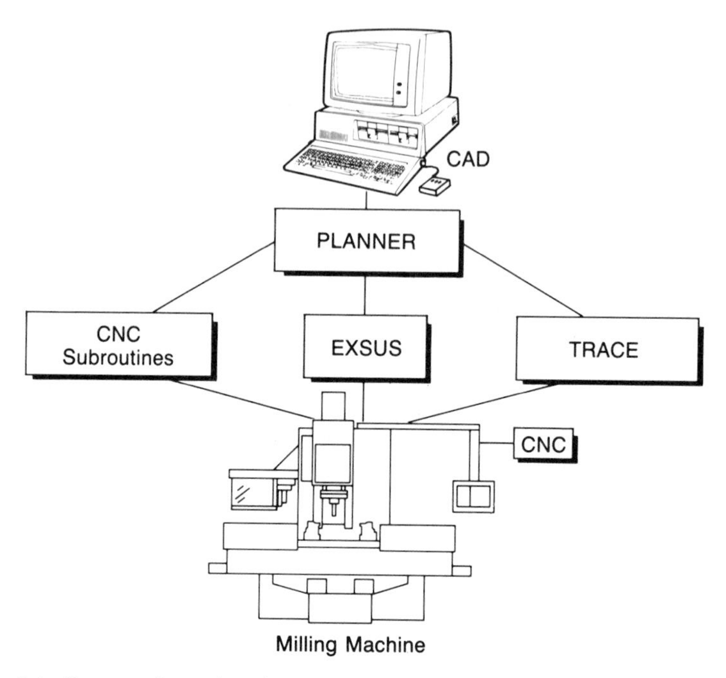

Figure 8.1 Outputs from the planning process include CNC programs, EXSUS, and TRACE. (Hayes & Wright, "Automated Planning in the Machining Domain," published by ASME, PED-Vol. 24, 1986. Reprinted with permission.)

the other side of the conversation, the machinist usually does not know anything about the design and how the part will be used, but he or she does have an intimate knowledge of the machine shop. Over a period of an hour or so, a conversation between these two people can usually solidify the design so that it can be easily made, while still satisfying the functionality required by the designer. Thus the automated system should also provide an environment that encourages this "spirit of compromise" between design and fabrication steps.

8.1.1 The System Design Specification

At the machining workstation, the planning process delivers the CNC subroutines in the usual way (see Figure 8.1). Once downloaded, these subroutines are triggered by following the expert setup sheet (see Figure 8.2). This setup sheet for the automated system is complete with information about tooling, fixtures, and stock materials. If required, the machinist or manufacturing engineer can also consult the setup sheet for details of the manufacturing process. In addition, if the human does not understand why a particular choice in the setup was made, the rationale can be displayed as a TRACE by pointing and clicking on the value in question. Figure 8.2 shows the human pointing at "2 flutes," thus questioning why this particular end-mill was chosen. The explanation can be displayed in stylized English or in a more verbose tabular display.

The setup sheet gives detailed information on the cutting tools, the stock size and preparation, fixture sizes and datum locations (i.e., geometrical origins), the ordering of setups and their relationships to individual CNC programs, the in-process quality control checks, and real-time comments on machining technology. In this latter category, there are occasions when a given fixturing procedure is risky, but there is no other way to machine the part. In this case, the machinist would be advised to reduce the feed rate to 10% of its usual value and to keep "on alert" (see Figure 7.7). The advice column in the setup sheet may then

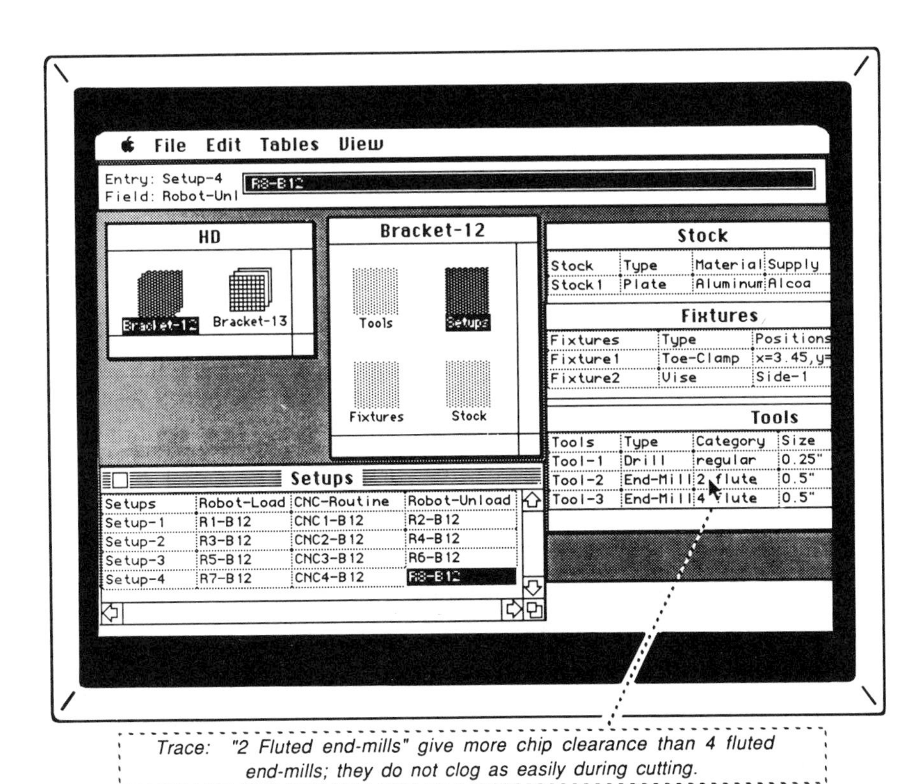

Figure 8.2 Features of EXSUS and TRACE.

say:

> If there is a lack of vibration and the surface finish looks satisfactory, you can cautiously increase the feed rate to the full 100% value.

Between passes, the machinist is also instructed to make measurements on the part to verify that everything is proceeding according to the plan. When the part has been completed, the machinist is supplied with detailed instructions for gauging and inspection.

In some cases, unexpected events occur during the machining operation. A tool may break or the machinist may be aware that the part is vibrating too much. At these times, the machinist can consult with the software system and explore alternative solutions to the current problem. As such, we can see how the system can be used by senior machinists who currently discuss problem situations among themselves.

In summary, the information in Figure 8.2 can help to train a novice machinist by simulating a more experienced man looking over the novice's shoulder. At the same time, the software is of considerable use to the skilled machinist—it allows faster setup, gives advice that improves the accuracy of the finished part, and generally improves the richness of the machining environment.

8.1.2 The First Thoughts of the Machinist During Setup Planning

To discover the starting point for creating a setup plan during knowledge engineering sessions, we analyzed the machinist's thinking patterns during the first 5 to 10 seconds of scanning over a new part drawing. The initial thoughts and questions tended to set the tone for the entire setup experience. Apparently the machinist used the strongest constraints almost immediately in order to eliminate many variations of setup that had no chance of working. Some of these constraints could be extracted from the video tapes and verbal transcripts:

- **Size:** What is the size of the stock in relation to the machine tool?
- **Size:** Does the stock fit in the available fixtures?
- **Machine Tool Characteristics:** Can the part geometry be made on available machine tools?
- **Tools:** Will the available tools be able to cut all the part features?
- **Fixtures:** Can the part be held in available fixtures, or do new fixtures need to be built?

- **Machinability:** Do the materials cause any unusual machining problems, especially considering tooling and fixtures?
- **Tolerances:** Do the tolerance requirements cause any unusual machining problems?

To simplify the proceedings, the expert system was initially written for parts made from rectangular stock, clamped in a parallel-sided table vise on a three-axis machine tool. We supplied the machinist with a piece of stock that was slightly larger than the final part to be made. The initial question to resolve was: Of the six possible orientations of this rectangular piece of stock in a parallel-sided vise, which was the best ordering of these orientations to obtain a block with all six sides machined? This ordering process was also mixed up with obtaining the features such as drilled holes and pockets. At first, the planning for these setups seemed to be arbitrary until some underlying principles were discovered.

8.1.3 Details of Stock Characteristics

Stock arrives at the machine tool in a variety of ways. It can be cut from a flat plate or from the end of a long piece of bar stock. In these two cases, there are already some smooth sides, which are cold rolled as an as-supplied condition. The machinist can often rely on these smooth sides for the initial referencing of the stock against the vise jaw. If the tolerances are not crucial, the machinist may even be able to use the dimension across two opposing cold-rolled sides as a final part dimension. In other examples, the machinist might receive the stock completely rough-sawn cut on all sides. In this case, all the sides must be machined to make them smooth. Here, the stock is generally 1/8 inch bigger all round than the final part size so that all faces can be cleaned up.

A key starting point for the production of any part is to ensure that three orthogonal sides are created in the stock. These sides can be used as reference planes for all features in the part as well as the opposing sides of the stock. In the first example, we present the simplest case where a machinist will end up with a fully orthogonal, six-sided, cube-like block into which features can be machined (see Figure 8.3). In practice, the experienced machinist may be able to start putting features into the block before the sixth step, because the features in one face will not be upset by subsequent squaring off operations on opposite faces. However, the steps shown in Figure 8.3 are the fool-proof, conservative way of making sure that the block is squared.

	Orientation or Process	Face Against Fixed Jaw	Notes
Step 1	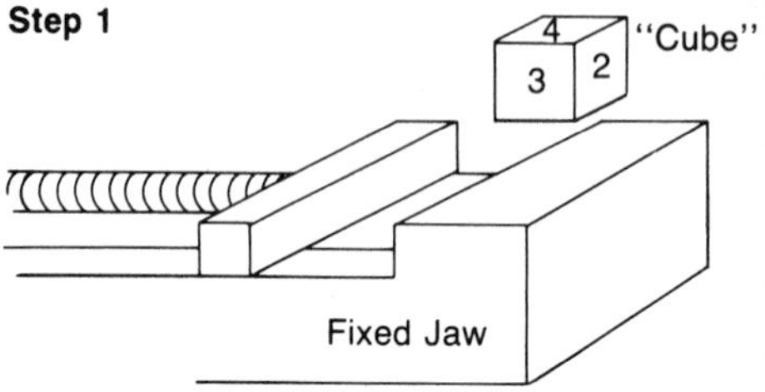	2	Put best side against the fixed jaw and prepare to face mill side 4.
Step 2	Machine side 4 (face mill)	2	First machine face.
Step 3	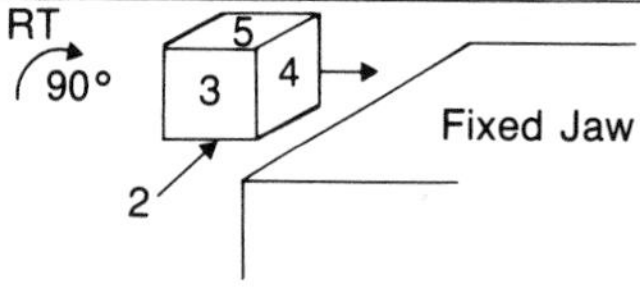	4	Turn by 90° to your right to put this well-machined face against jaw. Side 5 comes up.
Step 4	Machine side 5 (face mill)	4	Two machined sides
Step 5	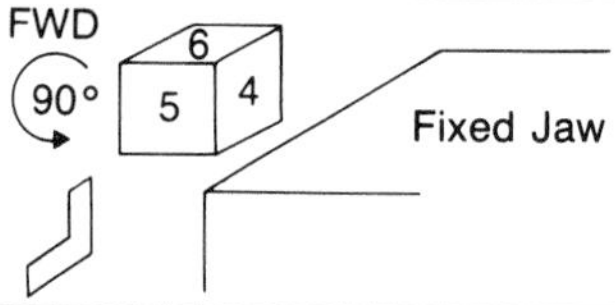	4 (still)	Flip toward you by 90° to bring side 6 up. But we must put a hand square against side 5 to make it vertical. (Side 3 is still saw cut.)
Step 6	Machine side 6 (face mill)	4	Three orthogonal planes.
Step 7	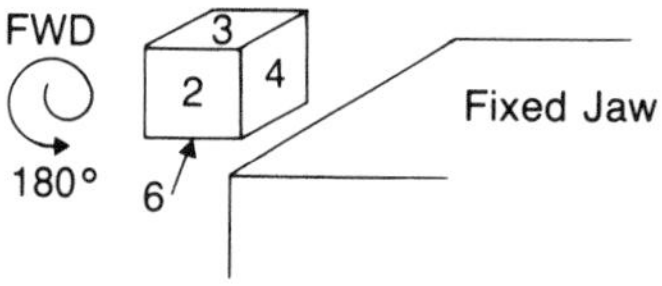	4 (still)	Flip 180° towards you with side 4 still against fixed jaw. Seat side 6 down (being a nice machined side) and prepare to machine sides.
Step 8	Machine side 3 (face mill)	4	Four machined sides.
Step 9	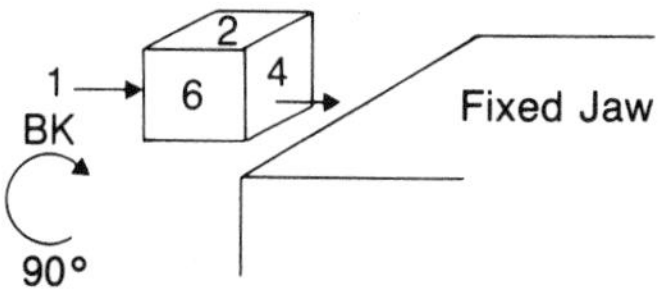	4	Rotate backward by 90° keeping side 4 against fixed jaw and side 5 on bottom.
Step 10	Machine side 2 (face mill)	4	Five machined sides.
Step 11	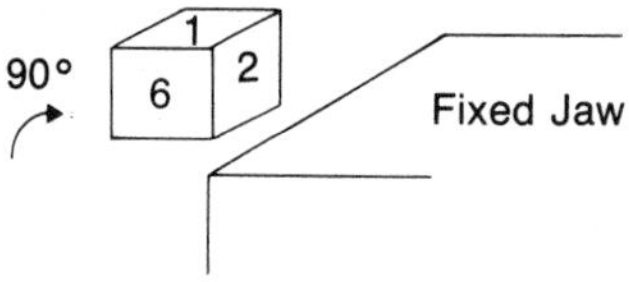	2	Rotate 90° to the right.
Step 12	Machine Side 1	2	All sides machined.

8.1.4 Case Study: Rules for Obtaining a Well-Machined Block

In this case study, we develop rules that a machinist must use in order to square up a piece of cube-like stock. This is the beginning of putting a machinist's intuition into a systematic framework. The reader may be concerned that this example is too simple, but these results have not been published in either academic or trade literature. However, these methods are well known to machinists and are commonly passed on by word of mouth when an apprentice enters the trade.

The procedure outlined in Figure 8.3 is especially important in skilled grinding operations. In grinding, the tolerances that an operator must achieve are even more exacting than in machining. Even if the stock comes partially smoothed from some previous cold-rolling operation, the cold rolling is still not good enough to keep faces aligned to the accuracies of 0.0002 inch commonly needed in grinding. Thus, this setup sequence is a widely used skill, and it is a good example of the type of information that must be in the setup sheet.

In this particular example, the stock is not allowed to overhang the sides of the vise, and there are no features in the final part. We just need to obtain six sides that are very flat and very square.

The machinist begins by scanning the part and decides which is the "best" of the six sides. He then positions this side against the fixed jaw of the vise. A machining vise generally consists of two jaws: one is fixed and permanently located against the bottom of the vise, and the other is movable. The fixed jaw (FJ), shown on the right of Figure 8.3, and the bottom of the vise are points of absolute reference. The machinist depends on the fixed jaw and the bottom of the vise being unmovable, never deflecting, and always remaining orthogonal. On the other hand, the moving jaw is not as reliable. This face is actuated by either a screw or hydraulics, but in both cases the moving jaw could twist slightly if the part is deliberately located off from the central axis of the vise.

In Figure 8.3, side 2 has been arbitrarily chosen to represent the best side of the six saw-cut sides. With side 2 against the fixture, the machinist then creates the first machined surface, which is side 4 in the example. He then puts this good machined surface, side 4, against the fixture and keeps it against the jaw as long as possible. This face is rotated several times while four other faces are machined. For this ex-

◀ **Figure 8.3** Setup procedures for an orthogonal block. (Hayes & Wright, "Automated Planning in the Machining Domain," published by ASME, PED-Vol. 24, 1986. Reprinted with permission).

ample, the machinist has to use a set square in one of the setups to make a machined face vertical to the bottom of the vise. The machinist achieves a major goal in step 6 in the diagram, where three orthogonal planes are obtained on the block. In fact, the subsequent steps become easier as the block becomes "squarer" and the exact ordering beyond step 6 is not absolutely crucial. However, the machinist knows that this sequence is conservative and guaranteed to succeed.

8.1.5 Rules for Setup: The Expert System

As the part dimensions, and hence basic stock geometry, begin to expand away from the simple cube-like stock, new constraints begin to influence the machinist's plans. The ease with which he can clamp across faces, and the stability of doing so, bring extra constraints into the previous case study.

As an example, consider machining a part that has the general dimensions of a thin paperback book. Its stock could be presented as either a "bread slice" or a "plate." These terms convey a mental image of the stock's appearance. The bread slice is rough cut on the two large faces, but smooth rolled on the four edge faces. By direct constrast, the plate is smooth rolled on the two large faces, but rough cut on the four edge faces. To square up such stock, the machinist is more guided by smoothness than by three-dimensional geometry. Thus, for the bread slice, the machinist first clamps across the edges and face mills a large face; for the plate, the machinist first clamps across the large faces and mills the edges.

Certain common ideas that are used over and over in setup situations are summarized as a set of definitions in Figure 8.4. Using these definitions, rules on setup have been established from the transcripts (see Figure 8.5). Each rule is classified according to whether it is a necessary condition for setup (required) or simply a good machining practice (preferred).

8.1.6 General Methods for Planning Stock Setup

The example for saw-cut, cube-like stock was carried out without overhanging the part from the side of the vise. In fact, the machinists will minimize the cleanup steps by clamping the stock so that it sticks out of the vise's side. In this way, a milling cutter can face-mill the top face and then side-mill the overhanging face in just one setup.

In the next set of experiments, the machinists were presented with

1. **Fittable**: A dimension that will fit within the jaws of the vise.
2. **Clampable**: Stock that is both "fittable" and will not bend or buckle when the clamp is tightened.
3. **Squaring Surface**: A side that is "clampable" and is at least 1/2 inch in all dimensions. Otherwise, the side is too short to guarantee accurate squaring.
4. **Flattest**: A saw-cut side is less flat than a rolled side, which is less flat than a machined side, which is considered the flattest.
5. **Major Reference Surface**: The side that is used to square the piece in two dimensions, usually the largest area surface.
6. **Minor Reference Surface**: The side that is used to square the piece in the one dimension that the major reference surface does not take care of, usually the longest remaining side.
7. **Reference Surfaces**: Depending on the tolerance needed, a reference surface is one that is flat enough for the task. If the tolerance needed is less than ±0.0015 inch, only a machined side is flat enough to be a reference side.
8. **Desirable Setup**: A desirable setup always has a major and a minor reference surface. Again, all squaring surfaces are either machined or rolled, but never saw cut. Major squaring surfaces must not be less than 1/2 inch in all dimensions. Minor squaring surfaces that are less than 1/2 inch thick must have a machined or rolled opposite side and they must be clamped between the jaws of the vise. Some desirable setups are shown in Figures 8.6 and 8.7.

Figure 8.4 Definitions for setup of prismatic stock. (Courtesy of Caroline Hayes, Robotics Institute, Carnegie Mellon University, 1987).

about 40 simple parts. These were all six-sided, generally rectangular parts, but the length-to-height-to-width ratios varied over the 40 samples so that very different types of starting stock were needed for the plan. Apart from the simple cube (see Figure 8.3) and the thin bread slice described earlier, stock was obtained in plate form (two large cold-rolled surfaces and four saw-cut edges) and in bar form (four opposing cold-rolled sides and two saw-cut ends like an extremely long bread slice). We studied the way in which the machinists handled these different stock geometries and surface finishes and expanded the rule set in Figure 8.5 accordingly.

After comparing all the finished plans, an important result emerged: the machinists used generic patterns of plans and these were termed

Required Rules

1. Obtain three adjacent orthogonal reference sides as the first subgoal.
2. If the stock is rough-sawn cut all over, then create the major reference side first by placing one of the largest area sides facing up in the vise for face-milling. Note that this face will next be rotated to clamp against the fixed jaw as step 3 in Figure 8.3.
3. If the largest area sides are rolled, then clamp them against the fixed jaw and use them as major reference sides.

Preferred Rules

1. If a hand-squaring operation is to be done, then do it as soon as possible.
2. If step jaws are used, then do not move them any more than is absolutely necessary.
3. If the major reference side is against the fixed jaw, then put the longest remaining side against the vise's bottom.
4. If the major reference side is against the bottom of the vise, then choose the longest, flattest remaining set of parallel sides to go against the vise jaws.
5. If a major reference side is available, then keep it against the same part of the vise (fixed jaw or bottom) for as long as possible.

Figure 8.5 Rules for setup of prismatic stock. (Courtesy of Caroline Hayes, Robotics Institute, Carnegie Mellon University, 1987).

"plan templates." Examples of these templates are shown in Figures 8.6 and 8.7 as plans A through F. Note that the cube-like stock (see Figure 8.3) with no overhang fits into plan B. The text along the top of the templates describes the kind of stock that fits into a particular plan. Each generic plan actually represents several possible plans; what is shown is a flexible sequence of setup steps. The arrows indicate a sequential transition of steps and the branches in the tree represent a set of setup choices, which can be made in any order. The awareness of the templates greatly simplified the problem: The basic rules were no longer used and it was necessary to only look at the stock to determine into which plan category it fit. The plan "construction" problem was transformed into a plan "recognition" problem. As discussed by Hayes and Wright (1986), this highlights one of the challenges in knowledge engineering. The expert knows how to do a task, but not on the basis of a

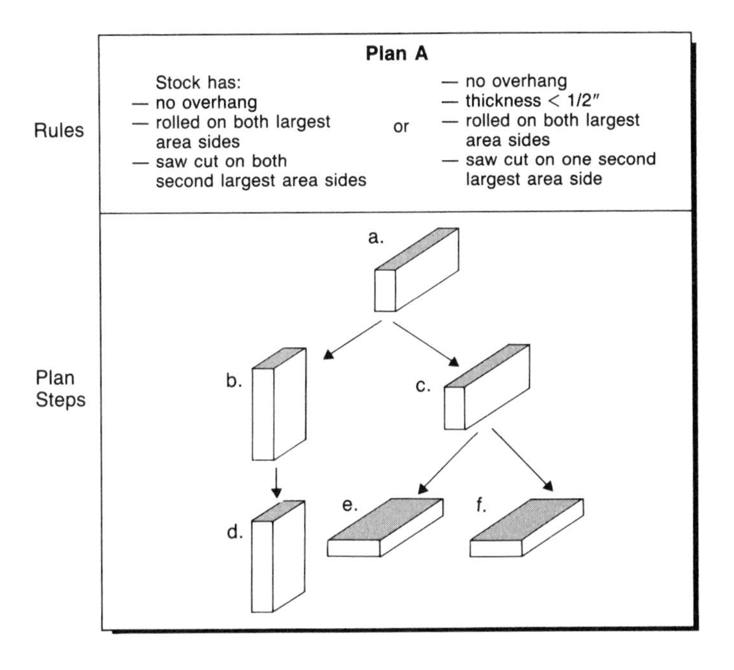

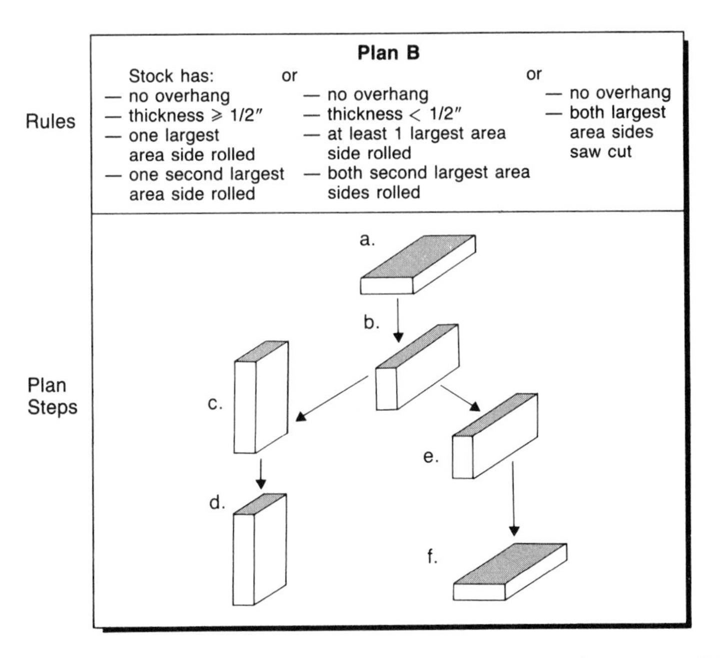

Figure 8.6 The structures of plan templates A and B. (Hayes & Wright, "Automated Planning in the Machining Domain," published by ASME, PED-Vol. 24, 1986. Reprinted with permission).

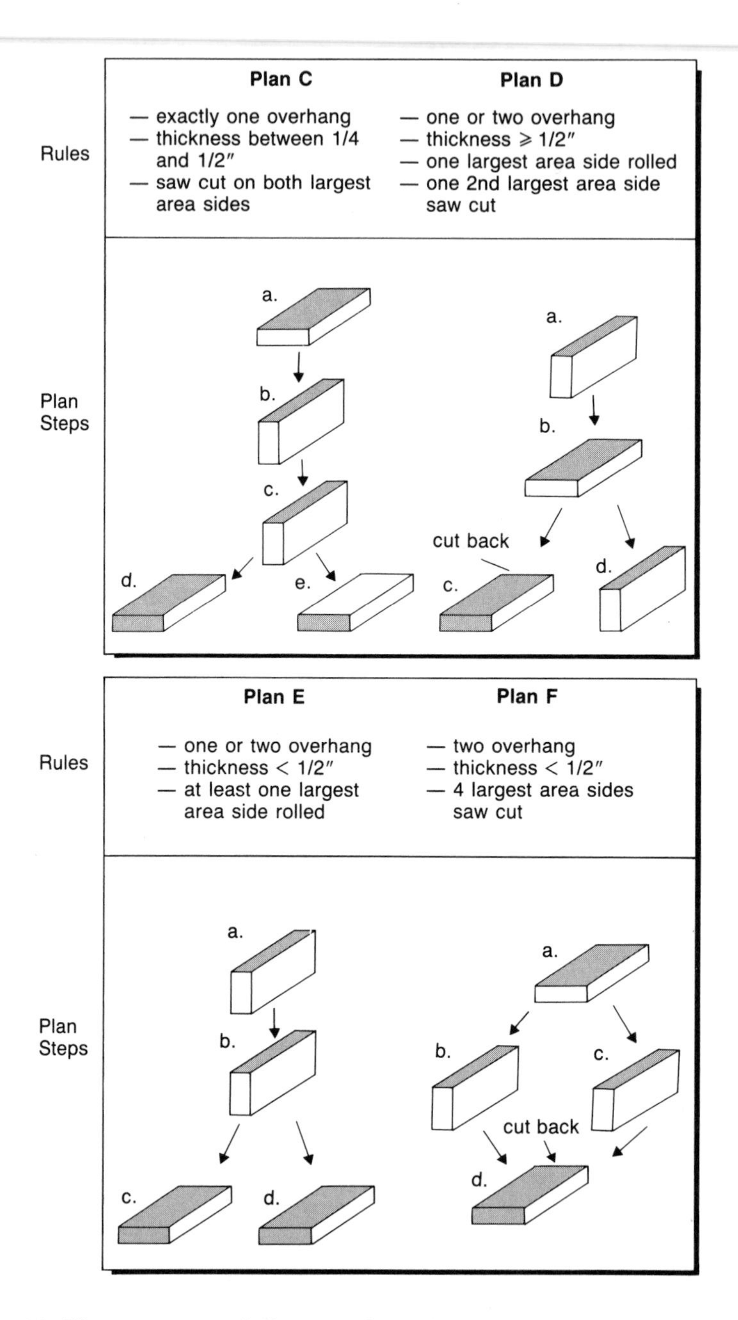

Figure 8.7 The structures of plan templates C, D, E, and F. (Hayes & Wright, "Automated Planning in the Machining Domain," published by ASME, PED-Vol. 24, 1986. Reprinted with permission).

logical sequence of rules; rather, the expert recognizes patterns and underlying forms (see the Pirsig quotation at the beginning of Chapter 1) that have been accumulated from many past experiences.

Ideally, knowledge engineering leads to creative schemes for knowledge representation, capturing the original expertise in a cogent, condensed form. Once in this form, the knowledge can be used in the various ways mentioned at the beginning of the chapter: education, benchmarking the science, communication, and building expert systems. Figure 8.8 is an example of how the rules for choosing the various setup plans can be conveniently condensed. This figure may be used to decide which setup plan to select, given the stock characteristics described along the top of the figure. The decision table contains "yes" (1), "no" (0), and "don't care" (-) responses to the questions. To use the table, consider the condition at the top and then see if there is a mismatch to the condition in the column below. If there is a mismatch, cross out that row, eliminating what would be an inappropriate plan. A don't care entry

Two overhang	At least one overhang	Narrowest dimension ⩾ 1/2 inch?	At least one largest area side rolled?	Both largest area sides rolled?	At least one second largest area rolled?	Both second largest area sides rolled?	Plan
—	0	—	—	1	0	—	A
—	0	0	—	1	—	0	A
—	0	—	0	—	—	—	B
—	0	1	1	—	1	—	B
—	0	0	1	—	—	1	B
0	1	0	0	—	—	—	C
—	1	1	—	—	—	—	D
—	1	0	1	—	—	—	E
1	—	0	0	—	—	—	F

1 = Yes
0 = No
— = Don't Care

Figure 8.8 The questions asked of the novice machinist, along the top of the table, lead to a decision to use a particular plan. (Hayes & Wright, "Automated Planning in the Machining Domain," published by ASME, PED-Vol. 24, 1986. Reprinted with permission.)

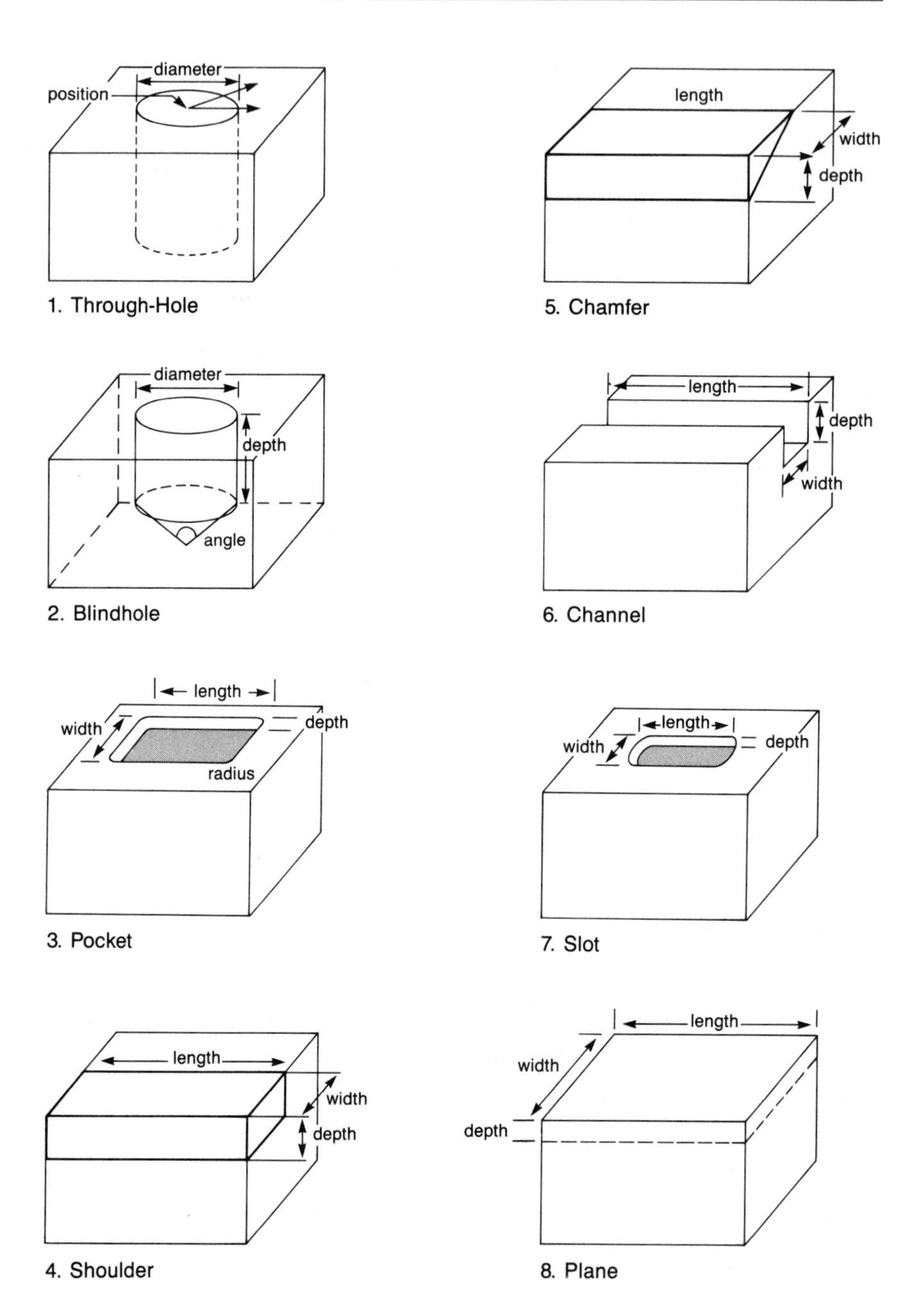

Figure 8.9 The primitive features. (Courtesy of Caroline Hayes, Robotics Institute, Carnegie Mellon University, 1987).

matches a yes or a no, meaning that a particular plan is independent of the question posed in that column.

Consider a specific example for a piece of bar stock that is 5 inches long and will be put into vise jaws that are 6 inches long. From the knowledge of the machining technology, we can allow one overhang during cleanup. The first column asks—are there two overhangs? Since we must reply no, we can cross out the mismatching yes entry and eliminate plan F. All other plans remain for consideration. The second column asks: Is there at least one overhang? Since we can reply yes, we can eliminate all the mismatching no's and eliminate the variants of plan A and plan B in the first five rows.

Since plan F is already eliminated, plans C, D, and E remain to be evaluated by other means. Suppose the stock is 1 inch thick. The third column asks: Is the narrowest dimension greater than or equal to 1/2 inch? Since the answer is yes, only plan D matches and it is selected as the most appropriate plan for this situation.

Hayes (1987) has implemented this decision making in a rule-based program. This program issues stylized English instructions on how to set up a part and also produces color graphics that a novice can use. This output looks like Figures 8.6 and 8.7. To reiterate, plan templates are for basic cleanup and to obtain orthogonality. The next two sections discuss how part features can be incorporated into the initial plan.

8.2 FEATURES

Features are primitive geometrical shapes that are cut into the surface of a block. Figure 8.9 shows the eight primitive features used in the experimental system. These features can be geometrically combined in various ways to describe different parts (see Figure 8.10). The concept of combining feature primitives is common in graphics and CAD research. In particular, we refer the reader to the work of Voelcker and colleagues Brown (1982) and Requicha (1980) who have developed the language PADL, which uses and combines primitive geometrical features. This language has become the basis of many "constructive solid geometry" CAD systems. More recently, Chan and Voelcker (1986) have also used PADL as the basis for constructing plans for machining. Their approach is exclusively driven by the part and tool geometry; our approach is similar, but it also takes into account heuristics that machinists have supplied. The reader is also referred to the geometrical approach to planning by Iwata and Fukuda (1986) and by Berenji and Khoshnevis (1986) who review similar systems.

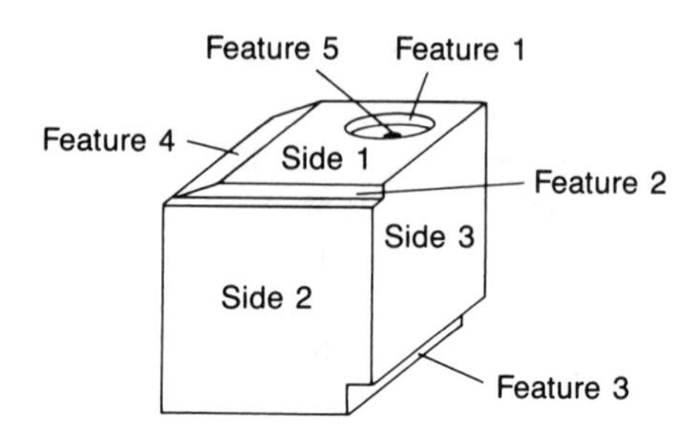

Figure 8.10 An example of specifying features in a simple part. (Hayes & Wright, "Automated Planning in the Machining Domain," published by ASME, PED-Vol. 24, 1986. Reprinted with permission.)

8.2.1 Part Features and Modifying Template Plans

Merging features into one of the basic templates shown in Figures 8.6 and 8.7 leads to a setup plan. The following examples illustrate some of the feature interactions.

The part with drilled holes shown in Figure 8.11 is going to be made from bar stock. Using the decision table in Figure 8.8, plan D is selected, as shown in Figure 8.12. Next, the features must be added to the plan. However, there are restrictions on where a particular feature can be added. A blind-hole needs three orthogonal machined reference sides from which the feature can be positioned. The *x*'s on the setups in plan D indicate the faces on which features may be cut, and they appear only at the point after which three orthogonal sides have been machined. To

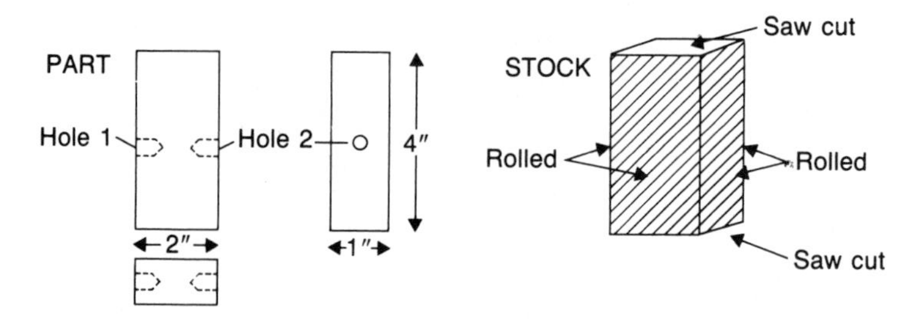

Figure 8.11 A part with two drilled holes and its initial stock. (Hayes & Wright, "Automated Planning in the Machining Domain," published by ASME, PED-Vol. 24, 1986. Reprinted with permission.)

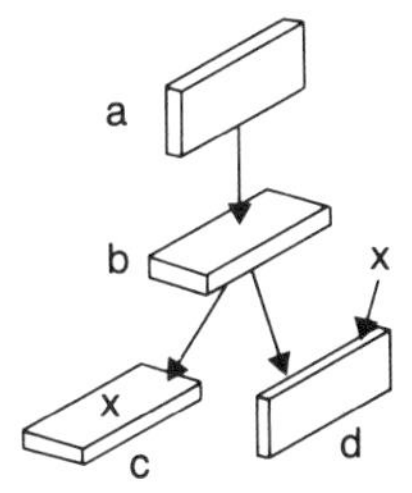

Figure 8.12 Plan D from the original Figure 8.7. (Hayes & Wright, "Automated Planning in the Machining Domain," published by ASME, PED-Vol. 24, 1986. Reprinted with permission.)

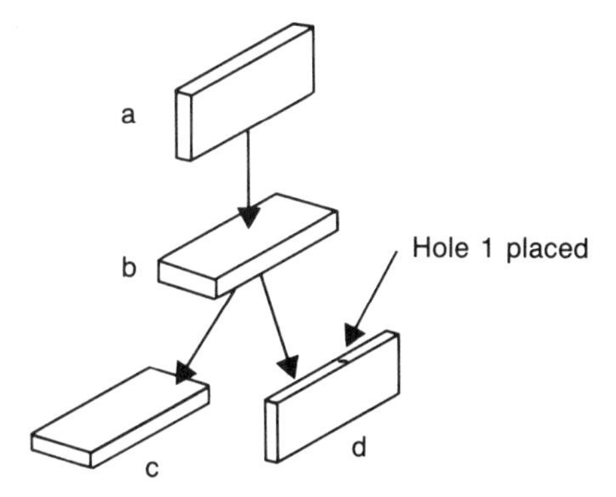

Figure 8.13 One feature added to plan D in the form of a drilled hole in the medium side at *d*. (Hayes & Wright, "Automated Planning in the Machining Domain," published by ASME, PED-Vol. 24, 1986. Reprinted with permission.)

place a feature in the plan, a setup must be found that has the correct side facing upward. Since both holes are cut in the medium sides (see Figure 8.13), we can place at least one of them on the lower right setup marked *d*.

Figure 8.13 shows how this first feature may be added. However, since there is no other setup in the original plan D that allows the other hole to be drilled, another setup must be constructed and added to the plan as shown in Figure 8.14. Finally, the setups are ordered, and the remainder of the sides are numbered, as shown in Figure 8.15.

In these examples, the templates are used to recognize the correct type of solution to a setup plan. In this case, plan D is selected. At the same

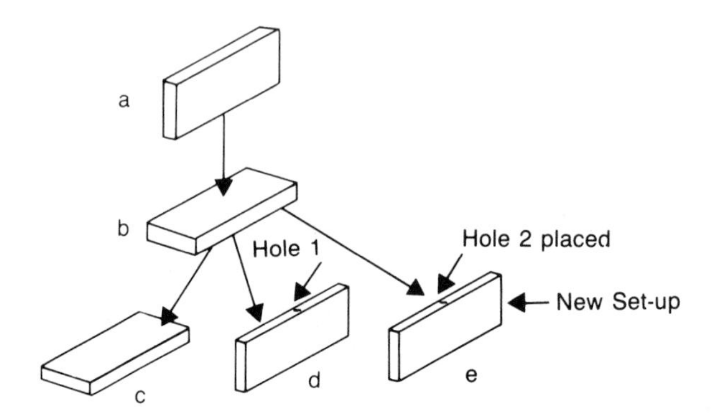

Figure 8.14 New setup added to plan D to allow the second hole to be drilled. (Hayes & Wright, "Automated Planning in the Machining Domain," published by ASME, PED-Vol. 24, 1986. Reprinted with permission.)

time, minor modifications are constructed and added to the plan in order to custom-tailor it to the particular features being put into a part.

8.2.2 Interacting Features

The next complication in the setup process is to see how interacting features are incorporated into a plan. Features can interact in two main ways: geometrical interactions and machining technology interactions.

In the first case, features may intersect with each other geometrically. For example, parts often require holes within pockets, as shown in the top of Figure 8.10. The correct ordering of such cuts helps to reduce burr formations. Burrs are rough edges at the side of a feature, which are thrown up by a cutting tool. Specifically, it is preferable to drill the hole after the pocket in Figure 8.10.

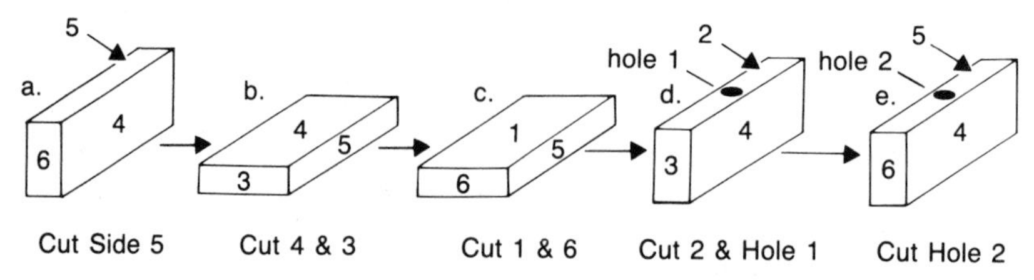

Figure 8.15 The ordered plan. (Hayes & Wright, "Automated Planning in the Machining Domain," published by ASME, PED-Vol. 24, 1986. Reprinted with permission.)

Otherwise, if the hole is already present, the milling cutter creating the pocket in soft aluminum will push a large burr into the already drilled hole. Such a burr will have to be removed in a subsequent operation. The skilled machinist knows how to minimize burrs by judicious planning when features intersect. Since rules governing such orderings are not absolutely crucial to the success of the plan, they have been embedded in our preferred rules in Figure 8.5.

In the second case, the rules governing the machining technology interactions are much stronger and must be included in the required rules (added to Figure 8.5). For example, a hole should not be drilled into a chamfered surface, or else the drill point will skid on entry and create an oversized hole. Thus, drilled holes should be integrated into the process plan at an early stage, while the block faces are still orthogonal. An exception to this rule may occur if dealing with complex forgings, castings, or parts with sculpted pockets. If a drilled hole is needed, at an angle to such surfaces, a special jig will be needed to guide the drill.

In other situations, the machining technology rules are concerned with clamping stability. For example, once a large feature has been formed in a part, it may be difficult to clamp for subsequent cuts. Again in Figure 8.10, the size of the shoulder at feature 3 could influence clamping stability. It would be appropriate to cut this small shoulder in the part during an early setup because this area would still be stable, in the bottom of the vise, while the machinist put the four features in side 1. However, if feature 3 were much larger, it might be better to cut it after the side 1 features.

As another example, Figure 8.16 shows a part with two shoulders and a hole. If the shoulders are cut before the hole is drilled, there will not be

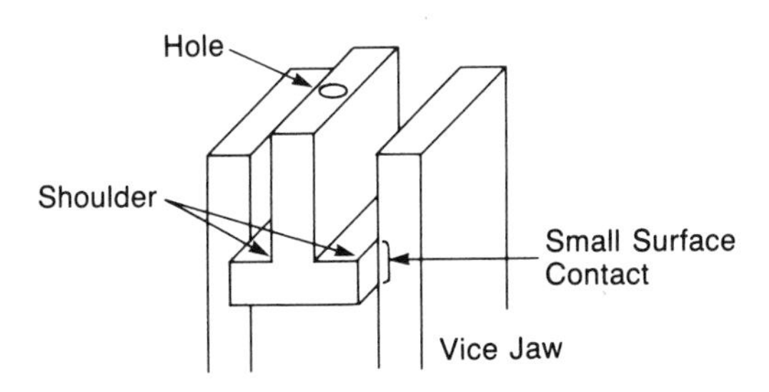

Figure 8.16 A part with interacting features where the hole and the pair of shoulders must be considered for clamping interactions. (Hayes & Wright, "Automated Planning in the Machining Domain," published by ASME, PED-Vol. 24, 1986. Reprinted with permission.)

enough surface area left in contact with the vise to grip the part firmly for drilling. The shoulders as a pair interact with the hole, and it is best to drill the hole first, while the block sides are large.

A skilled machinist brings a great deal of experience to these decisions especially when two or three features are interacting. The experts first look over all the features to identify their interactions and then pinpoint the major trouble spots. The problem areas of the part are then solved before the other features are put into place.

8.3 COMPARING HUMAN AND MACHINE EXPERTS

Several parts, one of which is shown in Figure 8.17, were arbitrarily selected for a comparison between the current expert system (Hayes 1987) and apprentice machinists. The plan generated by the expert system was based on rules from machinists with 15 to 20 years of experience (see Figure 8.18). This part can be made in seven setups, and the expert system took care to drill the holes prior to the chamfer. Recall that the rules specify that holes must preferably be drilled into orthogonal surfaces, rather than into chamfered surfaces where the drill point might skid.

The apprentices were not as efficient as the expert system in planning. For example, a three-year apprentice made a workable plan, but, adopt-

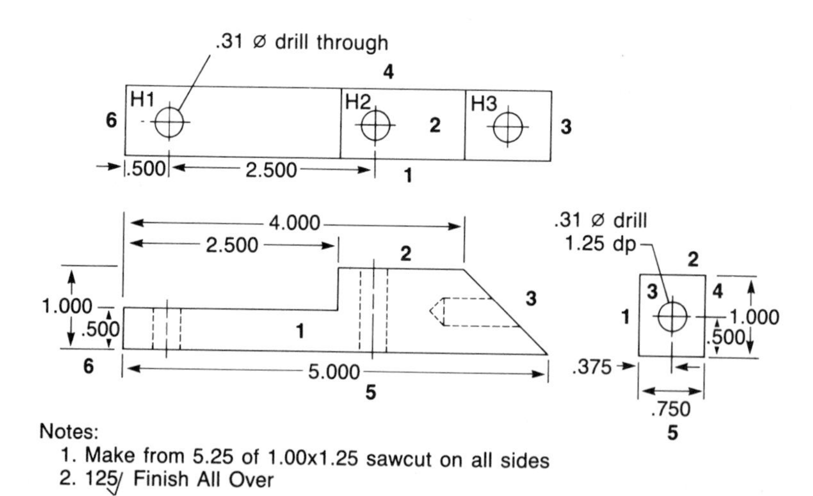

Figure 8.17 A part used to test EXSUS against two apprentices. (Courtesy of Caroline Hayes, Robotics Institute, Carnegie Mellon University, 1987).

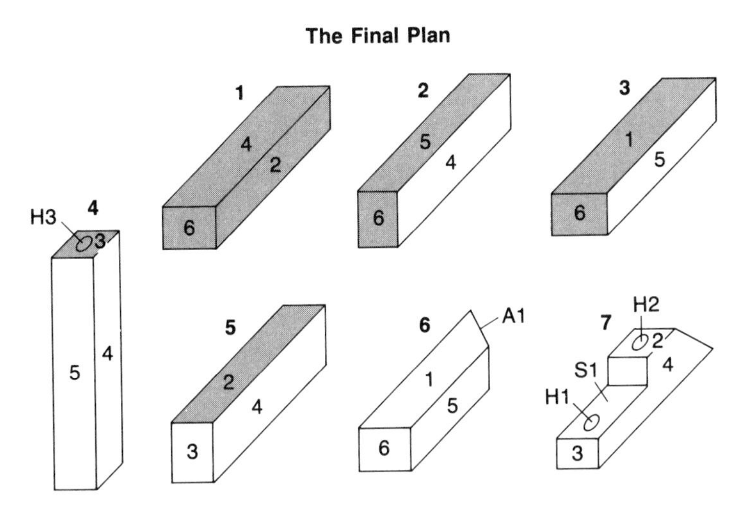

Figure 8.18 The plan generated by EXSUS for the test part in Figure 8.17. (Courtesy of Caroline Hayes, Robotics Institute, Carnegie Mellon University, 1987).

ing an overly cautious method, he cleaned up the stock first and then put in the features; this required more setups. A second-year apprentice made a plan with judgment errors. For example, the plan incorrectly specified that the hole would be drilled after the chamfer.

These comparisons between the expert system and apprentices need to be interpreted with caution. However, there is no doubt that the plan templates (Figures 8.6 and 8.7) and the decision table (Figure 8.8) have captured useful and applicable information for machine setup planning. It is also apparent that many setup activities depend on the effective fixturing and clamping of parts.

8.4 OPEN ISSUES AND CONCLUSION

The research into setup planning so far has established scientific principles, uncovered a useful methodology for machinists, and demonstrated relevant results. However, many challenges must be addressed before an expert system for general machining will be practical.

1. Complex part designs must be clamped with many kinds of special-purpose fixtures. Somehow, the fixture geometry must be combined with the part geometry to satisfy machining constraints.

2. Complex stock supplies are commonly used in industry. For example, the aerospace industry makes heavy use of both forgings and castings. The setup plans for these stock materials must be accounted for in a working factory system. However, note that the intermediate steps of any complex part face the same problems.
3. Aerospace materials (e.g., titanium, nickel alloys, and composite materials) pose unusually difficult constraints on machining. With these materials, there are often no ideal setup plans; there are only preferable setup plans.
4. Some part features are very difficult to cut, and again there is no way to cut them with high confidence. For example, very thin walls in aerospace parts are difficult to thin further because the wall actually bows during cutting. One solution is to cut a compensating path in the part assuming that the wall is going to bow, but this method is at best a trial-and-error procedure.
5. The optimal planning of interacting part features represents a major research opportunity for the manufacture of one-off or small-batch components. It also calls for a keen awareness of the machining technology. The previous discussion mostly dealt with clamping stability and interacting part features. But, the feature ordering also influences surface finish, chip disposal, machining vibrations, and even tool life.

The greatest benefit from the knowledge representations chosen in this work is not that we have a program for squaring up stock (see Figure 8.3) or creating a setup plan (see Figure 8.18), but that we have systematically organized the previously intuitive knowledge in this area. This can now be refined and compared with other projects. Although in its infancy, the science of setup has thus been given birth.

REFERENCES

Berenji, H. R., and Khoshnevis, B. 1986. Use of artificial intelligence in automated process planning. *Computers in Mechanical Engineering*, pp. 47–55.

Brown, C. M. 1982. PADL-2: A technical summary. *IEEE Computer Graphics and Applications*, Vol. 2, No. 2, pp. 69–84.

Chan, S. C., and Voelcker, H. B. 1986. An introduction to MPL—A new machining process/programming language. *Proceedings of the IEEE Conference on Robotics and Automation.*

Hayes, C. 1987. Planning in the machining domain: Using goal interactions to guide search. M.Sc. thesis, Carnegie-Mellon University.

Hayes, C., and Wright, P. K. 1986. Automated planning in the machining domain. *Proceedings of ASME Meeting on Knowledge Based Expert Systems for Manufacturing*. PED-Vol. 24, pp. 221–232.

Iwata, K., and Fukuda, Y. 1986. Representation of know-how and its application of machining reference surfaces in computer aided process planning. *Annals of the Confederation Internationale Pour Recherche de Production (CIRP)*, Vol.35, No.1, pp.321–324.

Requicha, A. A. G. 1980. Representations for rigid solids: Theory, methods, and systems. *ACM Computing Surveys*, Vol. 12, No. 4, pp. 437–464.

9 The Machine Tool's Hand: Jigs and Fixtures

You are to bind me with strong ropes and fasten me upright against the mast, so that I shall not be able to move. If I implore you and order you to set me free, you must tie me up tighter than ever.[1]

—Homer in *The Odyssey*

TO peel an apple, a human being clamps the fruit in one hand and the knife in the other. The interaction forces are low, and accuracy matters little. To shave a plank of teak with a wood plane, a cabinetmaker uses the vise on a shop workbench and puts both hands on the plane. The vise is used because, in comparison with the apple, the interaction forces are higher and accuracy is more important. To face-mill a large block of steel with an 8-inch carbide cutter rotating at 250 rpm, the experienced machinist at first pays considerable attention to clamping the part on the milling machine's bed. The machinist knows that the interaction forces will be several hundred pounds and that accuracies are being specified to a few thousandths of an inch.

The manufacturing hands in Chapter 6 tend to be more dexterous than powerful. This capability is often emphasized during part manipulation and can dominate a designer's solution. However, the machine tool's hand must be very powerful and rigid during cutting. This task constraint makes it difficult to propose a design that is both dexterous and powerful; thus, the power tends to dominate the design solution. In fact, a design that could combine both dexterity and power into a fixture would be very useful to the machinist. The added dexterity would be used to present more than one surface for cutting, which would effectively add to the machine tool's degrees of freedom.

Just like the manufacturing hand, the machine tool's hand must combine many sources of design information (see Figure 9.1). In this case, heuristic knowledge used by the machinist (e.g., pointed clamps can dent soft workmaterials) must be combined with hard physical facts

[1]*Homer's Odyssey,* translanted 1937 by W. H. D. Rouse. Reprinted with permission of the New American Library.

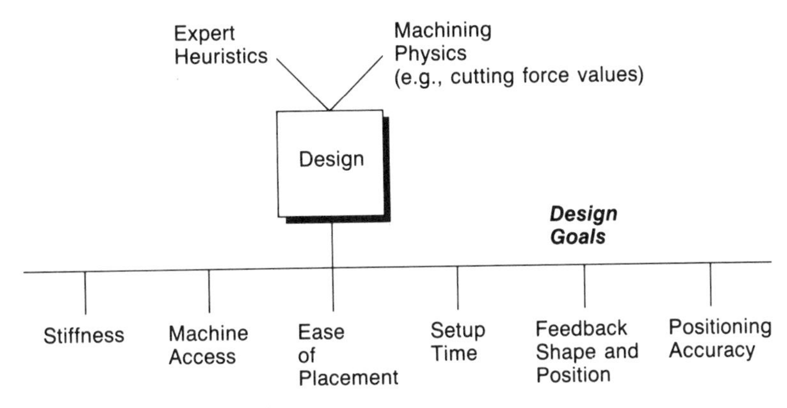

Figure 9.1 Designing the machine tool's hand.

(e.g., cutting force values). Recall that the design of the manufacturing hand was driven by the task. In machining, the task is to hold the part with absolute rigidity, while providing convenient access to the machine tool (see bottom of Figure 9.1 for more task requirements). This specific knowledge of the task should make the design of clamping mechanisms easier than the design of general robotic hands.

9.1 CLAMPS, JIGS, AND FIXTURES

In the machine tool community, the terms clamp, fixture, and jig are lazily interchanged and often paired (e.g., jigs and fixtures). In fact, there are some subtle differences between the three. A clamp provides a firm hold on a piece of stock so that it can be machined. A fixture also clamps the stock, but generally a fixture is a custom-crafted, one-off clamp that is made to hold an unusually shaped piece of stock (e.g., a casting of a ship's propeller or a forging of a turbine blade). Often, a fixture is the female shape of a part. A jig also clamps and also fixtures a part, but in addition, a jig has predetermined reference points in it for part alignment and/or tool alignment. With these reference points, a jig becomes a very accurate fixture or clamp. Jigs are particularly important in high-volume mass-production manufacturing. For example, as car engine cylinder blocks shuffle off conveyor belts into a multispindle drilling machine, a jig ensures that the blocks are guided to the correct location. Then, for machining, the jig may also contain hardened bushings that guide the drills to precise locations on the part's surface.

Rules

1. Positively locate the part and minimize movement in all the degrees of freedom, three translational and three rotational. This may be accomplished in a variety of ways depending on the types of locators or clamps used.

   - **Vise:** (a) The number of parallel sides must be greater than or equal to one; and (b) the number of points of contact with the bottom of the vise jaw must be greater than or equal to three, and the number of points of contact with the side jaw surface must be greater than or equal to two; and (c) the vise opening must be greater than or equal to the distance between the parallel sides of the part.
   - **Toe Clamps:** The number of points of contact between the part and the locating surface(s) must be greater than or equal to three in one plane and greater than or equal to two in another plane perpendicular to the first (Boyes 1982).

2. Apply a large enough clamping force to the part to immobilize it without damage.
3. Position the clamps and locators about the workpiece so that there will be no interference between them and the cutting tool.

Guides

1. Minimize the number of steps and the amount of time required to mount the fixture onto the machine tool bed. If there is a need for a specialized clamping device, keep its design simple.
2. Minimize the number of setups in the fixture for any part or assembly. The greater the number of setups required, the greater will be the chance for inaccuracies in locating and gauging. A good heuristic to minimize the number of refixturings is to maximize the number of cuts that can be made on the part for a given clamp orientation.
3. When multiple setups are a necessity, maintain the same locating surfaces for each setup. The engineering drawing should contain the necessary datum surface or geometrical origin. Error accumulations will be reduced if all operations are performed relative to these surfaces.
4. Position the clamps about the part so that the bending moments produced by the tool are minimized. That is, place the clamps as close to the tool path as possible. Avoid large overhangs of the part that would cause the cutter to induce large bending or torsional stresses in the workpiece.
5. Position the clamps about the part's axes of symmetry. Programming will be simplified, and the pressure exerted by the clamps will be more uniformly distributed so that part distortion will be reduced.

Figure 9.2 Initial set of rules and guides used for fixture placement.

There are many types of clamps for machining work. Although most of this book is concerned with milling, it is useful to mention the clamping process in lathe operations where the chuck represents the most straightforward kind of machine tool clamp. The chuck holds the rotating cylindrical workpiece while the cutting tools shape the part. Chucks come in various forms, the most common being 3-jaw, 4-jaw, and collet chucks (DeGarmo, Black, and Kohser 1984). All three of these actually correspond to the three-finger and four-finger grips in the lower left of the taxonomy for precision grips (Figure 6.12).

The intelligent machine tool concentrates on small-batch milling. In these situations, clamping is much more dependent on the human operator. For the rectangular parts discussed in Chapter 8, a parallel-sided vise is used; this is the simplest clamp for milling. As the geometry of the part becomes more complex, toe clamps and toggle clamps can be used. These fit into the tee slots on the bed of the machine tool and then grab onto the part at any convenient projection like a mountaineer's hand on a rock face. If the part geometry becomes very unusual or sculptured, a special purpose fixture may be needed.

In most automated systems, clamping is achieved offline by humans. They align stock in fixtures and then clamp the fixtures on movable pallets. The pallets move around the automated system on conveyor belts and then move in and out of different machines on branch lines of the main conveyor.

9.2 KNOWLEDGE ENGINEERING FOR FIXTURE DEVELOPMENT

As described previously, experts are being interviewed to determine the underlying principles of part setup. This knowledge will provide a basis for building automated machines that can make parts in small-batch manufacturing.

Machinists must satisfy certain rules for machine action planning and part setup, and they also must follow good machining practices. Figure 9.2 shows several rules and guides that were established in the machinists' interviews. If followed, these simple rules and guides can have many beneficial effects. For example, rule 1 can be used to maintain clamping stability, and rule 3 avoids interference between the cutting tool and the fixture. Also, the less stringent guides in Figure 9.2 can be used to achieve a minimum number of setups, while maintaining the integrity of the setup. This integrity will make it possible to machine accurately without causing unnecessary programming problems.

Figure 9.3 illustrates how a prismatic part with features can be cut while following guide 2. For this illustration, the tool database is limited to include mills and drills, and Chapter 8 showed the basic cut shapes available to the designer, for example, channel, shoulder, and plane. The first step is to generate a table of cuts and the sides from which they can be machined. Algorithms were developed to correlate the cuts that are necessary to make on each open side. For example, cut 2 in Figure 9.3 is a shoulder and it opens out on side 1 and side 2. However, when the

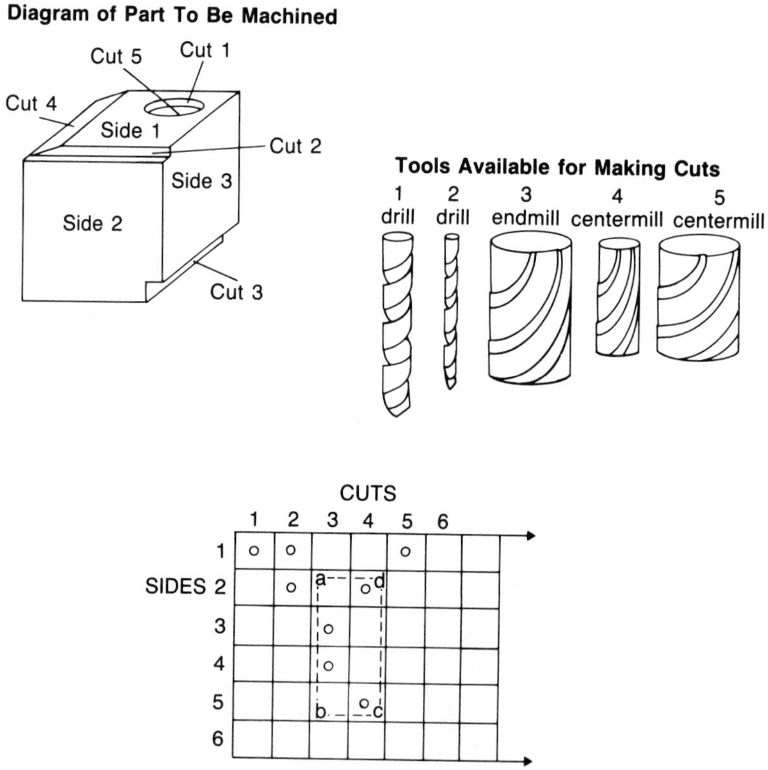

Diagram of Part To Be Machined

Cut 5, Cut 1, Cut 4, Cut 2, Cut 3; Side 1, Side 2, Side 3

Tools Available for Making Cuts

1 drill, 2 drill, 3 endmill, 4 centermill, 5 centermill

CUTS

SIDES	1	2	3	4	5	6
1	o	o			o	
2		o	a	o d		
3			o			
4			o			
5			b	o c		
6						

Figure 9.3 Preferences for ordering cuts in a prismatic part are determined by referring to the tool database and then filling in the table to show which sides the cuts can be made. The result shows that only three setups are needed to make the part. (Englert & Wright, "Applications of AI and the Design of Fixtures for Automation", Robotics & Automation Conference, San Francisco © 1986 IEEE.)

milling tool makes this cut, it will enter the part from side 3 and exit the part from side 6. Thus, there are four sides from which a shoulder cut opens out. A channel cut, however, only opens out onto three sides. Algorithms such as this can be used to uniquely define the different kinds of cut shape.

The table is then filled in by applying the set of algorithms and determining the allowable part orientations for each cut to be made in the block. In Figure 9.3, cut 1 is a round pocket, cut 2 and cut 3 are shoulders, cut 4 is a thru-prism, and cut 5 is a blind-hole. Cut 1, cut 2, and cut 5 can be made on side 1, and this could be the first side to be machined. This leaves cut 3 and cut 4 to be made. The smaller box highlighted in the center of the table, labeled *a b c d*, is the reduced matrix. It shows that the final two cuts can be made on side 2, side 3, side 4, or side 5.

In summary, only three setups are necessary to make the features on the part. Work continues to determine the preferred ordering of these three setups. For example, does the craftsman follow 1 → 3 → 4 or 3 → 4 → 1? Craftsmen often apply a rule that machines the most "featureful" side last in order to get the best accuracy. However, this rule may be modified if it causes an unacceptable problem with clamping stability.

9.3 CASE STUDIES: AUTOMATED FIXTURING

What can be done to automate clamping and fixturing for small-batch milling operations? To answer this question, the fixture designer must be fully acquainted with the "part family" that will be held in the fixture. However, at least two general approaches can be taken: conformable clamps and configurable clamps. The first choice is to design clamps that conform to the shape of the parts. Thus, these fixtures are only limited by the degree to which they can conform to different geometries without jeopardizing the critical rigidity requirements. Unfortunately, these fixtures tend to be quite complex and costly. The second approach is to configure the final fixture from a set of simple components. Since each of the components can be simple, they can be quite inexpensive. However, these configurable fixtures may require an extensive setup procedure, and the problem of where to place the fixtures on the part is magnified.

Not much research has been concerned with optimizing the design of flexible fixtures by combining elements of conformability and con-

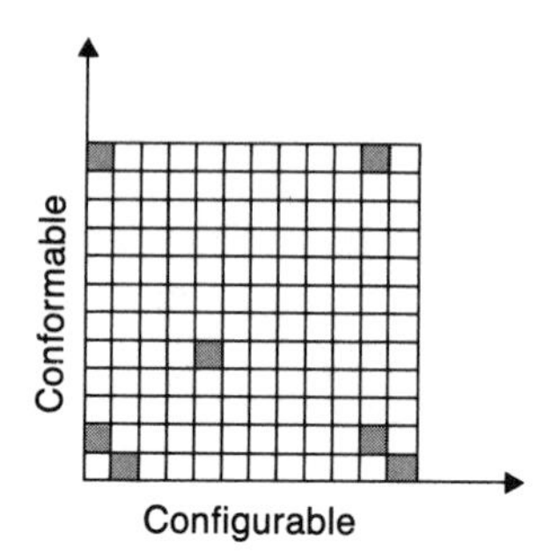

Figure 9.4 The configure-conform design space.

figurability. Figure 9.4 illustrates a range of designs that trade off between these two approaches to building flexible fixtures. Each square in this design space represents a possible design experiment, and the blackened squares represent the kinds of designs that can be tried without having to try all the possibilities. With this design philosophy, it is hoped that only a few design experiments need to be carried out in order to fully understand the design space.

Once the general design choice has been made for a fixture in the configure-conform design space, several other design decisions must be made. Figure 9.5 provides additional fixture design considerations that must be included so that the final task requirements are completely satisfied.

9.3.1 A Prototype Conformable Clamp for Turbine Blade Machining

In 1980, we began a conformable clamp pilot project, hoping to transfer it to a turbine blade machining facility (Cutkosky, Kurokawa, and Wright 1982). After closed-die forging, turbine blade airfoils of the type shown in the lower left of Figure 5.15 and in Figure 9.6 are finish-machined along their edges, around the lugs, and at the root. In the factory concerned with this case study, 6,000 different turbine blade shapes can, at one time or another, be fabricated. Depending on the machining operations, the fixtures can either be continuous hard profiles that rest in cradles and are the mirror images of the blade cross sections, or a low-melting-point alloy, cerubin, that is cast around the blade at certain cross-sections to allow machining in others.

DESIGN CONSIDERATIONS	TASK REQUIREMENT
Grasp Control	Securing Clamp to Shape
Kinematic Sensing	Shape and Position Digitization
Force Sensing	Monitoring Quality of Clamp
Rigid Mechanism	Avoid Vibrations during Cutting

Figure 9.5 Design and task requirements.

Both of these existing clamping methods have shortcomings. For example, the cerubin is very expensive, and the process of accurately aligning the blade and then pouring the alloy around it is very time consuming. At the end of machining, time must also be spent melting and

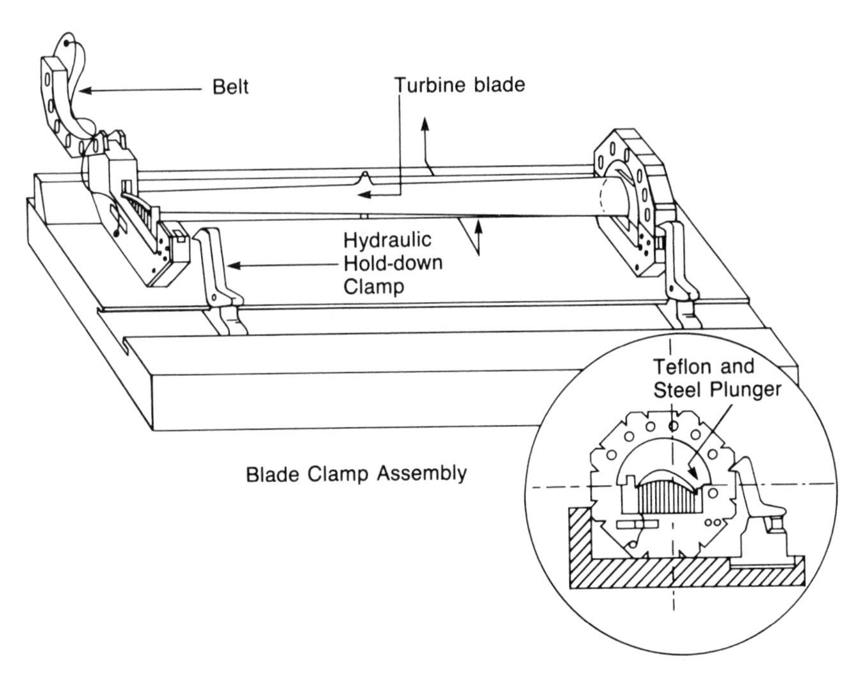

Figure 9.6 Turbine blade held in conformable clamps. (Reprinted courtesy of the Society of Manufacturing Engineers. © 1982, from Autofact 4 Conference Proceedings.)

retrieving the alloy. In the alternative method, the hard profiles must be custom made for each blade style, and this leads to massive inventories of different clamps for the 6,000 designs. The conformable clamps shown in Figures 9.6 and 9.7 have been designed and made in prototype form to address some of these shortcomings. It is hoped that they become a viable alternative in the factory. Note that their present design corresponds to a square close to the upper-left corner of Figure 9.4.

The conformable clamps consist of octagonal frames that are hinged so that they may be opened to accept a blade and then closed. The clamps

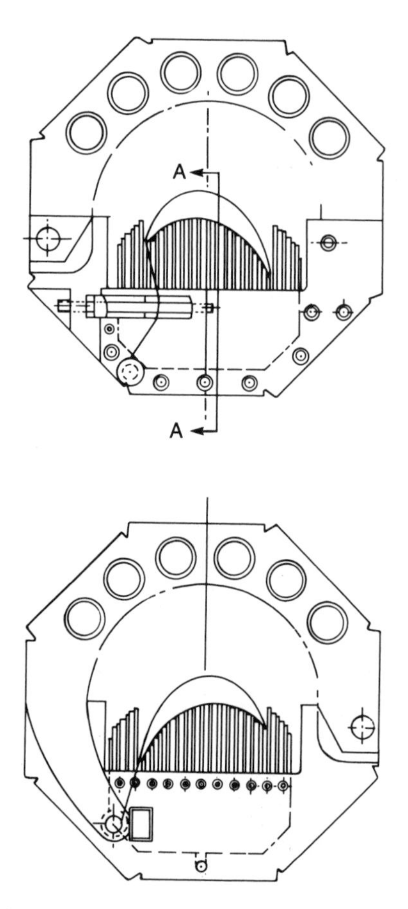

Figure 9.7 Conformable clamps (side view). (Reprinted courtesy of the Society of Manufacturing Engineers. © 1982, from Autofact 4 Conference Proceedings.)

may be placed almost anywhere along the turbine blade. Usually two or three clamps hold the blade rigidly. Using an octagonal shape, rather than a square, allows the blade-clamp assembly to assume more orientations about the blade's z-axis (length of blade). Since the blade cross-sections twist substantially from one end of the blade to the other, this is a necessary feature. The clamps could also be circular, but then holes or pegs would be required to establish alignment with respect to the z-axis.

The lower half of each clamp employs plates or plungers that, when released, are free to conform to the profile of the turbine blade as it is laid into the cradle. A similar technique has been used for programmable powder metal dies and for sheet metal forming (Wright and Holzer 1980). Air pressure forces the plungers against the turbine blade. Once a profile has been set, the plungers are mechanically locked in place and the air supply may be disconnected. A high-strength belt is then wrapped over the convex surface of the blade to force it against the plungers in the lower half of the clamp. The belt is tightened and the blade-clamp assembly is free to travel around a flexible manufacturing system. If the position of a clamp must be changed during the machining process, it is possible to allow a new clamp to conform to the blade in a new location, to tighten the new clamp, and then to remove the old clamp so that the orientation of the turbine blade is never lost.

After two or more clamps have been fastened to a turbine blade, the blade-clamp assembly becomes a pallet in the form of an octagonal prism. The assembly's outer dimensions are well defined and do not vary from part to part. A single fixturing arrangement can then be used on all the machine tools in the system. Furthermore, there is no need to modify the fixturing when the system changes to a new blade style. Figure 9.6 shows a blade held by two clamps mounted on a fixturing plate. Standard hydraulic hold-down clamps can be used to fix the assembly in place while it is being machined. If a robot were used to transport the blade-clamp assembly, it would also benefit from the standardized shape. The robot would not need to hold the assembly as rigidly as the machine tools, just firmly enough to keep the assembly from slipping during transportation. The robot could therefore use a simple gripping arrangement such as a rack with electromagnets.

Kurokawa (1983) showed that during machining operations the conformable clamps had good characteristics for force-excited vibrations. The basic harmonic component of the cutting forces acted far away from the resonance point of the blade-clamp assembly. General machining operations carried out on the blades as clamped in Figure 9.7 were suc-

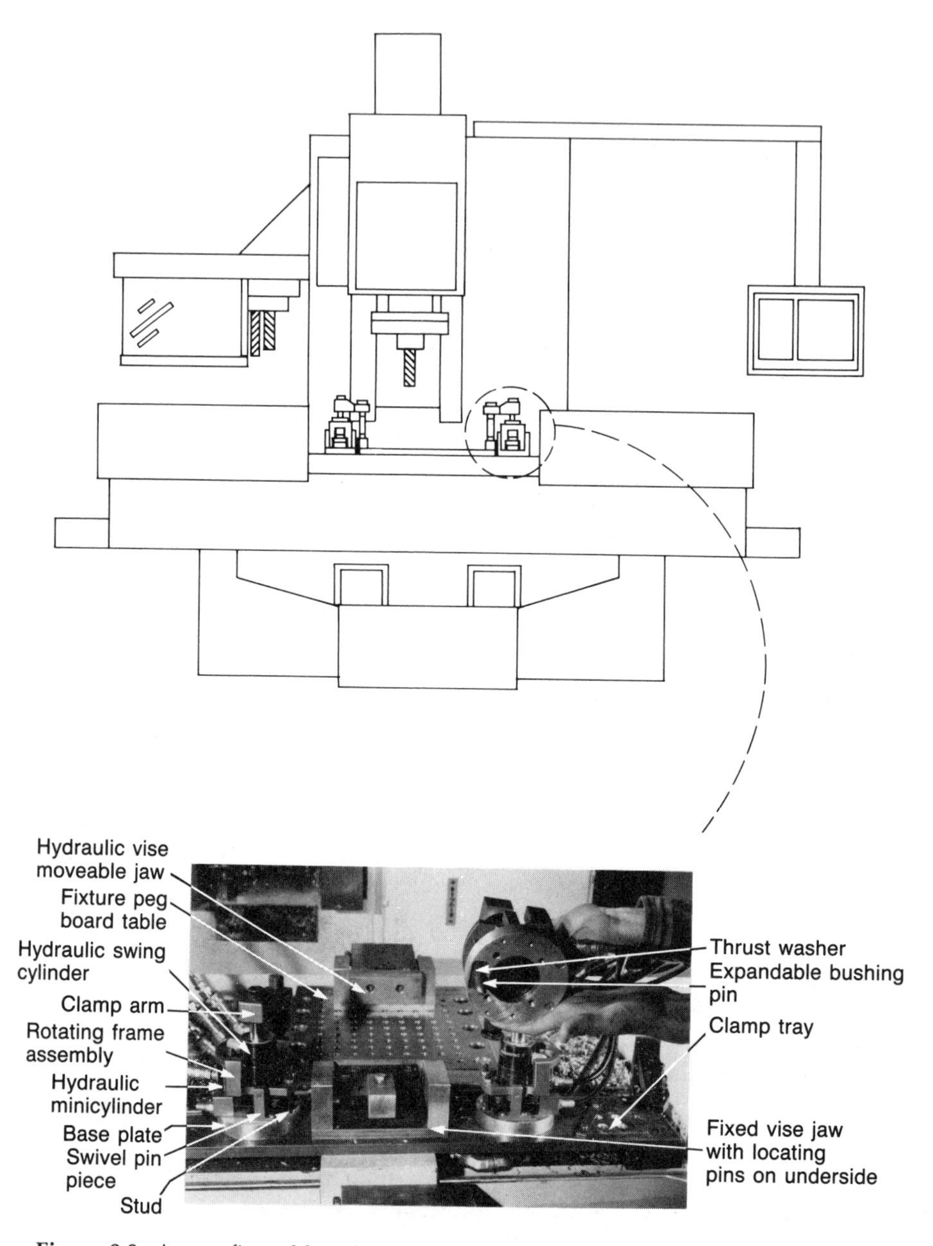

Figure 9.8 A configurable clamping device that can be secured under microprocessor control. (Englert & Wright, "Applications of AI and the Design of Fixtures for Automation", Robotics & Automation Conference, San Francisco © 1986 IEEE.)

cessful, and the prototype clamps are still being evaluated for possible use in production. Some preliminary ideas for achieving programmable control of the clamp plungers are discussed by Cutkosky et al (1982). However, these ideas have not yet been put into use. More recently, in order to demonstrate programmability in clamping devices, work has focused on rectangular workpieces (Chapter 8) using a hydraulic vise.

9.3.2 Configurable Clamps for Rectangular Stock

A configurable clamp has been developed that allows a fixture to be constructed by a robotic manipulator or by the machine tool itself (Englert and Wright 1986). The main components of the system are individual swing clamps that can be operated automatically from remote hydraulic controls. The principal subcomponents of one of the remotely securing clamps, shown in Figure 9.8, include: a hydraulic swing cylinder that turns 90° and pulls downward when powered, a clamp arm, a large clamp stud and leveling pad, an expandable bushing pin, a rotating frame assembly, a clamp base, and a hydraulic minicylinder. When the minicylinder at the rear of the base is energized, it pushes upward on the back part of the frame assembly and causes it to rotate about the swivel pin. This rotation causes two smaller studs at the front of the frame to move through slots in the clamp base and to push down on a thrust washer seated in a counterbore on the underside of the base. The resisting force from the thrust washer produces small radial expansions of the bushing segments, which lock the clamp into a fixture table. This fixture table is bolted to the machine tool table and consists of a matrix of through-holes. These holes accommodate the clamp's expandable pin. After the clamp has been loaded and locked in place, the main hydraulic cylinder is actuated to pull the arm, stud, and leveling pad down onto the part.

To confirm the feasibility of this device, a robot has been used to load a set of the remotely locking clamps onto the fixture plate, place a piece of stock in the fixture, signal the power source to actuate the clamps, and finally signal the CNC tool to begin the programmed machining operation. A portion of the sequence is shown in Figure 9.9. A small gripper has also been made to fit into the tool changer of the machining center so that the machine can prepare its own fixture without the robot. This appeals to the concept of self-reproducing and self-diagnosing machines that can function autonomously in the factory. In addition, work is being done to monitor the force readings obtained from strain gauges attached to the clamping elements, so that some of the "feel" or sensation, of the craftsman might be built into the system.

To quantitatively assess such "feel," we have equipped the machinist with a torque wrench. Sample parts, requiring slightly different clamping methods, have been investigated. The machinist has been asked to tighten down bolts to comfortable levels. This means that clamping forces are high enough to restrain the part during cutting, but not so high as to damage it. From these short, knowledge engineering experiments we have derived the range and median values of torque that the machinists commonly use. This is an example of nonlinguistic knowledge engineering.

Figure 9.10 shows a block diagram of the force controlled clamping system where desired force outputs from the clamps can be preset based on knowledge from the craftsmen. As the clamp pads make contact with the part, strain readings are generated in the gauges attached to the large clamp studs. When the compressive force in this clamp stud reaches a specified value, the pressure supply line is closed automatically. The desired clamping forces can be pre-programmed to suit the constraints of the particular part being machined. For example, if a machined part is

Figure 9.9 Flexible fixture clamping the stock. (Englert & Wright, "Applications of AI and the Design of Fixtures for Automation", Robotics & Automation Conference, San Francisco © 1986 IEEE.)

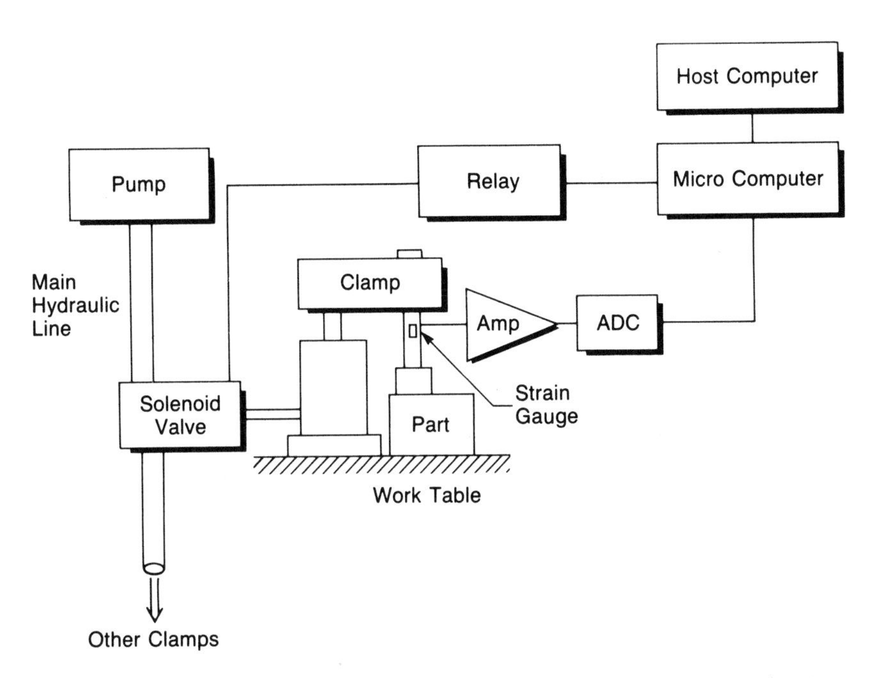

Figure 9.10 Force controlled clamping system. (Englert & Wright, "Applications of AI and the Design of Fixtures for Automation", Robotics & Automation Conference, San Francisco © 1986 IEEE.)

more likely to buckle on one side than another because a deep pocket has removed a great deal of metal, the individual clamps can be set to different levels to reflect part geometry. It is also easy to detect if the part is missing or at the wrong orientation if one or more of the clamps does not register a clamping pressure.

The use of toe clamps highlights rule 3 (Figure 9.2): "position the clamps and locators about the workpiece so that there will be no interference between them and the cutting tool." In setup 1 of Figure 9.11, the part is sitting on parallel bars to lift it up from the clamp base. Four of the configurable toe clamps restrain the part. In this first setup, side 3 and side 6 can be cleaned up by side-milling, and the majority of side 1 can be face-milled with the same tool. But the toe clamps obscure the corner areas of side 1. These obscured areas can, however, be finished off in setup 2. Thoughtful planning also allows side 2 and side 5 to be side-milled. In setup 3, the part is turned over to face-mill the remaining side, i.e., side 4. Once again the toe clamps obscure the corner areas and, to complete the cleanup, setup 4 is needed.

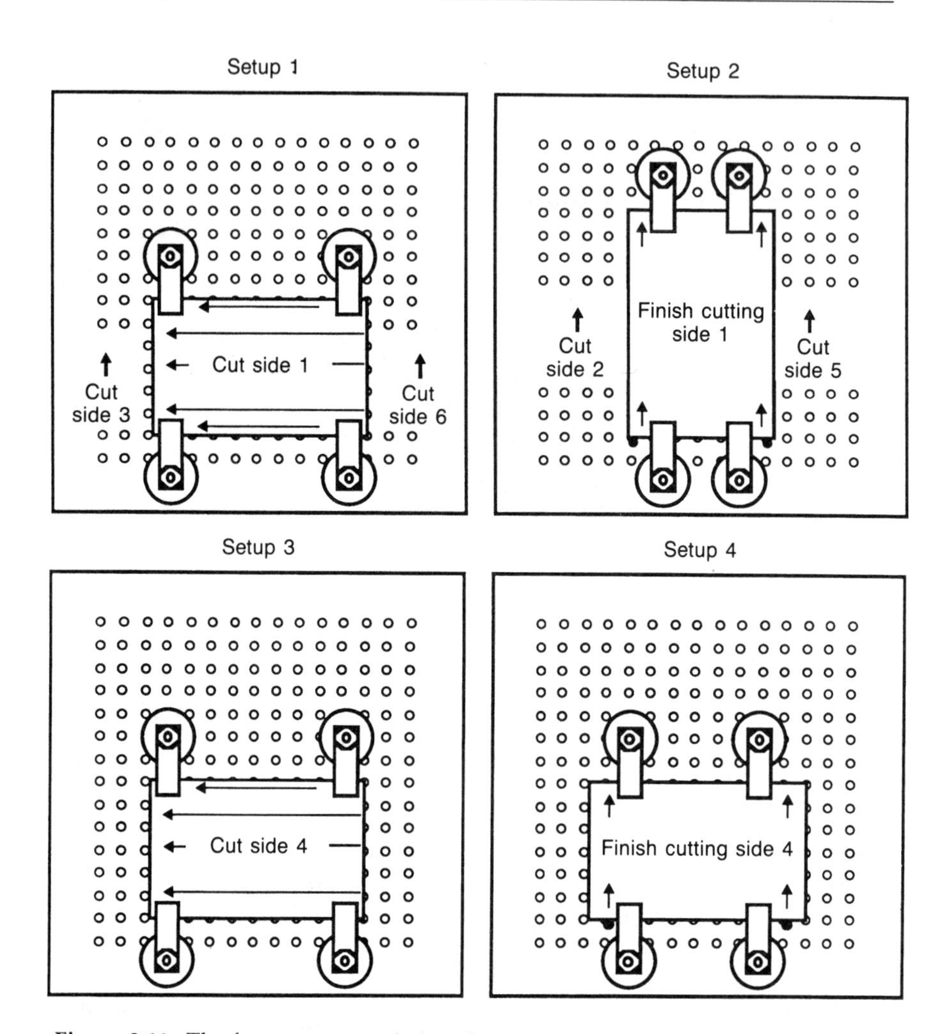

Figure 9.11 The four setups needed to cleanup rectangular stock when using toe clamps.

9.3.3 Experiments in Other Laboratories

Others have realized that flexible, programmable work-holding devices will improve the overall performance of automated manufacturing systems. Research scientists from the Hungarian Academy of Sciences and the Budapest Technical University (Markus, Markusz, Farka, and Filemon 1984) have developed a PROLOG system that configures clamps about a workpiece using CAD information and rules based on geometrical constraints.

Asada and By (1985) have implemented conformable clamps for assembly operations by using a magnetic table, a series of horizontal and vertical locators and clamps, a robot, and a CAD system all under computer control. The locators and clamps are configured by the robot using commands generated by the CAD system. In their case study, the commands are based on equations that define accessible positions for the clamps about the surface of the housing of an electric hand drill.

Ferreira, Kochar, Liu, and Chandru (1985) are developing a system known as AIFIX, which decomposes the overall fixturing geometry into independent subproblems and tries to solve each subproblem toward the final goal of work-holding.

Gandhi and Thomson (1986) have experimented with fluidized beds to hold unusually shaped parts. Here, the part is embedded in a vessel of fine powder which, when vibrated, behaves like a fluid embodying the part. When the vibration source is removed, the powder loosely holds the part. But, when a clamping plate is applied to the top surface of the powder, the whole assembly becomes strong enough to hold the part for machining.

The results from these previous studies have been encouraging. Although they show the feasibility of some hardware innovations, they have not revealed much information on where to put clamps around an arbitrary part. Consequently, new research is underway to consider the geometrical constraints in work-holding, partly from the viewpoint of traditional engineering mechanics and partly from the viewpoint of the more craft-oriented knowledge that is acquired from many years of experience. A tractable approach is to first limit the problem domain as in Chapter 8 and then concentrate on rectangular stock.

9.4 CONCLUSIONS AND OPEN ISSUES

The development of flexible fixtures from both a hardware viewpoint and a planning viewpoint is a research area that has only just begun. The hardware systems are in their early stages of development, and they can only cope with simple part geometries. The older textbooks on manufacturing (Boyes 1982) do not give many specific planning rules or guides for jigs and fixtures; rather, they describe many sample case studies for practicing machinists. Some open problems include:

1. Create new designs for swing clamps and automatic actuation systems to improve on the setups shown in Figure 9.9. The installation of sensors (e.g., strain gauges) that provide more meaningful

information on the quality of part clamping both before and during machining is also an important area to expand.

2. Develop analytical models for the stability of blocks of various shapes in different kinds of clamping systems (parallel-sided vises, toe clamps, or special fixtures). Besides static stability, also consider the buckling instabilities and vibrations that may occur in real-time machining.
3. Investigate correlations between these analytical models of part clamping and the rules-of-thumb extracted from the machinists for part setup (see top of Figure 9.1). Once relationships between these two bodies of knowledge have been found, this information should be embodied into a working expert system of the kind shown in Figure 9.2.
4. Develop a new kind of clamp that is dexterous enough to present different part surfaces, thus increasing the number of the machine tool's axes. This clamp must also be able to be locked into place so that it can satisfy the rigorous rigidity requirements in machining.

Current work is considering various situations where clamping stability is important. For example, the machinists intuitively know how far a block may extend beyond the vise before the pressure from a milling cutter or a drill will begin to deflect it. In other situations, they intuitively know when buckling or distortion of the part will occur if relatively thin cross-sections are clamped across the vise jaws. It is interesting that some of these situations resemble classical mechanics problems, where formulae have been developed to analyze the deflections of beams and buckling plates. In the authors' opinion, there is, at present, little to be gained from research that might look at some of these problems from a classical viewpoint. For example, finite-element analysis is not the best tool to understand part stresses and deflections in these part/vise configurations. The problem is that the boundary conditions for the finite-element model are impossible to describe. As a result, the calculations are extremely difficult to correlate with day-to-day practical results. The added practical constraints of cutting lubricants between faces and vibrating cutting tools make the problem definitions for stress analysis far from ideal. It is suggested that a fruitful area of research future is to develop heuristics that are commonly used by machinists, backed up with simple stress analyses that identify the working ranges for clamp pressures and clamp configurations.

REFERENCES

Asada, H., and By, A. 1985. Kinematic analysis of workpart fixturing for flexible assembly with automatically reconfigurable fixtures, *IEEE Journal of Robotics and Automation*, Vol. RA-1, No. 2, pp. 86–94.

Boyes, W. E. 1982. *Jigs and Fixtures*, 2nd Ed. SME, Dearborn, Mich., p. 340.

Cutkosky, M. R., Kurokawa, E., and Wright, P. K. 1982. Programmable conformable clamps. *AUTOFACT4 Conference Proceedings*, November, Society of Manufacturing Engineers, pp. 11.51–11.58.

DeGarmo, P. E., Black, J. T., and Kohser, R. A. 1984. *Materials and Processes in Manufacturing*, MacMillan, New York.

Englert, P. J., and Wright, P. K. 1986. Applications of artificial intelligence and the design of fixtures for automated manufacturing. *Proceedings of the IEEE Conference on Robotics and Automation*, San Francisco, Calif., April 7–10, pp. 346–351.

Ferreira, P. M., Kochar, B., Liu, C. R. and Chandru, V. 1985. AIFIX: An expert system approach to fixture design, *Computer-Aided/Intelligent Process Planning, presented at the ASME Winter Annual Meeting*, November 17–22, pp. 73–82.

Gandhi, M. V., and Thomson, B. S. 1986. An analytical and experimental investigation of adaptable phase-change fixtures. *Manufacturing Processes, Machines and Systems*, Society of Manufacturing Engineers, Dearborn, Mich., pp. 141–148.

Kurokawa, E. 1983. Flexible conformable clamps for a machining cell with applications to turbine blade machining, *Robotics Institute Report Series*, Carnegie-Mellon University.

Markus, A., Markusz, Z., Farka, J., and Filemon, J. 1984. Fixture design using PROLOG: An expert system, *Robotics and Computer Integrated Manufacturing*, Vol. 1, No. 2, pp. 167–172.

Rouse, W. H. D. 1937. *Homer—The Odyssey*, Mentor Books, New American Library, New York, P. 141.

Wright, P. K., and Holzer, A. J. 1980. A programmable die for the powder metallurgy process. *Proceedings Ninth North American Manufacturing Research Conference*, Vol. 9, pp. 65–70.

10 The Machine Tool's Cutting Sensors

Sensation is concrete perception of objects and people by means of our five senses. It provides the basic framework for our lives and in its unalloyed state renders us the experience of what we commonly regard as reality in its most direct and simple form. Our senses tell us what is.[1]

—Edward Whitmont in *The Symbolic Quest*

A machine tool is a dynamic system that must be operated in closed-loop control. The system cannot run open loop; otherwise, any operation that does not have a favorable outcome will prevent the machine from completing its task. Closing the loop involves the judicious placement of sensors and encoders that will continually report on the progress of an operation. These sensor reports are used to adjust the control of the overall system to keep an operation on track. In the case of the machine tool, sensors that monitor the cutting process are the most critical measurements that can determine whether or not an operation will be successful.

Our lives are filled with dynamic systems that operate in ways very similar to the machine tool. For example, an automobile is a system that runs partially in open loop if a brick is placed on the gas pedal. However, for standard operating conditions, the human is designed into the overall system as the feedback element that closes the loop and adjusts the overall system. The human uses eyes and ears to detect dangerous traffic situations and to make corrections with arms and feet. It is these senses, and others, that must be replaced when a system "drives" a cutter in a machine tool.

10.1 DIFFERENT MONITORING STRATEGIES

Society has imposed rules on automobile drivers to encourage safe conduct. For example, the national speed limit has been shown to reduce the number of fatal accidents. And, although many vehicles obviously can go faster and an alert, individual driver under perfect road conditions can complete the journey much quicker, the speed limit caters to the imperfect world with its unexpected events.

The "drivers" of machine tools are also faced with similar tradeoffs. Even with correctly chosen cutting speeds, the proper programs, and a good setup plan, the machining process is still unpredictable. For example, Figure 10.1 shows that natural inhomogeneities in workmaterials and cutting tools cause a wide variation in how much life can be expected from the tools being used. In this example, the variation in tool life is between 45 and 195 drilled holes. When such drills, or any other cutting tools are used in today's flexible manufacturing systems, the safest reaction to these variations is to respond to the worst possible case. In this situation, this would mean that every drill is replaced after the

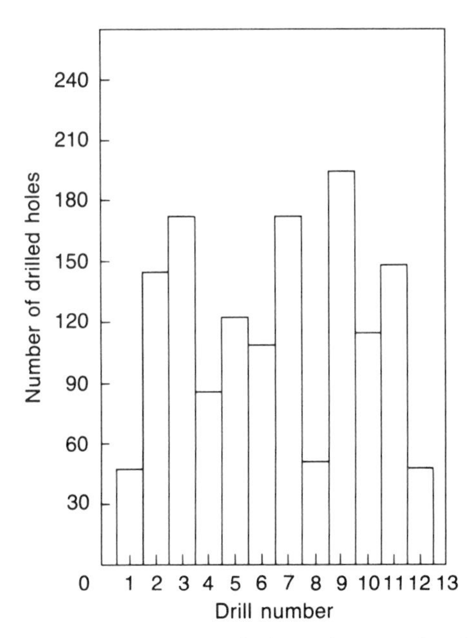

Figure 10.1 Variation of drill life for drills from the same batch.

Table 10.1 Use of Human Senses to Monitor Machining Activities with Possible Sensor Solutions

HUMAN SENSE	HUMAN MONITORING ACTIVITY	POSSIBLE MACHINE SENSORS
Vision	Establish datum points between incoming workmaterial stock, fixtures, and cutting tools.	Touch-trigger probe
	Watch for sudden breakage of cutting tools.	Machine vision and touch-trigger probes
	Carry out between-pass visual checks of trends in flank and crater wear.	Machine vision
	Detect tool temperature by studying chip color.	Thermocouples
	Watch for changes in part surface finish.	Machine vision
Hearing	Listen for excessive vibrations between tool and part.	Accelerometers and dynamometers
	Listen for sounds of tools near or at failure.	Accelerometers, dynamometers, and acoustic emission devices.
Touch	Feel excessive vibrations through floor or by touching fixtures.	Accelerometers
	Approximately sense excessive cutting forces by touching fixtures or tool-holders.	Dynamometers and strain gauges
	Touch surface finish and approximately gauge quality.	Stylus measurements of surface finish
Smell	Excessive tool temperatures sometimes change the smell of cutting fluids.	Chemical sensors
Taste		

forty-fifth hole. The remaining, potential life in the majority of drills would simply not be used. This approach prevents a cutter from breaking in the middle of a cut, but at the cost of discarding many perfectly good tools.

10.1.1 A Needs Analysis for In-Process Sensors

An automobile must be constantly checked to ensure that it is operating normally. The driver should monitor the instrument panel and be attuned to other signs that may indicate that the automobile is not acting normally. For example, if the car pulls to one side during braking, the brakes are either worn or the tires are not inflated properly. The machine tool is also a complex system that can have many malfunctions. It is often possible to notice a malfunction and to bring the machine safely to a stop without even damaging the in-process component. To accomplish this, the subsystems must be checked on a regular basis.

The role of today's craftsman is to watch for trends in the performance of the system and to change cutting speeds or switch tools before problems arise. How do the skilled machinists use their five senses and blend them together to make decisions? Table 10.1 lists some of the ways in which human senses are used. These monitoring and decision-making processes are more formally summarized in the block diagrams of Figures 10.2 and 10.3. The inputs from the engineering designer include a geometrical description of the part and a requirement for a low-cost, high-quality product. The expert machinist writes a CNC program, constructs a setup sheet, selects tools and fixtures, and consults a reference source (e.g., Metcut Research Associates 1985) for the appropriate cutting speeds and feed rates.

These selected rates of metal removal and the tool type control the physics of the chip formation "right down at the cutting tip." In particular, the selected rates of metal removal govern the stresses and the temperatures acting on the chip-tool contact area. These, in turn, directly control tool wear and expected life. The machinist watches and "feels" these processes using the skills in the middle column of Table 10.1. If unexpected workmaterial variations cause the tool to deteriorate, the machinist can either change tools or modify speed, feed rate, or depth of cut at the summing point in Figures 10.2 and 10.3. For automation, the quantitative measurements from sensors now replace the machinist's qualitative judgments (Figure 10.4.). Also, a reference model, based on the physical properties of the tool material, replaces the machinist's experience.

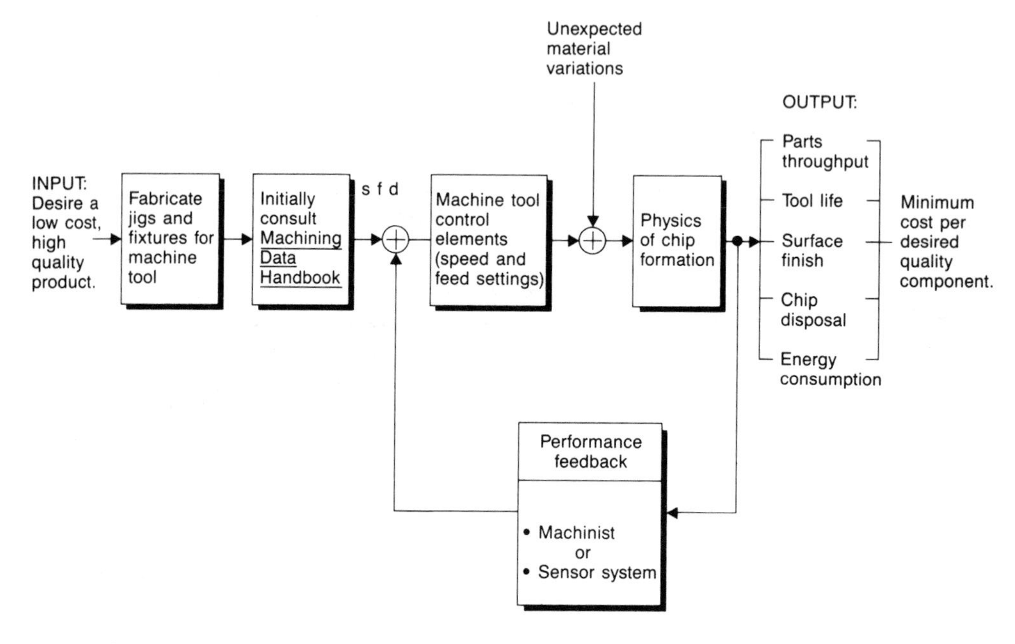

Figure 10.2 The feedback system, inputs and outputs in turning. (Wright, "Physical Models for Tool Wear for Adaptive Control," published by ASME, PED Vol. 8 *Computer Integrated Manufacturing*, 1983.)

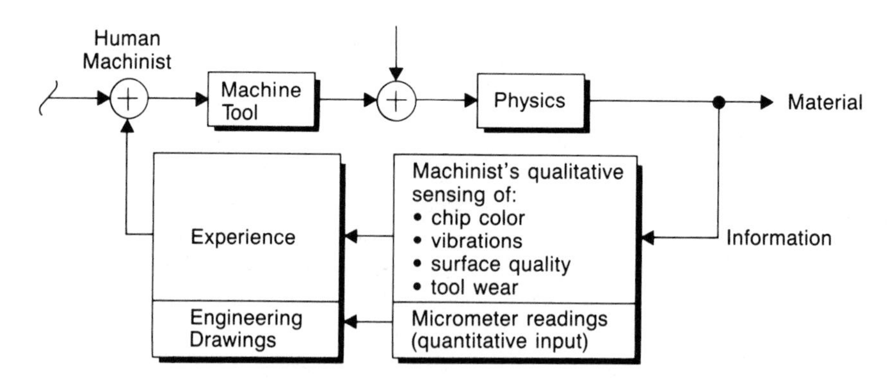

Figure 10.3 The performance feedback system for a human overseer. (Wright, "Physical Models for Tool Wear for Adaptive Control," published by ASME, PED Vol. 8 *Computer Integrated Manufacturing*, 1983.

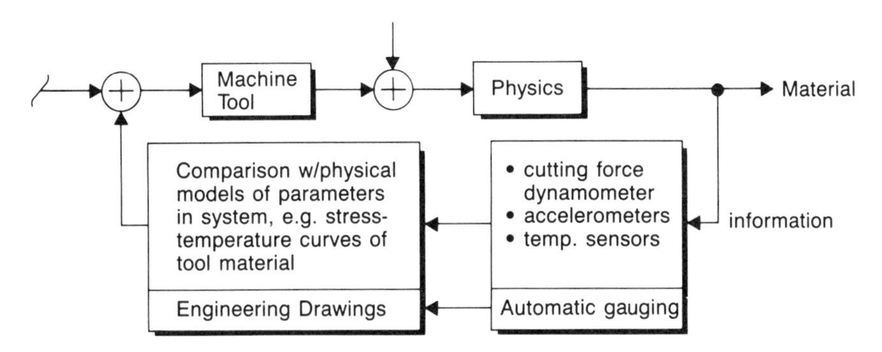

Figure 10.4 Sensor needs for closed-loop control and tool monitoring. (Wright, "Physical Models for Tool Wear for Adaptive Control," published by ASME, PED Vol. 8 *Computer Integrated Manufacturing*, 1983.)

Table 10.2 summarizes the needs analysis, i.e., the needed monitoring strategies, for machining with sensors. These are analyzed in more detail in the next three main sections of the text. The table compares the operations of a machine tool with those of an automobile, and it also acts as a road map to the remainder of the chapter. The general reader, with less interest in machining technology, could pass directly from the table to the next chapter without losing continuity. The interested student or manufacturing engineer should rewrite this table for another manufacturing process such as welding, the assembly of printed circuit boards, or textile production.

10.1.2 The State-of-the-Art for Post-Mortem Sensors

Today, simple sensors can be installed on machine tools to detect that a tool has failed. Many machine tool companies now supply, as standard features, touch-trigger probes (Tlusty and Andrews 1983), torque-controlled machining sensors (Schaffer 1983) and acoustic emission devices (Dornfeld 1983). For example, the touch-trigger probe (see Section 10.3.2) is installed on the machine tool's table, and it can be brought into the work area periodically to touch and measure the length of partially used drills to determine if rapid wear has occurred. Also, the probe can be used to check that a hole has been properly drilled by measuring its depth.

As another example, a torque-controlled machining sensor can be mounted in the machine tool spindle to monitor the torque on the drill.

Table 10.2 The Types of Monitoring Skills and Sensors Used for Machining With an Automobile Simile

MONITORING SKILL	AUTOMOBILE DOMAIN	MACHINING DOMAIN	SENSOR AND CHAPTER SECTION
Perform real-time adjustment of speed for smooth, efficient running.	Drive at cruising speed where car is smooth, fuel efficient, and powerful, but not over-revved.	Machine at a speed feed and depth of cut that is safe, but productive. The stress (σ) and temperature (*T*) on the tool will be high, but no so high as to cause failure.	Monitor cutting stresses in real-time with piezoelectric dynamometer and measure temperature with thermocouples. Keep σ and *T* just below critical values based on metallurgical data (Section 10.2).
Keep alert for sudden-death situations where unexpected failures arise (despite good monitoring above) beyond the operator's control.	Occasionally glance at oil pressure gauge or warning light to make sure engine is performing well.	Occasionally glance at tools and drills to make sure they are not broken or excessively worn. Also monitor chip color and surface finish of part for sudden changes.	Use dynamometer to register sudden "spike" force from worn-out tool or touch-trigger probe to check for very worn or broken drills (Section 10.3).
Carry out routine maintenance between usage and estimate time needed before service work is due.	Check oil, water, and tires after long journeys or weekly; e.g., How much life is left in tires?	Check cutting tools after each machining pass or between parts; e.g., How much life is left in tool?	Monitor flank wear land between passes, or parts, using computer vision. Then estimate percentage of life left in tool (Section 10.4).

But, presently, it is not sensitive enough to detect the gradual wearing of the drill. At best, it can reveal a sudden "spike," followed by a zero torque reading, to indicate that the drill is broken. All three of the simple sensors just mentioned only provide post mortem information. If a drill or tool has indeed broken, the problem still remains to stop the machine and retrieve the partly machined component that has the shattered drill shank still stuck in a half-drilled hole.

10.1.3 The State-of-the-Art for In-Process Sensors

A manufacturing engineer might say that these previous sensors report, after the fact, on the "sudden death" of a drill. But, to continue this medical metaphor, we prefer to monitor the cutting tool's health during the course of its life and take remedial action before its sudden death.

More advanced sensors that can monitor tools in process have been developed in laboratories (Barth, Blake, and Stelson 1980; Brecker and Shum 1977; Mathias 1978; Schaffer 1983). However, the machine tool users have not been able to take advantage of them (Kegg 1983). On the factory floor, it seems that these advanced sensors for predicting in-process trends need too much calibration and retuning. While the research engineer is on site nurturing the new creation, adjustments can be made to account for such factors as new tooling types, variations in fixtures, and unpredictable workmaterials. However, after the engineer has returned to the laboratory, no one on the production staff can, or has the time to, keep up the nurturing process.

Despite this lack of success, if factories are to be unattended, machine monitoring and failure prediction activities must be automated. This is the subject of the following sections.

10.2 CLOSED-LOOP CONTROL WITH IN-PROCESS SENSORS

Closed-loop control is concerned with monitoring the in-process cutting conditions and making adjustments to account for unexpected disturbances. These disturbances may arise from an external source, for example, a new batch of workmaterial with a different heat treatment. Or, they may arise from an internal source, for example, a faulty fixturing element that causes the tool and part to vibrate excessively. The sensor system must first detect how these unexpected disturbances are influencing the tool's integrity. Second, comparisons must be made with

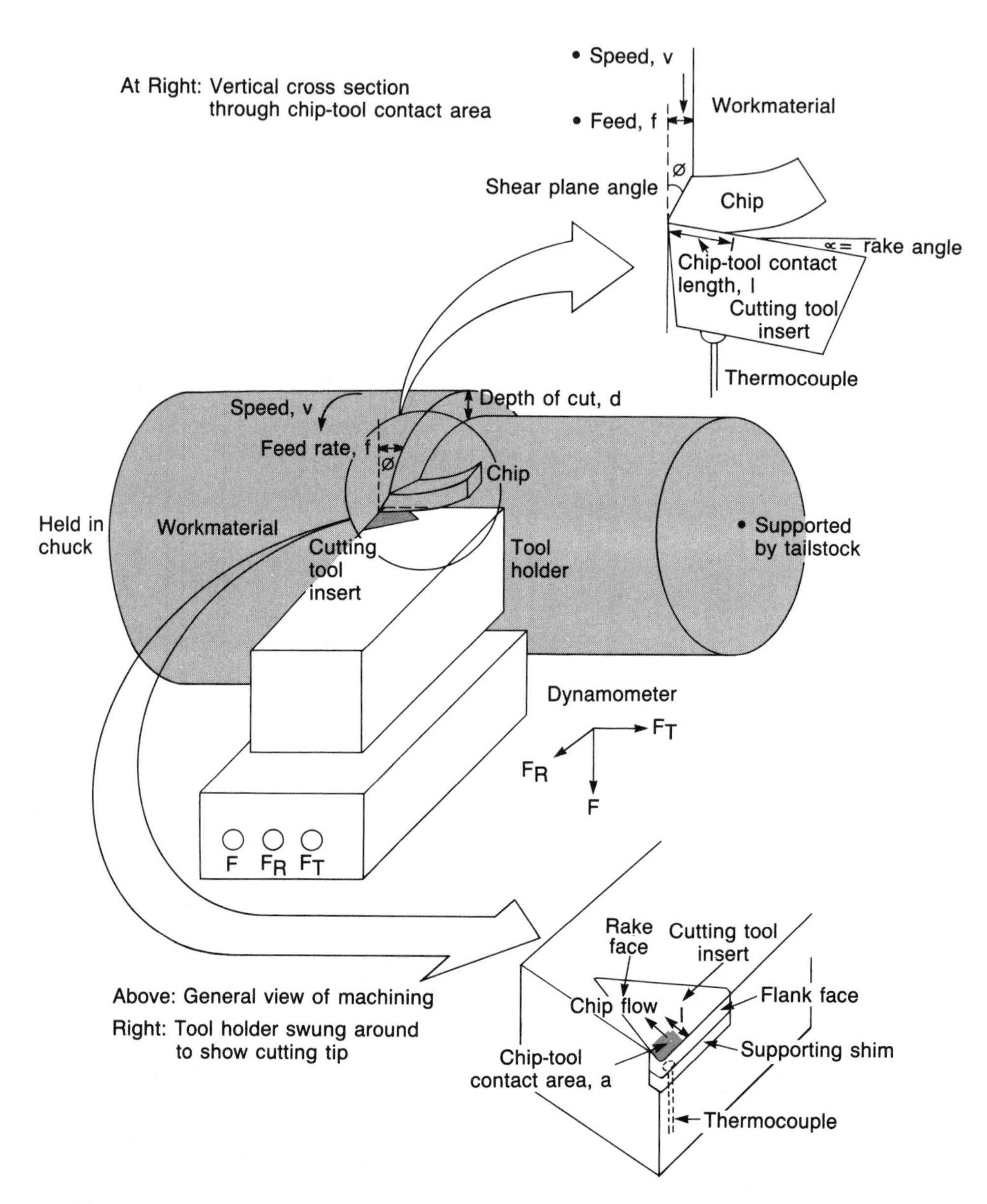

Figure 10.5 A general view of the turning process including a vertical cross-section through the forming chip and an enlarged view of the tool insert. While this is the simplest geometry to analyze, all cutting processes (drilling, milling, and so on) can be modeled by the two-dimensional cross-section shown at the top.

a reference model of an "ideal" system to see how far the working system has deviated from the ideal reference source. Third, corrections must be made to return to normal operation.

Manufacturing engineers often, but wrongly, refer to this as adaptive control. Unless the reference model is updated over time and effectively "learns" about machining, the system is not, strictly, adaptive control (Takahashi, Rabins, and Auslander 1970). We thus use the term closed-loop control for the application in the following sections.

10.2.1 A Reference Model for Closed-Loop Control

A new approach to the closed-loop control of machining is presented. Tool temperatures and stresses are monitored simultaneously and modified on the basis of a reference model. This model is rooted in the metallurgical properties of the tool material being used for cutting. This approach is believed to be worthwhile because it is more sympathetic than earlier schemes (e.g., Mathias 1978) to the physics of tool wear processes. By being based on the metallurgical properties of the tool material, it is also independent of the particular cutting process.

The geometry of the simplest cutting process, turning on a lathe, is shown in Figures 10.5 and 10.6. Actually, all cutting operations have fundamentally the same geometry (Trent 1984). Consider the common, hardware store drill. A vertical slice, or cross-section, taken normal to one of the cutting edges at the end of the flutes reveals the same geometry as that shown at the top of Figure 10.5. The same is true for all milling cutters, when sectioned at the appropriate angle.

For effective closed-loop control, there must be a quantifiable relationship between the three levels of the problem shown in Figure 10.7. Viewing this figure top-down, the selected cutting speed and feed rate generate particular temperatures and stresses in the tool, which, in turn, cause wear. Thus, looking bottom-up, the wear rates must be controlled by keeping the temperature and stress below critical values, which, in turn, must be controlled by a judicious choice of speed, feed rate, and depth of cut.

The goal of the control system is to establish physical relationships between the three levels in Figure 10.7 and then, for practical machining, adjust the control variables to obtain the maximum rate of metal removal while minimizing wear.

If perfect relationships concerning the physics of chip formation were available and the tool geometry remained unaffected by wear, the preceding scheme could be implemented without sensors. It would be a good open-loop control scheme. However, to accommodate the inaccuracies in the modeling, random workmaterial variations, and the change in tool shape, sensory devices that monitor temperature and stress are necessary for feedback. Referring to Figure 10.6, the three tool wear processes that have been analyzed are:

- **Plastic collapse of the cutting edge region:** This occurs when the local normal stress on the edge exceeds the elevated temperature yield strength of the tool material in this region. When viewed

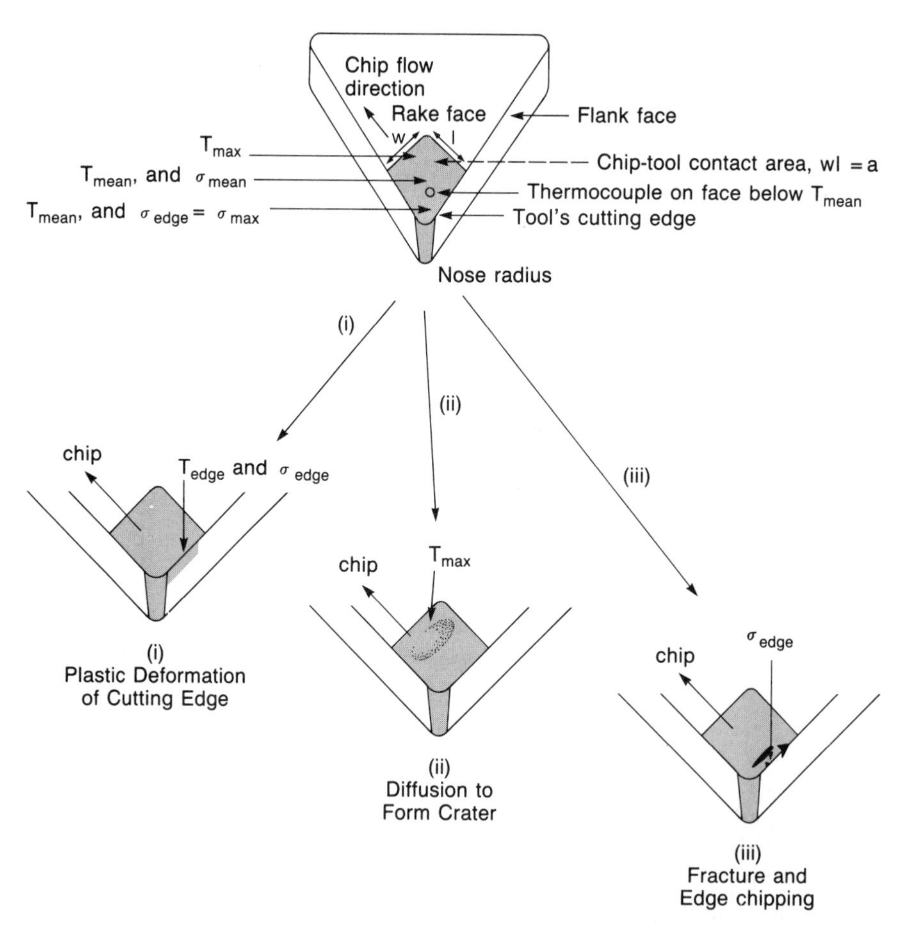

Figure 10.6 Three types of tool wear separated as "intrinsic images" (Barrow and Tenenbaum 1978) from a tool showing generalized wear (see Boothroyd 1985 and Trent 1984).

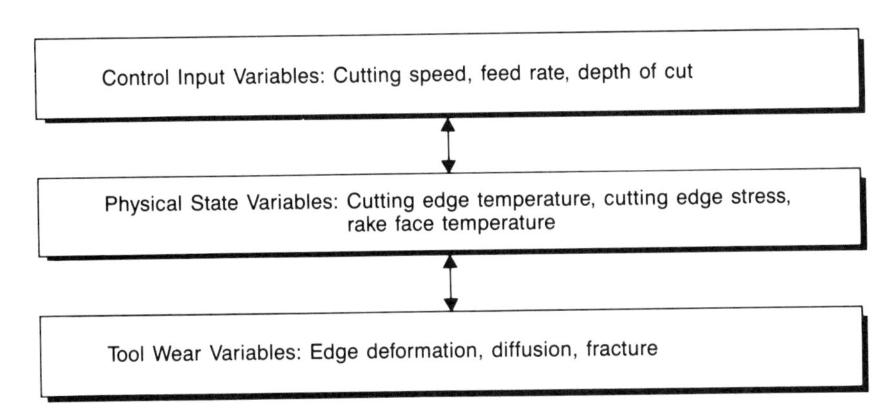

Figure 10.7 Levels of problem analysis which need to be translated into a reference model for closed-loop control. (Wright, "Physical Models for Tool Wear for Adaptive Control," published by ASME, PED Vol. 8 *Computer Integrated Manufacturing*, 1983.)

under a stereomicroscope, the edge appears plastically depressed or pushed down.

- **Cratering of the tool rake face:** This occurs by the diffusion and solution of tool material in the chip. Trent (1984) observed that this process is accelerated above a critical temperature. In this case, the worn surface is very smooth because the solution process resembles a gradual, high-temperature erosion.
- **Brittle fracture of the tool emanating from cracks in the contact area:** This occurs when the fracture criterion for the tool material is exceeded by the combination of normal and tangential stresses acting on the tool face. When viewed under the stereomicroscope, the cracked areas appear crystalline and specular, similar to a broken rock sample.

In some cutting operations, just one of these wear mechanisms will be particularly dominant. For example, when machining aerospace alloys, wear by plastic deformation of the cutting edge tends to be dominant because the high strength of these alloys imposes intense temperatures and stresses on the tool. In other cutting operations, all three (and other) wear mechanisms will occur simultaneously with a similar intensity. For example, if the stock material is an unusually shaped casting or forging, uneven cutting loads will be imposed on the tool. This may cause local fracture and chipping of the tool edge in addition to the plastic deformation. And, if cutting speeds are high, a crater may additionally be worn into the rake face by diffusion.

When viewing such a worn tool under a stereomicroscope, it is easy for the trained human eye to distinguish the different wear modes. Indeed, the wear modes have been illustrated in this way in Figure 10.6. Such a separation evokes the same concept used for creating intrinsic images from a general visual scene, as shown in Figure 5.6. Recall that these "intrinsic visual images" can be separated, with sensors tailored to extract each intrinsic image, which then greatly simplifies the analysis. Likewise, "intrinsic wear images" can be separated and analyzed individually. First, for example, the acoustic emission sensor (Dornfeld 1983) can be used to isolate wear due to fracture and chipping. Second, temperature sensors can be used to ensure that the tool rake face is below the temperature at which diffusion wear becomes especially rapid. Third, a combination of temperature sensing and stress sensing can be used to detect and control the wear by plastic deformation. This third case is amplified in the next section.

10.2.2 Case Study: Wear by Plastic Deformation of the Cutting Edge

To illustrate the control scheme, one of the wear processes, plastic deformation of the cutting edge, is now analyzed in more detail. Figure 10.8 presents laboratory compression tests on different tool materials. These plots show the relationship between the yield strength and temperature. However, Figure 10.8 should be viewed as more than just the physical properties of the tool materials presented in a way commonly found in metallurgical databooks. It also represents a "fail-safe" diagram for the edge of a cutting tool made from one of the materials. A point in the lower left space, below the curves, represents a safe condition where the tool edge will not yield and plastically deform. Conversely, a point in the top right area, above the curves, represents a condition where the material will yield and deform. Ideally, the cutting speed and feed rate need to be adjusted so that the temperature and stress values of the tool's cutting edge are in the safe region. However, operating conditions cannot be "too safe." To machine and produce parts quickly, the speed and feed rate, and consequently temperature and stress, must be pushed into the shaded region, just under the curves in Figure 10.8.

Figure 10.8 shows that the metallurgical integrity of the tool material is controlled by both temperature and stress. Although there is no direct

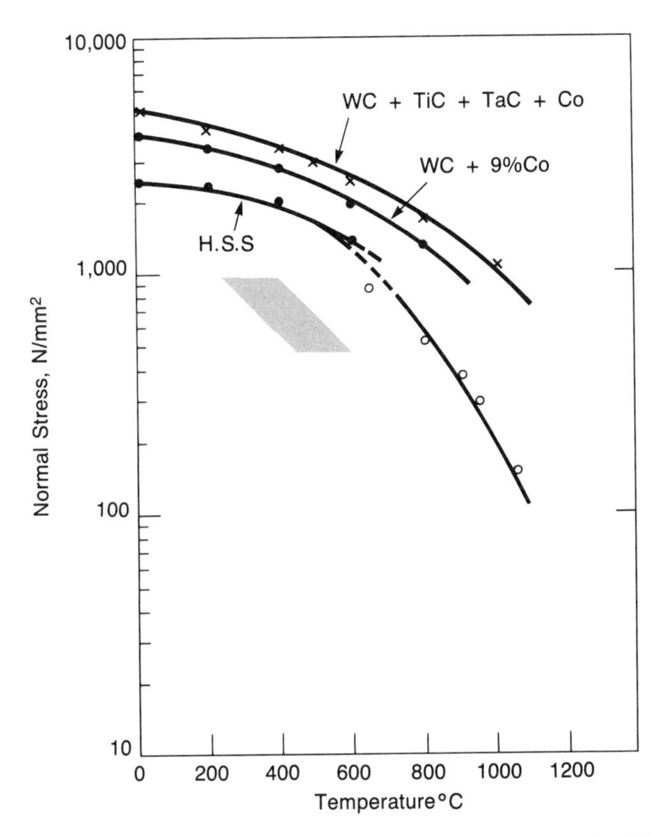

Figure 10.8 Elevated temperature yield strength of tool materials. The shaded area is the "safe-but-productive" region for temperature and stress, when machining with high-speed steel. The top two curves are for cemented carbide tools made from tungsten carbide (WC), titanium carbide (TiC), and tantalum carbide (TaC), cemented together with cobalt (Co). The lower curve is for high-speed steel (HSS.) (Wright, "Physical Models for Tool Wear for Adaptive Control," published by ASME, PED Vol. 8 *Computer Integrated Manufacturing*, 1983.)

physical correlation between temperature and stress, an increase in either will push the state of the tool material toward a failure condition. All elevated-temperature, engineering designs face this issue: the integrity of the component in use is governed by its operational temperature and stress states. For example, the turbine blades of an aircraft engine operate at high temperatures under the influence of centrifugal stresses. The life of a turbine blade is thus influenced by its alloy composition, by its shape, and by the temperature and stress state in which it operates. Analogically, the life of a cutting tool is influenced by the same factors. But, while the alloy composition and shape are designed in by the manufacturer, it is the end user of the tool who determines its temperature and stress state. This state is controlled by the cutting conditions that the machinist or manufacturing engineer selects. We now describe this relationship.

For a constant depth of cut d, Yen and Wright (1983) presented functional relationships between the control input variables speed s and feed rate f, and the physical state variables temperature T and stress σ.

$$T_{edge} = \text{function} \frac{s \; f \; \alpha \; \tau}{K \; \phi} = AT_{mean} = BT_{max} \tag{10.1}$$

$$\sigma_{edge} = \sigma_{max} = C \; f = 1.75 \; \sigma_{mean} \tag{10.2}$$

Where α is the tool rake angle in Figure 10.5; τ is the shear strength of the workmaterial being cut; K represents the thermal properties of the tool material; ϕ is the shear plane angle in Figure 10.5; and A, B, and C are constants.

The empirical analysis of tool wear by plastic deformation of the cutting edge requires a knowledge of the particular stress σ_{edge} and particular temperature T_{edge} in this region. The preceding equations have thus been written to begin with these terms. Experimentally, it has been found that the stress on the tool's edge is also the maximum stress σ_{max} on the tool. These values are shown, also experimentally, to be 1.75 times higher than the mean stress σ_{mean}, acting over the complete chip-tool contact area in Figures 10.5 and 10.6 (Zorev 1963; Bagchi and Wright 1986). The chip-tool contact area supports the cutting forces. This area varies according to cutting speed, the tool-work combination, and any lubricant used. It also fluctuates during a cut and its dimensions cannot be predicted a priori.

Contrary to first intuition, the cutting edge is generally cooler than the remainder of the chip-tool contact area in Figures 10.5 and 10.6 because, as the chip moves along the chip-tool contact area, the frictional heating is cumulative. Thus, the mean temperature T_{mean} approximately in the center of the contact area is higher than the temperature at the edge. In turn, the temperature toward the rear of this frictional, planar heat source is highest of all, T_{max} (Trent 1984).

The solution of Equations (10.1) and (10.2) is aided by some practical heuristics. It is well known that high-speed steel tools deform very rapidly, under any cutting load, once their secondary hardening temperature is exceeded (at approximately 650° C). For cemented carbide tools, the temperature at which rapid deformation occurs is 1000° C to 1150° C, depending on composition (Trent 1984).

Thus, to establish a control strategy that avoids wear by plastic deformation of the tool's edge, we begin by selecting a safe temperature for T_{edge}, e.g., 800° C for a cemented carbide tool. A corresponding stress value is then obtained from Figure 10.8, and then appropriate values of speed and feed can be calculated from Equations (10.1) and (10.2).

Minor adjustments can be made on the basis of the sensor readings we describe in the next sections, a thermocouple for T_{mean} and a dynamometer for σ_{mean}. For example, suppose the strength τ of the workmaterial being cut is higher than originally anticipated. The sensor readings would, in turn, show higher than expected values. The cutting speed and feed rate could then be adjusted, according to Equations (10.1) and (10.2), to return to safe, but productive conditions. Refer to Trent (1984) for a general review of tool wear and to Yen and Wright (1983) for a deeper discussion of how the other two major wear mechanisms, diffusion and fracture, can be controlled.

10.2.3 Real-Time Monitoring of Temperature and Stress

To implement the control scheme, a remote thermocouple is needed to monitor temperature and a dynamometer is needed to monitor cutting stress. These are well-established, relatively inexpensive sensors. For the turning operation in Figure 10.5, the dynamometer can be used to support the tool-holder and to directly measure the three cutting force components. The principal cutting force F is divided by the chip-tool contact area a to give the mean stress on the tool. The other two force components are the radial force, F_R, and thrust force, F_T. The remote thermocouple is placed between the cutting tool insert and the supporting shim, directly below the chip-tool contact area.

These two sensors provide the raw data shown as the first step in Figure 10.9. The second step is to convert this data, in real time, to the mean temperature and stress values. Third, these mean values can be converted, again in real time, to the localized temperature and stress values at the very cutting edge of the tool.

Temperature Monitoring

In Figures 10.5 and 10.6, the remote, chromel/alumel thermocouple is 6mm away from the chip-tool contact area. This dimension reflects the typical thickness of a cemented, carbide, cutting tool insert. The mean

STEP	TEMPERATURE MONITORING	STRESS MONITORING
1. Obtain raw data.	Measure temperature at remote thermocouple on bottom of tool insert.	Measure principal cutting force (F) with dynamometer.
2. Calculate mean stress and mean temperature.	Solve heat-diffusion equations to render T_{mean} on top, rake face in center of contact area.	Use a tool with controlled contact area (a) and calculate mean stress, $\sigma_{mean} = F/a$.
3. Calculate localized stress and temperature of tool edge. Also calculate the maximum temperature toward the rear of the contact area.	Use generic-temperature profile to determine both the temperature of the tool edge T_{edge} and the maximum temperature T_{max}.	Use generic-stress profile to determine σ_{edge} (which is also σ_{max}).

Figure 10.9 Summary of methodology that converts raw sensor information into more meaningful stress and temperature information.

temperature T_{mean} on the top of the tool in the center of the contact area can be found by solving the three-dimensional heat diffusion equations for the complete cutting tool and its holder (Boothroyd 1975; Tay, Stevenson, and deVahl Davis 1974; Chow and Wright 1987). This analysis treats the contact area a as a frictional, planar, heat source lying at the center of a sphere. Heat diffuses from this source to the remote thermocouple.

A reverse solution allows T_{mean} to be calculated from the remote thermocouple reading. When machining steels, at cutting speeds typical of commercial practice, the remote thermocouple readings are in the range 100–150° C. Chow and Wright's (1987) thermal analysis shows that these thermocouple readings should be multiplied by 6.25 to obtain T_{mean}. Mean temperatures in the range 625–950° C, depending on the chosen cutting speed, have also been obtained by Chao and Trigger (1959) using work-tool thermocouples, by Tay et al. (1974) using finite element analysis and by Wright and Trent (1973) using the metallographic method described below.

The mean temperature can be used to estimate the edge temperature T_{edge} and the maximum temperature T_{max} toward the rear of the contact area. The simplest way to do this is to use information from a loss-of-temper metallographic method (Wright and Trent 1973). This involves a post mortem analysis of cutting tools where changes in microstructure are correlated with the temperatures that existed in the tool during use. This cannot be used online, but it gives a detailed map of the temperature distribution around the tool edge and rake face area. From this, the generic-temperature profile along the chip-tool contact length can be found (Figure 10.10). For one particular work-tool combination, the general shape of this profile is the same over a wide range of cutting speeds (Wright, McCormick, and Miller 1980, Trent 1984). While the absolute values of temperature change with speed, the profile's shape remains the same.

The T_{mean} value in Figure 10.10 has been calculated from the area under the curve. Data from several samples have then been averaged over a range of cutting conditions. Thus, for cutting steels at commercial speeds and feeds, we find that a good algorithm to use is that $T_{edge} = 0.76\ T_{mean}$ and that $T_{max} = 1.18\ T_{mean}$. These constants can thus be used

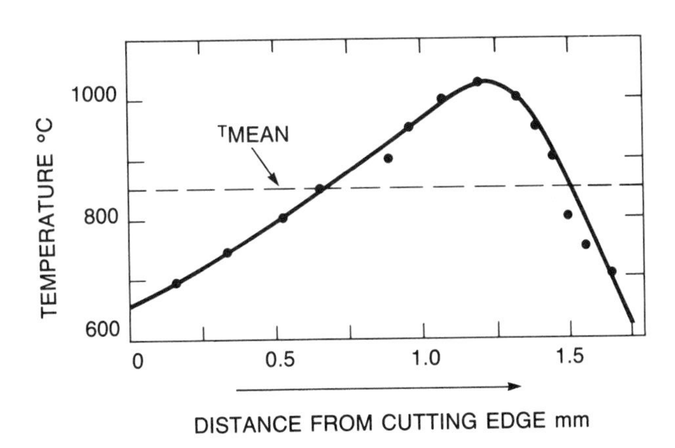

Figure 10.10 Generic-temperature profile from post-mortem metallographic method, and T_{mean} value calculated from area under curve. Machining AISI 1008 Steel. (Wright, "Physical Models for Tool Wear for Adaptive Control," published by ASME, PED Vol. 8 *Computer Integrated Manufacturing*, 1983.)

to convert the remote thermocouple reading (see its position in Figures 10.5 and 10.6) directly to the edge or maximum temperature. For an industrial application, these constants should be reevaluated for the particular tool and workmaterials used by a practitioner. The cutting-edge temperature is used to control wear by plastic deformation (see Figure 10.8), and the maximum temperature used to control wear by diffusion.

Stress Monitoring

Piezo-electric dynamometers and strain-gauge dynamometers are, today, the most practical devices for obtaining the raw, cutting force data (Powell, Cosic, Erickson, Herko, Kline, Mayer, and Varma 1984; Pearson, Syniuta, and Cook 1980). Although it is relatively easy to measure the forces, there is no direct relationship between them and the physics of the chip formation process (Wright 1983; Williams, Smart, and Milner 1970). To exploit the forces, the chip-tool contact area a must be specified, and then the stress can be calculated ($\sigma_{mean} = F/a$). Actually, this is the same as in any other engineering design situation. The simple tensile test is the best illustration that forces are meaningless unless they are related to the area under load.

To correlate forces with mean stresses during real-time machining, we have been using controlled contact area tools. These predetermine the value of the area that bears the load (Figure 10.6). These tools also provide other practical advantages such as chip-shape control and the overall reduction in cutting temperatures (Chao and Trigger 1959).

To implement the closed-loop control scheme, we have converted the mean stress to the maximum stress on the tool edge. This allows wear by plastic deformation to be controlled as described by Figure 10.8. Wear by fracture is also strongly influenced by the maximum stress acting on the tool edge (Loladze, 1975; Yen and Wright 1983).

The relationship between σ_{mean} and σ_{max} has been of interest since the early work of Zorev (1963). He proposed that stress was a maximum at the edge, falling parabolically along the chip-tool contact length. Amini (1968) confirmed this proposition in photoelasticity experiments with soft, epoxy-resin cutting tools (Figure 10.11). Bagchi and Wright (1986) also confirmed the result in photoelasticity experiments with single crystal, aluminum oxide tools. But, with this hard tool material, stress results could be obtained very close to the edge. The results show that for many cutting conditions and workmaterials, $\sigma_{edge} = \sigma_{max} = 1.75\ \sigma_{mean}$. This result has been used in the real-time determination of the stress acting on the edge of the tool.

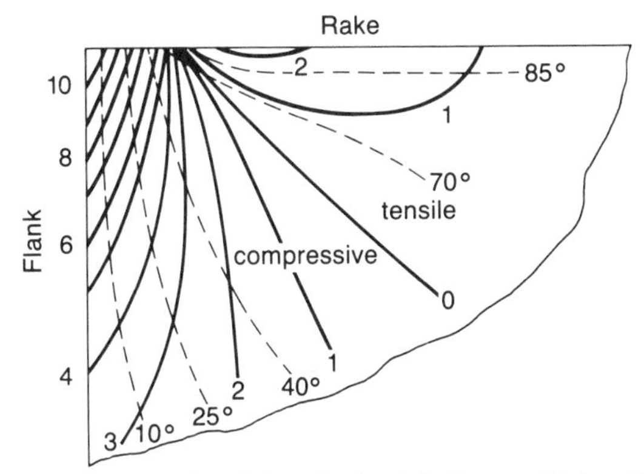

Composite isochromatics and isoclinics obtained during machining 12L14 steel at 75 m min^{-1} and uncut chip thickness of 0.381 mm. Solid lines, isochromatics; dashed lines, isoclinics.

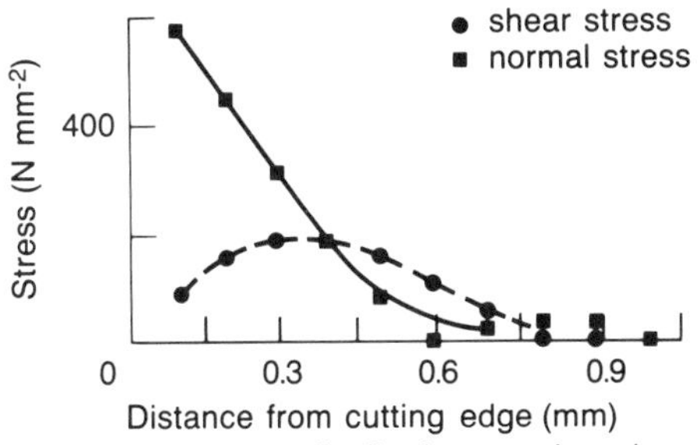

Normal and shear stress distributions on the rake surface

Figure 10.11 Isochromatic lines in single crystal aluminum oxide tool used to empirically obtain $\sigma_{max} = 1.75\ \sigma_{mean}$. (Bagchi & Wright, "Stress Analysis in Machining with the Use of Sapphire Tools," *Proceedings of the Royal Society of London A*, Vol. 409, 1987. Reprinted with permission.)

10.2.4 Summary

To reiterate, this closed-loop control scheme establishes the "cruising speed" for cutting where rates of metal removal are safe, but productive. The closed-loop control can been illustrated by analyzing one of the major wear mechanisms, plastic deformation of the cutting edge. Thermocouples and dynamometers can be used for real-time machining to ensure that operations remain safe, but productive (see shaded area of Figure 10.8). In this way, the inevitable approximations in the modeling and the unpredictable variations in actual machining can be accounted for by the sensor feedback (see Figure 10.4)

10.3 MONITORING OF FINAL TOOL FAILURE

We now consider the second monitoring strategy in Table 10.2. Vigilant diagnosis by closed-loop control should ensure that both automobiles and cutting tools are not overstressed and that they lead a long productive life. However, automobiles and cutting tools are bound to wear out eventually. If they did not, we would not be making the best use of them.

The inevitable wear of an automobile engine can be intuitively sensed by an alert driver and more evidently displayed by sudden fluctuations in the temperature and oil pressure gauges. The attuned driver may well be able to stop the car before a fatal seizure of the engine. In the worst case, the driver may be able to disengage the clutch before the seizing engine fatally damages the transmission. The sensors described in the next two subsections, by analogy, provide the same responses for machining. In the first case, rapidly rising force signals, with fluctuations superimposed on them, indicate that the tool or drill is so worn that it is seizing against the workmaterial. In the second case, the touch-trigger probe can be used to detect a worn out or broken off cutter that should definitely not be used; otherwise, the part and even the fixtures and machine tool will be at risk.

The inevitable wear on a cutting tool's flank face occurs by a combination of the wear mechanisms: some local plastic deformation of the irregularities on the edge, some diffusion, and some wear by local fracture known as attrition. Trent (1984) and Boothroyd (1975) show empirically that the end of useful tool life occurs when the flank wear is greater than ≈ 0.03 inch. Then, cutting temperatures and stresses begin to rise sharply because the tool is blunt, and it ploughs rather than cuts the workmaterial. Figure 10.12 shows the three well-documented stages of

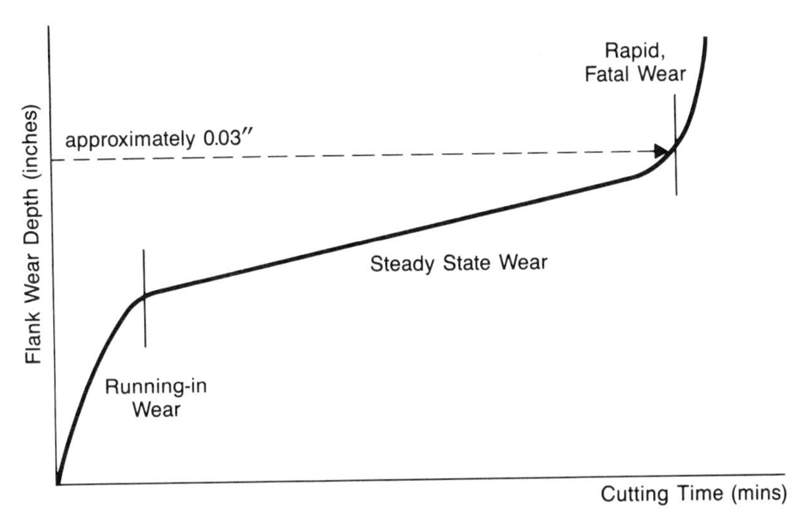

Figure 10.12 The three stages of tool wear over time. Although flank wear depth is shown on the ordinate, cutting temperatures and stresses generally follow the same trend (Trent 1984; Boothroyd 1975).

tool wear and hence corresponding stages of rising temperature and stress: running in, steady state, and rapid.

10.3.1 Force Measurements and Final Tool Failure

In our dynamometer readings, it has been possible to recognize the sharp increase in force that occurs as the steady-state gradient begins to "take off" into the rapid gradient. This development is shown at the top, right corner of Figure 10.12 beyond 0.03 inch of flank wear. Powell et al. (1984) have used the same trend for their tool condition sensor. It is important to replace the tool at this point before more advanced failure occurs.

Thangaraj and Wright (1987) have also used this trend to predict the incipient failure of a drill and retract it before significant damage occurs. It was found that the gradient of the drilling thrust force, calculated using a digital filter with the necessary frequency specifications, exhibited a sharp increase several seconds before any serious failure. This is shown in Figures 10.13 and 10.14. This sharp increase was correlated with a stick-slip process that began, at this stage, at the drill's outer margin. As a comparison with Table 10.1, this could be heard as a

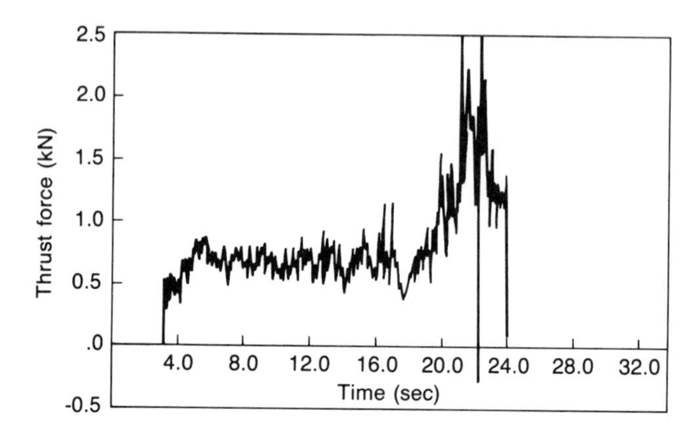

Figure 10.13 Variation of thrust force with time.

high-pitched squeal. Depending on drill size and workmaterial, this incipient failure was detected somewhere between 2 and 7 seconds before rapid failure of the drill. Using appropriate software, a signal could arrest the feed rate of the machine tool in less than 1 second, showing that this is a viable production scheme to protect drills and, more importantly, the part being machined.

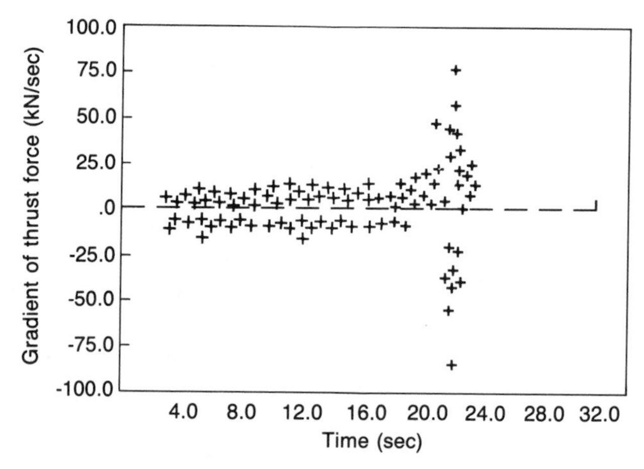

Figure 10.14 Variation of the gradient of thrust force with time.

10.3.2 Touch-Trigger Probe Measurements and Final Tool Failure

The industrially well-proven touch-trigger probe (Figure 10.15) provides the last resort, in an unattended system, for detecting a totally worn out or broken off cutting tool. The touch-trigger probe can be mounted off to the side, on the machine tool bed, and then brought into the work zone periodically to check the lengths of drills and other tools. An alternative strategy is to mount the touch-trigger probe in the machine tool spindle, as is done for part indexing. In this setup, the probe can be brought in to verify that a hole in the part has been drilled to the correct depth. Or, if necessary, it can check that a broken drill is not stuck in a hole prior to a threading operation with an expensive tapping tool.

The touch-trigger probe can also be used for dimensional checking of a partially or completely machined component. For completeness, we also mention that the touch-trigger probe replaces the older style, edge-

Figure 10.15 Touch probe for part dimensioning and simple tool inspections. (Courtesy of Renishaw, Inc.)

finding tool for locating a piece of stock that has just been clamped in a fixture. By touching the top face and the two adjacent sides of a block in a fixture, the probe ascertains the precise location of the top corner, which becomes the block's origin. This *xyz* coordinate is then used as an offset to the coordinates in an NC program so that machining can begin.

10.4 MACHINE VISION FOR THE INSPECTION OF CUTTING TOOLS

We finally consider the third monitoring strategy in Table 10.2. In the automobile domain, proper maintenance in between uses can preempt potential problems during a journey. The experienced driver periodically checks oil and water levels, tire pressures, and brake pads. The experienced machinist makes the same kinds of periodic checks.

Our video-taped protocol analyses of machinists at work reveal that, to check for wear, machinists frequently glance at the tool point, especially between passes and during tool changes. The machinists have explained that not only are they checking for the current quality of the tool edge, but also attempting to estimate how much life is left in the tool. This is important because they want to plan ahead for when to change the tool and, almost at any cost, they want to avoid changing the tool in the middle of an engaged cut. An in-cut tool change involves minor but frustrating readjustments and always leaves a rub mark or slight step in the machined surface. Now that machine vision has become a proven factory device, it is appropriate to ask if it can be used to make this between-pass check for wear and if, in addition, it can help estimate the amount of remaining tool life. For the prototype system being developed and described below, it is imagined that the cameras for the machine vision unit will be installed immediately adjacent to the tool change carousel. Then, tools indexing around the carousel can be inspected between passes.

In a bench-top experiment, worn cutting tools were examined with a CCD camera coupled to a machine vision unit. Images were obtained by viewing orthogonal to the worn rake and then orthogonal to the worn flank surfaces. The picture resolution averaged 0.0014 inch per pixel. We analyzed the kinds of images that were obtained from a tool with a chipped cutting edge, or a broken nose radius, or a worn flank.

For example, Figure 10.16 shows encouraging results for the vision system's ability to monitor the development of flank wear during the

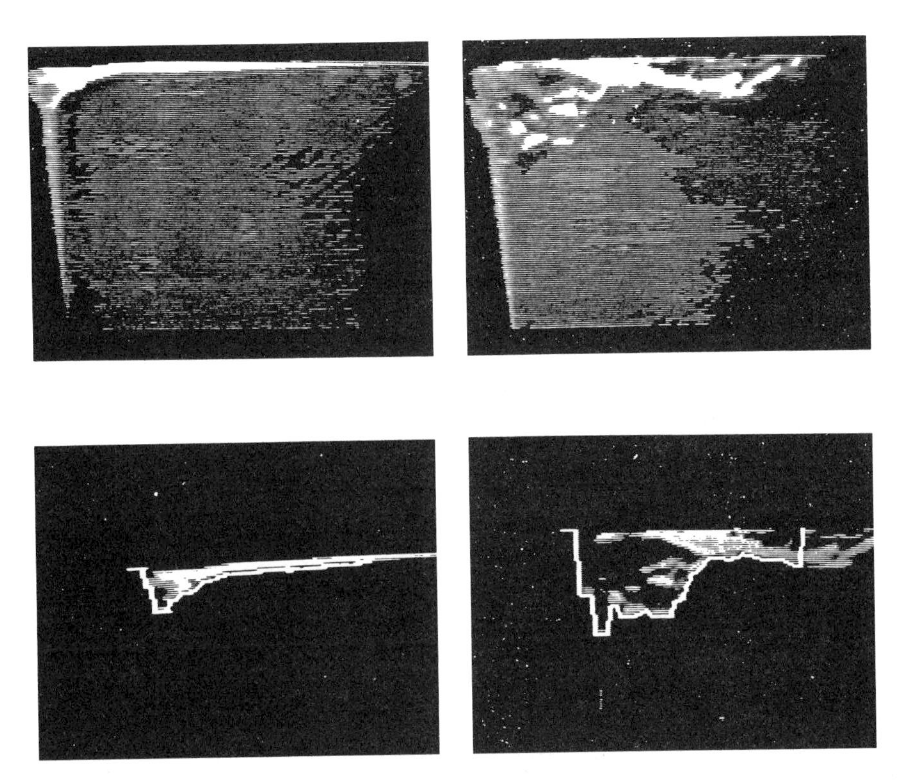

Figure 10.16 A cutting tool viewed at two points in its life after minor use (at left: with 0.008 inch mean flank wear), and at the end of useful service (at right: with 0.03 inch mean flank wear). Each view illustrates binary values produced from a threshold value.

service of the tool. Shown on the left is the connected flank wear areas for a tool that was used for only a few cuts. The maximum depth of the flank wear is 0.025 inch around the nose radius, and the mean land depth is 0.008 inch. After more service, the tool appeared as shown on the right, where the mean flank wear land is 0.03 inch. Experience has shown that it is wise to replace the tool once the mean flank wear land depth reaches this value (Figure 10.12). These results provide the starting point for an expert system that would answer the question: How much life is left in the tool? The appropriate responses for the two views of the tool in Figure 10.16 are 73% and 0%, respectively.

10.5 SUMMARY AND OPEN ISSUES

Even for the most modern batch-production factories, a substantial gap remains between the concept demonstrations of sensors in research laboratories and what is practically feasible in production. In real production settings, there are so many random variables in day-to-day operations that closed-loop control sensor schemes are difficult to tune. It is also clear from Tables 10.1 and 10.2 that several types of sensor-based strategies will be needed to eventually replace the overseeing machinist. In the approach described in this chapter, simple thermocouple and force measurements have been advocated for closed-loop control. These may be augmented by touch-trigger probes for detecting final tool failure and machine vision for between-pass inspections of flank wear.

Future work must be devoted to the refinement of these existing sensor schemes. The models such as Figure 10.8 must be carefully documented for a wide range of commercial cutting tools. In addition, the hardware for thermocouples, dynamometers, and machine vision must be ruggedized and tailored to the geometrical constraints of various machine tools. Before closed-loop, unattended machining of small batch sizes can become a reality, many challenges must be addressed:

1. As mentioned, aerospace alloys pose particularly challenging tool wear situations. Indeed, when machining these alloys, the temperatures and stresses at the tool's edge are so intense that the tool's shape wears away constantly during the cut. To obtain a component of the correct size, sensor schemes that continuously monitor tool length and/or part dimensions would be very valuable.

2. Acoustic emission sensors that detect tool fracture and chipping have been successfully demonstrated in laboratories. Generally, this wear process indicates that an inappropriate grade of cemented carbide tool has been chosen for the cut. However, acoustic emission devices have not yet found wide acceptance in industry. A great deal of on-site tuning seems to be needed. Research into the development of general-purpose acoustic emission devices is believed to be worthwhile.

3. Machine vision has not yet been greatly exploited for monitoring cutting operations. The example shown in Figure 10.16, however, indicates that it has great potential especially for checking tool

wear between passes. Machine vision may also be used to check the surface finish of components and verify machined part features. When finer resolutions are available, machine vision may also provide an alternative to the touch-trigger probe for dimensional measurement.

Note that the recently available frame-rate vision systems may be useful for general monitoring of machining. For example, a machinist is particularly on-alert (see Chapter 7) when making the first cuts on a new part. The machinist checks for correct tool operations, proper fixturing, effective lubrication, and good chip clearance. Frame-rate vision systems may be able to provide this service in unattended systems.

4. It is reemphasized that if workmaterials and toolmaterials are more homogeneous and predictable, many in-process problems can be avoided. Even if these materials remain variable from their manufacturing source, it may be possible to qualify them immediately before entering the machining system. Sensors for preprocess hardness testing of workmaterials and sensors for a priori tool inspections, e.g., high-resolution touch sensors or machine vision units, are suggested as being potentially useful. Such preprocess checking could also be used to automatically tune the main machine sensors such as the acoustic emission devices and dynamometers.

The open issues must include not only the development of new sensors, but also the ability to use existing sensors in better ways in order to effectively understand the physics of chip formation. We must also combine the outputs from such sensors into a software architecture that more closely models the human machinist's ability to decide what to do next, given the current state of the manufacturing process. This area is developed in Chapter 11.

REFERENCES

Amini, E. 1968. Photo-elastic analysis of cutting tools. *J. Strain Analysis*, Vol. 3, pp. 206–213.

Bagchi, A., and Wright, P. K. 1986. Stress analysis in machining with sapphire tools. *Proceedings of the Royal Society*, Vol. A409, pp. 99–113.

Barrow, H. G., and Tenenbaum, J. M. 1978. Recovering intrinsic scene characteristics from images. In *Computer Vision Systems,* edited by A. R. Hanson and E. M. Riseman. Academic Press, New York, pp. 3–26.

Barth, C. F., Blake, D. E., and Stelson, K. E. 1980. Adaptive control through on-machine sensing. *Computer Applications in Manufacturing Systems,* ASME Special Volume at the Chicago Winter Annual Meeting, Vol. PED-2, pp. 53–70.

Boothroyd, G. 1975. *Fundamentals of Metal Machining and Machine Tools.* Scripta, Washington, D.C..

Brecker, J. N., and Shum, L. Y. 1977. Reducing tool wear with air gap sensing. *SME Paper,* #TE77-333.

Chao B. T., and Trigger, K. J. 1959. Controlled contact cutting tools. *Trans. ASME, J. Engineering for Industry,* Vol. 81, pp. 139–151.

Chow, J. G., and Wright, P. K. 1987. Sensor development for on-line monitoring in machining. *Trans. ASME J. Engineering for Industry.* In press.

Dornfeld, D. A., and Lan, M-S. 1983. Chip form detection using acoustic emission. *Proc. of the 11th North American Manufacturing Research Conference,* Vol. 11, pp. 386–389.

Kegg, R. 1983. Discussion of papers, *10th NSF Conference on Production Research and Technology,* SAE, Detroit, Mich.

Loladze, T. N. 1975. Nature of brittle fracture of cutting tools, *Annals of the Confederation Internationale pour Recherche de Production (CIRP),* Vol. 24, No. 1, pp. 13–16.

Mathias, R. A. 1978. New developments in adaptive control. *SME Technical Paper,* #MS78-217.

Metcut Research Associates. 1985. *Machining Data Handbook,* Cincinnati, Ohio (2 volumes).

Pearson, J. F., Syniuta W. D., and Cook, N. H. 1980. Development of an instrumented tool holder transducer. *Computer Applications in Manufacturing Systems,* ASME Special Volume at the Chicago Winter Annual Meeting, Vol. PED-2, pp. 71–80.

Powell, J. W., Cosic, J. E., Erickson, R. A., Herko, F. M., Kline, W. A., Mayer, J. E., Jr., and Varma, A. H. 1984. Sensing and automation for turning tools, *SME Paper,* #MS84-909.

Schaffer, G. 1983. Sensors: The eyes and ears of CIM. *American Machinist,* Special Report 756, July, pp. 109–124.

Takahashi, Y., Rabins, M. J., and Auslander, D. M. 1970. *Control and Dynamic Systems.* Addison-Wesley, Reading, Mass., p. 554.

Tay, A. O., Stevenson M. G., and deVahl Davis, G. 1974. Using the finite element method to determine temperature distributions in orthogonal machining. *Proc. of the Institution of Mechanical Engineers*, Vol. 188, No. 55, pp. 627–638.

Thangaraj, A., and Wright, P. K. 1987. Computer-assisted prediction of drill-failure using in-process measurements of thrust force. *Trans. ASME J. Engineering for Industry*, Vol. 109. In press.

Tlusty, J., and Andrews, G. C. 1983. A critical review of sensors for unmanned machining. *Confederation Internationale pour Recherche de Production (CIRP) Proceedings*, Vol. 32, No. 2, pp. 563–575.

Trent, E. M. 1984. *Metal Cutting*. Butterworths, London.

Whitmont, E. C. 1978. *The Symbolic Quest*. Princeton University Press, Princeton, N.J., p.143.

Williams, J. E., Smart, E. F., and Milner, D. R. 1970. The metallurgy of machining. *Metallurgia*, Vol. 81, pp. 3–94.

Wright, P. K. 1983. Physical models of tool wear for adaptive control in flexible machining cells. *Computer Integrated Manufacturing*, ASME Special Volume at the Boston Winter Annual Meeting, Vol. 105, pp. 31–38.

Wright, P. K., McCormick, S. P., and Miller, T. R. 1980. Effect of rake face design on cutting tool temperature distributions. *Trans. ASME, J. Engineering for Industry*, Vol. 102, No. 2, pp.123–128.

Wright, P. K., and Trent, E. M. 1973. Metallographic methods of determining temperature gradients in cutting tools. *Journal of the Iron and Steel Institute*, Vol. 211, No. 5, pp. 364–368.

Yen, D. W., and Wright, P. K. 1983. Adaptive control in machining: A new approach based on the physical constraints of tool wear mechanisms. *Trans. ASME, J. Engineering for Industry*, Vol. 105, pp. 31–38.

Zorev, N. N. 1963. Inter-relationship between shear processes occurring along tool face and on shear plane in metal cutting. *Trans. ASME, International Research and Production Engineering*, New York, pp. 42–61.

11 The Machine Tool's Team of Experts

So may a thousand actions once afoot
End in one purpose, and well borne
Without defect.

—William Shakespeare in *Henry V*, Act I, II, 211.

TO make the first part right the first time requires the expertise of a team of engineers working together. If the fixture designer fails to tell the tooling engineer that a side cut needs a particular, geometrically shaped, end mill in order to avoid a collision with the fixture, the part will not be made correctly the first time. If the fixture selection adversely affects the part material because it dents the surface, the part will not be made correctly the first time. For every combination of expertise, there is a potential error situation. Unfortunately this can lead to a combinatorial explosion where, if there are ten experts, there are ten factorial ways to make team assignments.

To avoid this uncontrollable situation, a foreman (human or expert system) must strongly bias the result toward the solution. This means that the foreman must have a good idea what all the solutions are before the process even begins, and that the foreman must be able to resolve conflicts between team members. In the worst cases, this solution could fail in two ways: the foreman could take forever to solve the problem or the foreman could make mistakes.

11.1 THE EXPERT SYSTEM TEAM

A roster of the intelligent machine tool's team and a brief accounting of each member's duties follow:

1. **Design Verification and Advice**: This team member validates the manufacturability of a part and communicates manufacturing problems to the designer in a language most suitable to the designer.
2. **Part Setup Planner**: From a part design, this member collects miss-

ing information that is required for manufacturing and make an overall process plan for the part fabrication (see Chapter 8).

3. **Machine Specialist**: This member understands the machine's coordinate system, peculiarities, language, error messages, and other physical and electronic capabilities.
4. **Material Selection**: This team member keeps an inventory of available materials and suggests which materials are best for particular part functions.
5. **Tool Selection and Monitoring**: This team member keeps an inventory of available tools, including their condition and expected life, and suggests which tools are best suited to make particular cuts.
6. **Fixture Selection and Placement**: This team member designs a fixture that will firmly hold a part during manufacture, while not getting in the way. Once the fixture is designed, this member also suggests how the fixture should be placed on the machine (see Chapter 9).
7. **Robot Grasp Selection and Path Planning**: From a description of the raw stock, the part design, or the fixture, this team member suggests a way to grasp the object and a robot path that avoids collisions with objects in the area of the machine tool (see Chapter 6).
8. **Sensor Monitoring**: This team member monitors the machining process and reports back on either error situations or situations that could lead to errors in the near future (see Chapter 10).
9. **Visual and Tactile Inspection**: Once cuts have been made on the part, this team member measures the part and compares it with the expected dimensions (see Chapter 5).
10. **Program Construction and Verification**: Each step of the process plan involves a program segment. This team member constructs it in the language of the machine tool and watches for conflicts that would only appear in the program.
11. **Error Diagnosis and Recovery**: If and when the whole team fails, this member is responsible for containing the error so that only the minimum amount of damage is done. After the error is contained, this member makes a plan that will at least reset the machine for another try (this may take the whole team to solve).

This team roster is not complete. In some cases, specialists must be brought in to solve difficult problems. These specialists can be used to solve an ad hoc problem before control returns to the "first team."

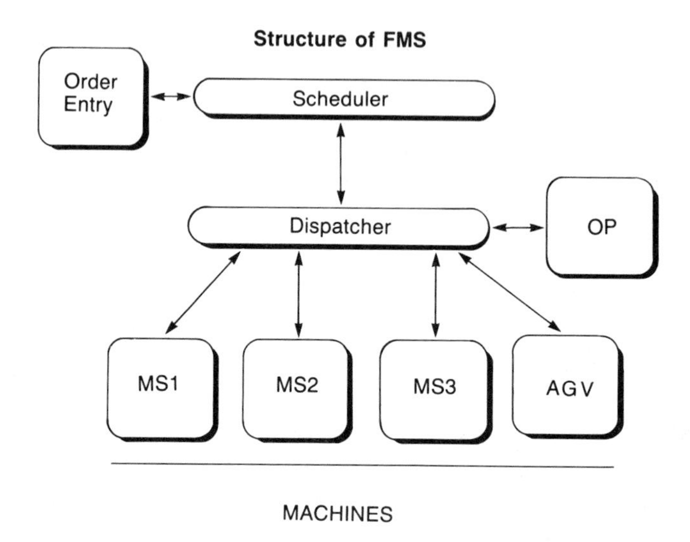

Figure 11.1 The shape of a typical cell control system.

11.2 MULTIDOMAIN ARCHITECTURES

In both a flexible manufacturing cell and system control, we have used a team of "in-software" specialists, where each specialist knows how to communicate to its machine (see Figure 11.1). At the top, there is a scheduler that makes strategic decisions about what work the team will execute. It also provides the first guesses to the solution to avoid a huge computational expense. The dispatcher uses these guesses and the work order to sequence commands that satisfy the work order and avoid conflict between the specialists (e.g., a robot crashing into a furnace in Figure 2.5.)

A similar team arrangement can be used to implement the intelligent machine tool. At the top of Figure 11.2, the monitor makes initial suggestions on how to solve the problem. It also monitors the behavior of the system to make sure that it is converging on a solution. The system blackboard is where information can be shared among team members, but it also has the responsibility for translating goals and suggestions between subsystems. The blackboard should also detect redundant messages and other errors in judgment caused by single team members.

At first, it may seem surprising that a flexible manufacturing cell control and a multidomain expert system can use the same system architecture. But, further examination reveals that the cell control is coordinating machine specialists into an expert system that knows how to interpret a range of commands; this is identical to a multidomain expert.

11.3 SYSTEM PERSONALITY

The personality of the team leader strongly determines how the system will operate. The team leader can be greedy, jumping on the first partial solution that comes along without looking at the complete situation. In relatively simple situations, with only one maximum point in the search space, a greedy algorithm generates optimal solutions very quickly. However, if there are many local maxima in the search space, a greedy algorithm can lead the system away from the best solution.

At the passive extreme, the team leader does not have any decision-making powers, leaving all the decision making to the local experts. In this case, the experts must have enough intelligence to know when they should be involved in the planning process and when they have no valuable information to add.

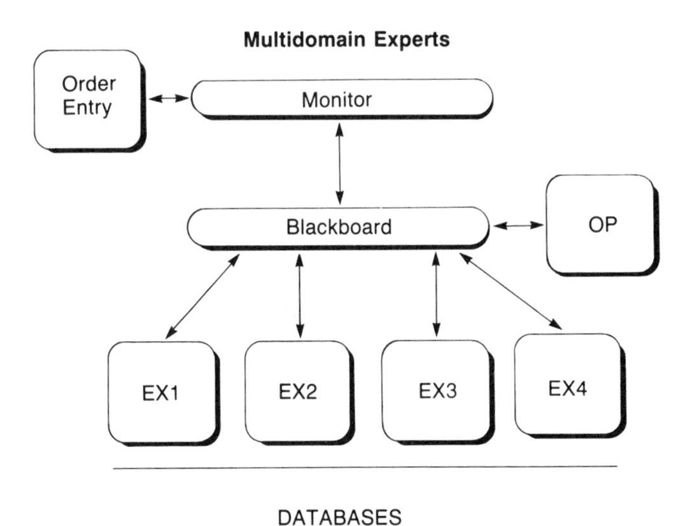

Figure 11.2 The shape of a multidomain expert system.

Figure 11.3 shows that the personality dimensions should be combined in a flexible way. An ideal system is one that automatically tunes its personality to different situations. Quick decisions can be made when the problems are known to be easy, whereas a more lengthy systematic approach can be taken in difficult situations. There has been very little research on how to balance different control strategies.

Berliner has tried one approach in the game of backgammon, where the program sticks with one game strategy until there is an overwhelming amount of evidence that the program should switch, at which time it completely changes its strategy (Berliner 1980). (In backgammon, the first player can stay back waiting to strike the second player's pieces and send them home, or the first player can play aggressively to get all his or her other pieces off the board.) This avoids an awkward situation where the program is caught between two extreme positions, not being able to capitalize on either one. For this reason, the software for the intelligent machine tool intentionally skews the personality space to suit the particular demands of a manufacturing application. This avoids crosses between multiple optimization strategies.

Another dimension in the personality space is depth versus breadth. In the first case, a systematic search tries to completely resolve one expert's position and only then move onto the next expert. This is a depth-first analysis. In the second case, a little time is given to each expert so that they all have an opportunity to contribute to the solution, thus avoiding false trails. This is a breadth-first search. An interesting combination that has been used in some systems (Michie 1967; Michie and Ross 1970) combines the extreme positions of the depth dimension and the intel-

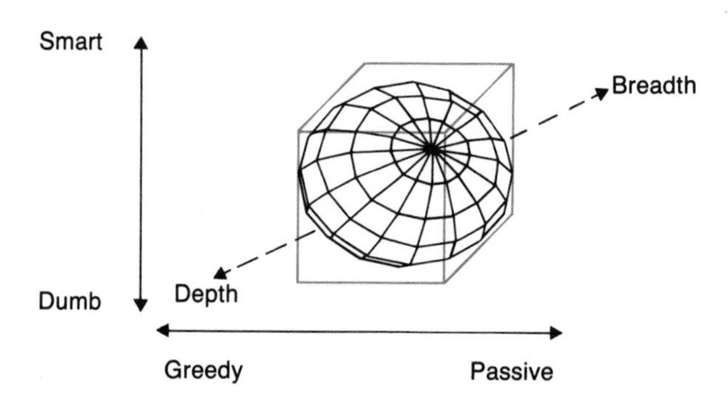

Figure 11.3 Personality space of team leader.

ligence dimension. In this case, each partial solution is evaluated and only the best candidate is pursued. This strategy has been coined a best-first search.

11.4 A TYPICAL ASSIGNMENT FOR THE TEAM LEADER

Consider the precision manufacturing of a one-off component. A skilled craftsman never attempts to obtain the final dimensions and tight tolerances of $\pm$ 0.001 inch during the first milling cut. The craftsman knows that there will be unpredictable deflections in fixtures and tooling, which may lead to inaccuracies in the component. Therefore, a few roughing cuts are made. Measurements are taken between such cuts to see how the total system is behaving. During these trial cuts, the machinist has an opportunity to learn about the specific conditions of the job in progress. As a result, subtle adjustments can be made to fixtures, tools, and machine tool accessories to allow for slack, distortions, and thermal movements. This is all completed prior to making the final commitment to a finishing cut that produces the final dimensions.

Given such uncertainties in the machine tool domain, an expert system solution must encompass this cut-and-test iterative procedure. Our planning software (see Figure 8.2) has thus been designed to learn about the machining environment during the creation of the initial outer-envelope roughing cuts. The CNC subroutines first generate a prismatic part that is slightly larger all around than the desired end product. After one of these roughing cuts, when the stock has been reduced by a prescribed amount, machine vision and tactile probes are used to verify that the expected amount of material has been machined away. In other words, the software makes a "hypothesis" about how much metal should have been removed and then "tests" to see if this hypothesized reduction has been obtained.

What happens if there are noticeable differences between the hypothesis and the measurements? With an expert craftsman in attendance, the craftsman can monitor individual system components and make corrective adjustments to tools, fixtures, and the machine itself. The intelligent machine tool's software will be designed to make the same decisions as the human would. It exploits reliable sensor information, domain experts, and programmable hardware in such a way as to create an autonomous system.

Even the skilled human craftsman may not always be in a position to properly fix all the problems. Suppose one of the slideways of the machine tool has a 2% inaccuracy and that this is discovered in one of the early roughing cuts. Obviously, corrective maintenance ought to be done, however, there may be no time to do it. A good craftsman can cope with this dilemma by making ad hoc compensations to overcome the less-than-ideal equipment. We have designed the software so that the expert system's team leader also has this responsiveness.

Successful management of less-than-ideal conditions is the hallmark of the best team leaders, whether human or expert system. An important open problem is to conduct protocol analyses with machinists in order to characterize their problem-solving strategies in these less-than-ideal circumstances. Rules and knowledge bases that mimic this skill of minor-crisis management will constitute a major step in the creation of an autonomous system for small-batch machining.

REFERENCES

Berliner, H. 1980. Computer backgammon. *Scientific American,* Vol. 242, No. 6, pp. 64–85.

Michie, D. 1967. Strategy building with the graph traverser. In *Machine Intelligence 1,* edited by N. L. Collins and D. Michie. American Elsevier, New York, pp. 135–152.

Michie, D., and Ross, R. 1970. Experiments with the adaptive graph traverser. In *Machine Intelligence 5,* edited by N. L. Collins and D. Michie. American Elsevier, New York, pp. 301–318.

Part Four

AUTONOMOUS MANUFACTURING

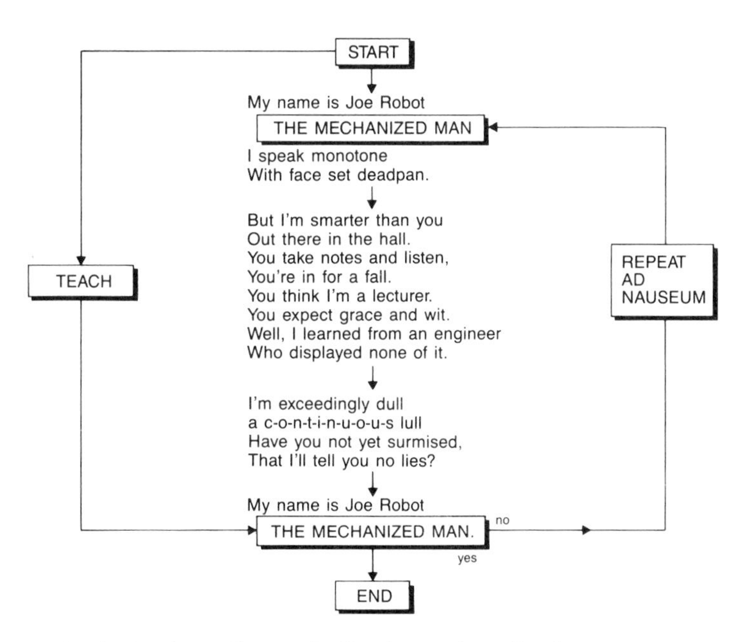

—Paul Wright in *Poetry for Engineers from Engineers*

Part Four makes several predictions about the state of manufacturing in thirty years. In addition, a glossary has been provided to give working definitions for many of the technical terms that have been used.

12 Manufacturing in the Year 2020

20/20 hindsight is always easier to achieve than 20/20 foresight.
—Proverb

ACADEMIC training strongly determines how people think about the future. Because one of the authors is trained as a mechanical engineer and the other as a computer scientist, we juxtapose their views of the future, rather than trying to integrate them. The truth probably lies between the views rather than at their extremes. How would you answer these questions? Your answers will in part determine the future.

Question 1: What generations of machine tools will we see in the next thirty years?

Mechanical Engineer: In the next thirty years, unattended manufacturing will advance in the large corporations. Based on the experience with NC, it will take five to ten years (by 1997) for a knowledge-rich controller to be commonly used by these large organizations. Other research on programmable hardware moves more slowly because rugged sensors and feedback methods take longer to build and need more trial-and-error cycles of design. To be useful, the hardware also has to interact meaningfully with an automated knowledge-rich setup sheet. It will take almost the thirty-year period to achieve a fully comprehensive diagnostic package that monitors and successfully reacts to emergencies in small batch manufacturing of aircraft parts to be operational in large companies. In the meantime, the communications between machines will be solved within five to ten years by schemes such as the GM-MAP project, so at least the developments in offline programming at CAD stations can be put to use. Some developments in machine tool design such as composite/granite structures and high-speed spindles will take place. Table 12.1 summarizes the predicted position of large corporations:

When posed with such a question, it is often useful to look back and consider the changes that have occurred over an equal time period, in

Table 12.1 Upcoming Generations of Machine Tool Technology

TECHNICAL ADVANCE	MEDIAN YEAR
Knowledge-rich controllers	1997
Effective machine-machine communications	1997
High-speed spindles and modified structural elements	1997
Accurate, automated loading and fixturing for small batches	2000
Error recovery diagnostics for small batches (unattended)	2010

this case, from approximately 1960. At that time, NC machines had been developed in research laboratories for nearly a decade, and the early versions were in use in larger companies that had the resources to experiment with new technologies. Now, we have seen noticeable improvements in programming methods moving from paper tape to CNC to DNC. The NC machine has become easier to program, the technology of the servo-drives and feedback elements has been refined, and the machine structures are heavier.

However, the machine tool industry is rather conservative, certainly in comparison with the computer industry. As a result, the research and development laboratories of the machine tool companies have not capitalized on all the new developments in microcomputers, programming methods, and programming languages. It is amusing to watch the raised eyebrows of the dyed-in-the-wool computer scientists being introduced to the programming environment of an NC machine for the first time. A paper-tape reader is still an obvious feature of CNC machines and many manual inputs and corrections need to be punched in at the machine console, reminiscent of old computers programmed in machine language. But, the computer scientists may not consider the economic problems that frequent model changes would cause in the machine tool industry. It is more the market for the machines that is conservative and slow moving.

Recognizing this lag in the machine tool industry, the U.S. Air Force has issued contracts to investigate how small businesses can be encouraged to use CNC equipment and unattended machining centers. Despite this, there are sprawling industrial areas in Chicago and New Jersey where scores of small machine shops make a healthy living today without one NC machine on their premises. At the same time, some of the industrial giants are experiencing cut-backs and diversifying. It is also possible that in the next thirty years the American automobile industry will become even more depleted by Japanese and Korean competition, as some economic forecasts suggest. It is clear why the machine tool industry operates with an uncertain future and therefore caution. Based on the current spectrum of the machine tool user group, even in thirty years some smaller machine shops will still be using manually operated milling machines.

Computer Scientist: A factory is only modernized once every fifteen years. This seemingly slow modernization process is due to several factors:

1. Incredibly high capital investments are required.
2. The mentality of the workforce is geared to maintain the status quo.
3. Unions prohibit the speedy introduction of new technology by imposing work rules.
4. Many factories are based on fully mature technologies that are not advancing very quickly.

This trend of fifteen year modernization cycles could start changing because the new enhancements to the factory are based on computer technology, which is on a three- to five-year product life cycle. However, the first three of the four "conservative" factors have not changed.

One result of these unbalanced technology cycles is that senior management is discouraged from modernizing old facilities because all the necessary "new" technologies never seem to be ready at the same time. In fact, this is true just within the computer industry. For example, once the "best" computer hardware is chosen for a project, it often turns out that the network facilities are not ready for market; and once the network facilities are ready, some key software is not ready; and once the software is ready, the hardware is obsolete. This dilemma is much worse when the technology groups are broader and include products from "traditional industry" (e.g., machine tools).

In the next thirty years, these trends will continue. However, there will be one new element in the equation. As computers become more sophisticated and take on more of the workload, much smaller companies will be able to build a new factory. Essentially, these factories will be the "mom and pop" shops of the industrial communitity and will cater almost strictly to local clients. The way in which computer-based technology is used will also change. By 1995, the machine tool controller will be no different from the computer in a travel agency and not much different from the personal computer at home. This will be the single largest shift because today's industrial machine tool builders spend fortunes developing their own proprietary controllers, and they are outdated by the time they reach the factory floor.

The second biggest change will be the widespread use of knowledge-based systems (by 2000). These systems will be sold by the current industrial giants to small units, much the way Burger King sells hamburger franchises. The result will be that the small units will be able to benefit from the accumulated knowledge of many people, although only a few would be needed to actually run the shop. Finally (by 2010), new sensor technology, coupled with knowledge-based systems, will make it possible to automate the diagnosis of routine machine failures. Again, it will be much easier for small groups of people to operate their own shops.

Question 2: What role will people play while working with machines thirty years from now?

Mechanical Engineer: In thirty years time, the factory floor of the larger corporations will be unattended and machines will operate largely autonomously. Initial setup and long-term maintenance will need human intervention, but day-to-day operations of machines will be unattended. The small machine shop catering to custom needs and very rapid turn-around will be DNC, and there will be little need for humans to watch the machines.

Computer Scientist: People will remain in the design and sales roles, but they will work in close proximity to an autonomous manufacturing station. If we consider the textile industry, it will be possible to go to a clothing designer's shop to see a color three-dimensional display of the fall line. Picking out clothing sizes will no longer be an issue because a laser scanner will take a customer's exact measurements. The design system then regenerates the design for a particular person's taste or need. Further modifications can then be made based on the materials in stock, fastener types, and desired accessories. As soon as the customer is satisfied, the designer asks the system to fabricate the dress on the spot. The people who would own and operate such a shop would be a cooperative group of designers who collectively purchased the autonomous dress fabrication system (ADFS). This is actually a combination of "reverse" engineering (deducing the design from the shape of the person) and "forward" engineering (the usual case).

Question 3: What is the biggest problem in manufacturing today and how will we think about this problem thirty years from now?

Mechanical Engineer: We have begun to come to grips with the machine-machine communication problem, and solutions for this will be increasingly available to any size of industry over the next five years. This leaves the setup aspect of metal manufacturing—batch manufacturing in particular—as the major source of frustration and economic loss in day-to-day factory operations.

Thirty years ago, it was common to see machinists laboring over setups and adjusting the handles of the manually operated machines. Today, in many situations, the view is much the same, except that they are frowning at the hieroglyphics on the controller and pushing buttons (the artist who thought of Figure 1.5 obviously saw the same thing). The research needed to solve the setup problem has been started and Chapter 8 described the initial results. In thirty years, the problem will not evaporate entirely: there will still be bugs in CNC programs, errors in fixtures, and random fluctuations in workmaterials and tools, but the number of such occurrences will be less and the setup will be smoother. Obtaining a "good first part the first time" will be possible and expected.

Computer Scientist: System integration in manufacturing is the biggest problem today, but this will not be the case in thirty years. Large factories will lose out to smaller job shops as well as to specialty cooperatives, which will operate a single autonomous manufacturing station. There will be no need for economies of scale.

Society will be so computerized that science will start to view mechanical and scientific problems within a computational paradigm. For example, there will be a "universal machine tool" in the sense of a "universal computer." This universal machine tool will be able to make every part within the limits of the time and space of the machine. This includes operations that would replace: casting, forging, extrusion, machining, grinding, deburring, assembly, and whatever other operations would be necessary for a complete fabrication system.

Question 4: What will the controller of a machine tool look like in thirty years?

Mechanical Engineer: Very small. At one stage in our laboratory, we used a vertical machining center of 1971 vintage. In its lifetime it gave exceptional service, yet its distinguishing feature was that the controller took up as much floor space as the machine tool. The 1980s vintage machine currently in use has a significantly smaller controller. In thirty years, we will not see a physical controller separate from the machine. There will be a maintenance/operator interface that can display, if necessary, a color graphics, three-dimensional version of the part being machined in real-time, together with the EXSUS and TRACE information for trouble-shooting. But even this facility will be more for maintenance and trouble-shooting rather than for an operator, in the 1980s' sense. Gradually the operator will not be needed, just as research mechanical engineers have been distanced from their minicomputers.

In the 1970s, with minicomputers invading the mechanical engineering laboratories, faculty members and graduate students were excited to rub shoulders with the hardware and, in addition to programming, adjust the disks and boards. But, now, they would rather have the machines in a central resource area and speak to them from a terminal. So, ideally this is how machine tools will be eventually: spoken to from a distance. The control functions will be inside the machines, next to the servomotors like the direct-drive robots in today's research laboratories (Kanade and Schmitz 1985). And the maintenance/operator interface will be an information terminal, only used for emergencies and difficult setup procedures.

Computer Scientist: The control software will be an off-the-shelf autonomous system with special knowledge about machining provided for the installation. In fact, the control software will be virtually identical to the control system for the home robot and to the autonomous family transport system. Each system will have general knowledge of the other applications, but specialist knowledge will continue to be reserved for individual expert systems.

The control hardware probably will also be an off-the-shelf component, again virtually identical to the other devices. The technology will certainly be radically different. Most likely, the device will be photonics (as opposed to electronics). This area is being actively investigated today, but it probably will not become practical for another fifteen years. Most research efforts in photonics aim to mirror the mechanisms found in contemporary computers—essentially looking for a mechanism that is equivalent to a transistor, but that works for light (Abraham, Seaton, and Smith 1983). It is likely that once a practical photo-transistor is invented and used to build full-scale computers, the next step would be to run millions of virtual computers through the same optical mechanisms—all at different frequencies. The result would be a very small parallel machine. Technically, it would be a single-instruction, multiple-datapath machine (SIMD). This machine would be similar to the famous ILLIAC-IV, but with each machine working on the same device at the same time.

Question 5: What will the mechanical part of a machine tool be like in thirty years?

Mechanical Engineer: Current trends to high-speed spindles, higher accuracy lead screws, and improved feedback devices with laser scales will continue. Tool-changers will become faster and more selective, also enhanced by tool-wear sensing and replacement devices. Vision and other sensors will be evident inside the machine. Chip control and disposal techniques will need to be improved to suit nearly unattended manufacturing.

In terms of machine tool design, we have seen or are now witnessing the main structural change. In the early days of CNC controls, the machine tools looked much the same as throughout this century: the machine was still designed for human ergonomics, with the CNC control uncomfortably stuck out to one side or in a floor panel adjacent to the machine. Indeed, in the late 1960s, CNC machines were deliberately designed to still have hand wheels, in case the machinist felt a need to do something, as well as the CNC control. Thus, many of the CNC small milling machines are still designed with the table at waist height on a vibration-prone knee support so that a person can intervene. This design has already begun to disappear and most new machines take advantage of the removed operator by having heavy tables close to the ground for maximum support and vibration resistance. The vibration damping, resistance to warping, and thermal stability of machine tool tables can also be improved by making them from composite materials. For example, granite powders made of carefully mixed ball sizes for close packing and cemented with epoxy resins are typical new materials with these properties.

Computer Scientist: The unpredictable wear behavior of cutters and drills will encourage researchers to eliminate them from machinery as soon as science allows it. Even today, there are very sophisticated water-cutting and laser-cutting machine tools (La Rocca 1982). It can be expected that the advent of research from the Strategic Defense Initiative (SDI) will result in considerable improvements in lasers and related (ion beams) technologies. This technology will certainly trickle down into manufacturing. This kind of technology propagation has been part of our history; huge arrays of technology are a direct result of World War II and the space race.

Devices like computer assisted tomography scanners or nuclear magnetic resonance detectors that have the power to etch, instead of just image, internal structures into a solid piece of material can be imagined. This would be done by focusing enough energy onto small internal spots. Of course, a problem is that even if it is possible to locally melt internal material, it still has to be extracted. One possible solution to this problem is to develop workmaterials with a honeycomb structure or controlled inner porosity. Then, after local inner melting and resolidification, a size-controlled cavity would be created within the component. An additional challenge will be to control the location of the newly solidified material. It may be possible to control the liquid flow before solidifying in simulated gravity fields (e.g., a certifuge), possibly in gravity-free space.

Question 6: What are the parts and components that will be made in thirty years, and how will these effect the machine tool?

Mechanical Engineer: The automobile and aircraft industries will still dominate manufacturing designs, methods, and parts. The trend toward lightweight, high-strength alloys will continue. Both industries will also use easy to machine composite materials. To cut these at productive rates, machine tools will need to move with faster table feeds and higher spindle speeds. Quality control will be more exacting, and machines will need to be more accurate and stable. Machine sensors will be needed to maintain thermal and vibrational stability. This question is strongly influenced by global economic trends as well as by technology. A main theme of this book has been the trend toward batch rather than mass production. This has naturally led to the need for faster setup, keener quality control methods, and the ability to get "a good first part right the first time." Even in mass production these goals are still desirable.

Computer Scientist: The dominant applications in thirty years will be machining very large and very small components. The problem of machining very large components has so far been avoided by building parts that are then fitted together, but at the price of reducing the structural integrity and adding weight to the final part. The problem of machining very small components has been barely addressed. This problem is different from fabricating integrated circuits because miniature mechanisms are inherently three dimensional, instead of the 2 and 1/2 dimensions that are needed to build circuits. Currently, three-dimensional micro-components are very simple planar structures etched from silicon like materials (Kaminsky 1985). Some researchers have also attempted to laminate these together, but have found it prohibitively hard to align the laminates.

The large components will be needed for ultralight, but strong systems (e.g., planes, boats, spacecraft). And, the small components will be required to build sophisticated arrays of sensors (e.g., robot skin), output devices (e.g., flat display panels), and actuators (e.g., muscle like motors from fibers). The manufacturing processes that will be needed to construct these kinds of objects will have to be considerably different than those of today.

Question 7: Will remote repair stations and outer space change the way we think about manufacturing?

Mechanical Engineer: Yes. This question focuses even more attention on unattended daily operations with self-diagnosing and self-correcting machinery. In addition, any type of remote manufacturing facility needs to be self-sufficient and able to operate on minimum inventory whether it is in space, in a remote repair station on Earth, or on-board an aircraft carrier in the middle of the Pacific Ocean. In the "days of plenty," the 1950s and 1960s, it was possible for a manufacturing facility to spread out physically, hoard workmaterial stock, build up large inventories of special-purpose fixtures, and rely on custom-made tooling for many ad hoc applications. Generic fixtures, generic tooling, and generic raw materials, all of which can be accommodated in outer-space or remote location factories, are the requisite solutions to reduce inventories.

The research area of generic fixtures was pursued in Chapter 9 where a main goal is to be able to use the programmable fixture device for many different and varied components. The outcome of such experiments should be a reduction in the number of custom-made fixtures in a manufacturing plant. Of course, in addition to designing generic fixtures, it is desirable to create part designs that make parts easier to fixture in the first place.

In the automated factory, generic tooling inventories should be minimized and, ideally, at one particular machine tool, the same tool set should be able to produce a wide variety of parts. In the design room, the engineers should thus aim to design all new parts for both the generic fixture and the generic tooling philosophy.

It is also relevant to mention the concept of generic raw materials (Wright and Holzer 1981). In outer space in particular, it will not be possible to store large quantities of exactly the right bar stock as is common in a 1980s machine shop, i.e., various sizes of round, rectangular, and hexagonal bar. This makes way for near net-shape forming of powder metals. Imagine in a repair shop in outer space how much easier it would be to store barrels of powder that could, to suit each ad hoc repair part, be pressed and sintered into a preform prior to machining. It would eliminate the need to carry the wide variety of bar stock sizes and geometries. In preliminary work, we used movable plungers in the powder-metal female die to create die cavities of different shapes. In this way, the same die could produce several preform styles within a similar family. The products had good mechanical properties, but they exhibited surface irregularities from the discrete plunger sets that were 0.25 inch square in cross-section. However, this is not a serious problem, because the part is finish-machined.

Computer Scientist: Manufacturing technology will contribute to the technology of space vehicles (and vice versa), because many of the problems encountered in manufacturing are also encountered by autonomous vehicles. Self-diagnosis, contingency planning, and creative use of resources are but a few of the solutions that will contribute to autonomous space vehicles.

On its own, space offers many possibilities for the manufacturing process itself. For example, growing crystals in the correct shape is certainly less difficult without the encumbrance of gravity. There also may be interesting results in areas such as chip removal because simple mechanisms could be devised to get the metal chips moving in the right direction. This most common of processes (machining) on Earth may not be considered a very good solution in space. See the answer to Question 5.

There are also interesting possibilities in space computing. Today's super-computers' biggest problem is heat removal, so elaborate schemes have to be used to increase the surface area of chips (e.g., gold leaf wings). When absolute zero is approached, electrons flow faster. In state-of-the-art super-computers, a factor of two speed increase is obtained by super cooling the CPU board. Space will provide some inexpensive solutions to this problem, making super-computers more cost effective than they are on Earth.

Question 8: What will the automated factory look like architecturally?

Mechanical Engineering: The automated village for the electronic era, Chrysalis, was conceived by Ringeride and Wright (1984) to suit the remote repair station. The factory model shown in Figure 12.1 can be applied to a number of situations. At one extreme, for an industrial park, the upper areas would comprise the management, design, and human facilities contained in present factories; at the other extreme, for lunar manufacturing, full facilities for urban living would be provided. The lower floors have been designed for the hardware. In the lowest level, the machine tools and conveyor systems are densely arranged to take advantage of nearly unattended manufacturing. Architecturally, this is desirable to maximize the percentage of square-footage devoted to manufacturing. The central control room, on the manufacturing mezzanine, has visual access to the hardware for a small group of monitoring engineers. Galleries on the mezzanine provide for emergency access to the hardware with cherry-picker-like landing stations. The factory operates in just-in-time mode, with minimum goods-in and goods-out locations on each side of the access bay.

Computer Scientist: Architecturally, the factory of the future will look more like a doctor's office of today than a factory because the customer support, the design facility, and the fabrication facility will all be brought together into one location. This answer assumes that the product is customer oriented; in this example, we describe a fabrication facility for clothing.

Figure 12.2 shows the way a factory layout might look. A customer enters the facility at A, signs in with a receptionist at B, and sits down to wait at C. A receptionist escorts the customer into the measurement room at D and helps the person through whatever steps are required to get a full digital body measurement. The customer then goes into the sales room at E to select a new wardrobe using animation facilities to model clothing based on the customer's exact physical makeup; all with the help of a professional fashion advisor. After the selections have been made, the customer can go into an observation area at F to watch the selected clothes be constructed in N. Finally, the clothes are wrapped and deposited for pickup at O, and the customer exits at G. The cost of the garments are automatically charged against the customer's bank account.

The designers' office area is on the other side of the facility. It is likely that these designers would also be the principal owners of the shop. At H there is a traditional office receptionist, in front of a series of design

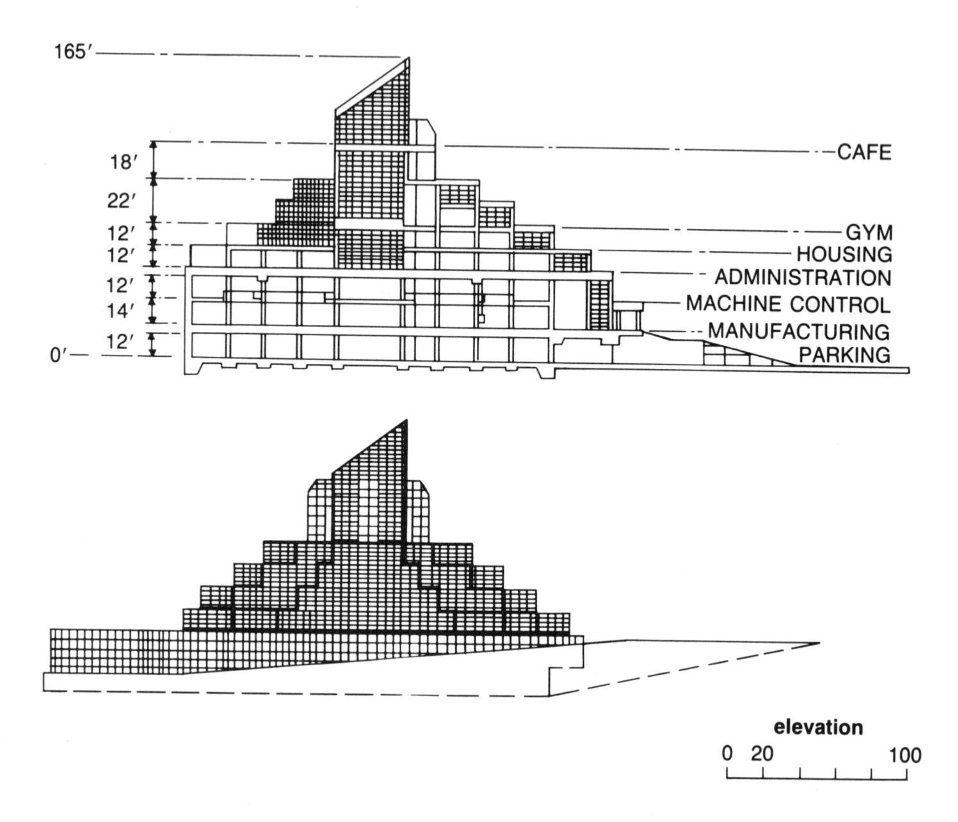

Figure 12.1 Chrysalis—The automated village for the electronic era. (Courtesy of Diane Ringeride.)

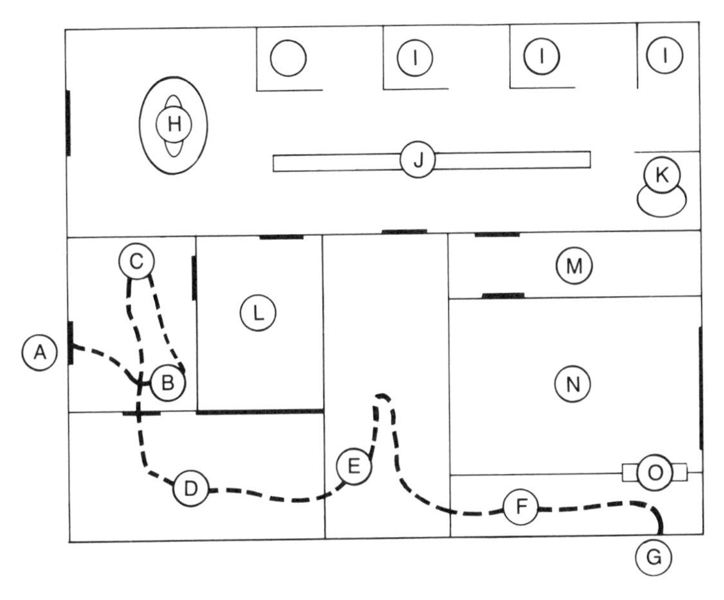

Figure 12.2 Floorplan for factory of the future.

workbenches I that are equipped three-dimensional color stations. There will probably still be need for common resources (e.g., a digital library at J). And, to keep the noise level down in the design area, an air space M is used as a noise buffer, and it doubles as a maintenance area for the fabrication system.

Beneath this facility a whole floor is dedicated to inventory, and the fabrication facility N is fed from the bottom by an automated storage retrieval system. Outside of the facility, a satellite dish communicates with other designers world-wide.

Question 9: What will be the state of manufacturing science in thirty years?

Mechanical Engineer: At present, so much of manufacturing is an art, and there are very few processes or methods that can be regulated with equations. Between the 1940s and 1960s, some analytical techniques for metal working operations were applied industrially to processes like rolling, extrusion, and wire drawing. The reader is referred to a good comprehensive text in this area by Rowe (1977). For example, optimum die angles for wire drawing can be predicted; temperature fields in extrusion can be predicted in order to assess how the heat treatment of the billet is affected; for many two-dimensional extrusion, rolling, and closed-die forming operations, the working loads can be calculated as an aid to equipment design.

For other processes such as polymer extrusion or sheet metal forming, engineers now can explain parametrically what can be done to improve product and process quality. The analytical techniques may not be fully predictive, but, like the forming limit diagram in sheet forming (Kalpakjian 1984), they at least help the engineer to understand trends and to experiment with different lubricants or die designs in a parameter-controlled sense, rather than in only a craft sense.

Ironically, it is the process most discussed in this book, machining, that is the most rebellious analytically. At present, tool forces, tool temperatures, tool wear, component surface finish, and ease of chip disposal are all impossible to predict a priori with any certainty and/or without a large number of quasi-empirically obtained constants. In comparison with other processes like extrusion, machining is an unconstrained metal working operation. Simply put, as a tool cuts a layer t from the top of a block, depending on workmaterial strain-hardening characteristics and the amount of friction on the tool face, the resulting chip can be either slow moving (Figure 12.3a) or fast moving (Figure 12.3b). Although the cases near the extremes are shown in Figure 12.3, in fact, the chip thickness $t_0 \rightarrow t_1$ can be anything in between the cases shown in the diagram. This complete unpredictability in chip thickness and hence the angle ϕ leads to the inability to calculate a priori tool forces, tool temperatures, wear and chip type.

It is doubtful that an engineering mechanics solution to this issue will be found in thirty years because the roots of the problem are in other unpredictable phenomena such as friction and strain-hardening. This is why Part Three advocates an approach that combines the heuristics of the craftsmen and the empirical data collected on machining over the last eighty years since the work of Taylor (1907). In one sense, such an approach can hardly be called a science. However, it is leading to better control of machining processes. And, in the turbine blade forging application, the feedback from the vision system will lead to high-quality forgings, as information on hammer positions provides for better control. For processes such as open-die forging, we probably will not be able to fully predict the behavior of the open-loop system. But, if the feedback techniques—both hardware and software—are increasingly im-

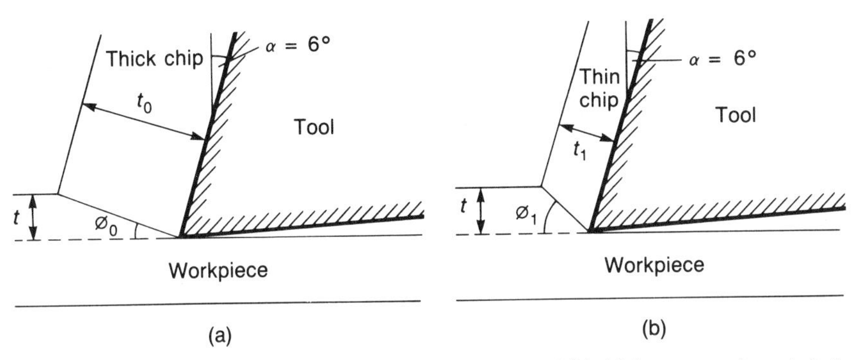

A tool removing a layer of material of: (a) thickness t_0 and (b) thickness t_1, where $t_0 > t_1$

Figure 12.3 Unpredictability in the mechanics of cutting, where, for a fixed value of t, variations in tool face friction and workmaterial strain-hardening can lead to any value of chip thickness between t_0 and t_1. (Figure by Paul Wright, from *Encyclopedia of Materials Science and Engineering*, M. B. Bever, ed. Copyright 1985, Pergamon Books Ltd. Reprinted with permission.)

proved over the next thirty years, at least the approach to manufacturing will become more rigorous. We will be able to reliably change parameters and control the effects on the manufacturing process.

Computer Scientist: A science for manufacturing currently does not exist. Many thick books on manufacturing fall between two extremes. First, there are introductory texts that are extremely descriptive of machines and processes. Such books also have guides and a few parameterized relationships, but no science. By contrast, the more advanced texts and handbooks contain thousands of analytic equations for manufacturing conditions that rarely occur. Even if they did apply, the equations are usually so complex that someone working in the factory probably would not understand them. Industrial people tend to ignore both types of books and instead to make up rules that seem to work most of the time (i.e., heuristics). The problem here is that no one writes them down, and they are carried around in a person's head until retirement.

Thirty years should be enough time for expert system technology to mature and to be included as a built-in component of a machine tool. Some heuristics will be encoded by today's expert and some will be automatically deduced by the controller. In any event, these heuristics will codify experience and will make it possible to systematize these ideas into a heuristic theory of manufacturing. Actually, a better term is "constraint theory" because these rules would circumscribe the working parameters of manufacturing. Thus, they would outline regions of safety, danger, and disaster.

Question 10: Will manufacturing be economically viable in countries with a large traditional manufacturing base?

Mechanical Engineer: Experience shows that this question is more influenced by investment, management, and government policy than by the new technology itself.

Today's management policies in the United States cannot continue without disastrous consequences. Executives are preoccupied with takeover bids and divesting for profit; industrial managers are consequently obliged to favor short-term returns instead of long-term reinvestments in modern equipment. By contrast, competitors in Japan, Korea, and Taiwan are continuously remodernizing and dedicating themselves to "quality circles." And their direct labor costs are much lower. The production of a metric ton of steel contains labor costs of $23 in Korea, $132 in Japan, and $164 in the United States (Kraar 1987). It is no counter-argument either to say that these differentials are only relevant in mass production. It is true that in small-batch production the factory floor labor costs are generally a smaller percentage of overall costs. But, there are still large mid-level labor costs, (for manufacturing engineers, schedulers, maintenance engineers, and designers) that can be pruned by sound investments in computerized equipment and by thoughtful resource planning.

Large-scale domestic manufacturing can prosper in the next thirty years, but only under the following conditions:

1. The government's support for bold remodernization of industry through low interest rates and favorable tax policies for manufacturing investment.
2. The board of directors' awareness, in all companies, that investments in "intelligent" machines must be made at all levels of the factory hierarchy in Figure 1.2. This is the only way to increase quality and decrease labor costs.
3. The industrial managers' awareness that the "intelligent" machines must be supported by good housekeeping in, for example,
    a. Inventory reduction.
    b. Efficient use of plant and warehouse space.

c. Reductions in manufacturing lead time.
d. Reductions in machine setup time.

History provides plenty of warning signs. In just four generations, the greatness of Victorian industry has been destroyed by myopic executives in the city of London who had no interest in modernizing the northern industrial towns and who just prayed that the Empire markets would last forever. In just one generation, the greatness of the American steel industry has been eroded by a similar lack of remodernization and a hope that a tame domestic market would continue to buy overpriced, mediocre-quality, late-delivered products.

There are some signs of a reawakening. The domestic automobile industry is now making large investments in "intelligent" manufacturing equipment, designing progressive looking cars, and making high-quality products. Also, the government is now aware that it must support and protect vital source producers such as the semiconductor industry. But these are just small steps in the right direction: a national awareness of these remodernization and good housekeeping issues is vital.

Computer Scientist: In ten to twenty years, the prospects for the traditional manufacturing economies will greatly improve. Rather than sending manufacturing tasks to developing countries, it will be advantageous to keep the design and manufacturing base close to the customer base.

As batch sizes decrease (see Figure 1.1) to satisfy the demands of customers, the design departments and manufacturing departments will become much more tightly coupled. This coupling will be made even tighter as manufacturing facilities begin to run unattended. In this case, designers will be anxious to see the results of their work, and it will be economically expedient to have the designer bear the administrative burden of both the design and manufacturing facilities.

The net result will be that new manufacturing technologies will begin to stabilize the current trade imbalances that countries are facing all over the world. Local townships and regions will begin to take control of their product designs, rather than depending on world-wide suppliers. This local specialization of designs will begin a new market for design specifications that will be purchased and transferred over world-wide networks. This new type of trading will, in turn, cause brand new problems. But, on an optimistic note, it does not seem as if the world is going to become a bland universe where everyone wears the same clothes, eats the same food, drives the same cars, and plays the same games.

The decentralization of the manufacturing base proposed in this answer is similar to the decentralization that in underway in computing environments. Here again, there is a very strong push to decentralize the main mass of computing into local workstations and only leave costly components (i.e., large file servers) in centralized locations.

REFERENCES

Abraham, E., Seaton, C. T., and Smith, S. D. 1983. The optical computer. *Scientific American*, Vol. 248, No. 2, pp. 85–93.

Kalpakjian S. 1984. *Manufacturing Processes for Engineering Materials.* Addison-Wesley, Reading, Mass., Chapter 7.

Kaminsky, G. 1985. Micromachining of silicon mechanical structures. *Journal Vacuum Science and Technology B*, Vol. 3, No. 4, pp. 1015–1024.

Kanade, T., and Schmitz, D. 1985. Development of CMU Direct Drive Arm II. *Proceedings of the American Control Conference*, Boston, Vol. 2, pp. 703–709.

Kraar, L. 1987. Korea's big push has just begun. *Fortune Magazine*, March 16, pp. 72–78.

LaRocca, A. V. 1982. Laser applications in manufacturing. *Scientific American*, Vol. 246, No. 3, pp. 94–103.

Ringeride, D., and Wright, P. K. 1984. *Chrysalis One: An Automated Village for the Electronic Era.* Department of Architecture Report, Carnegie Mellon University.

Rowe, G. W. 1977. *Principles of Industrial Metalworking Processes.* Edward Arnold (Publishers) Ltd, UK.

Taylor, F. W. 1907. On the art of cutting metals. *Transactions of the American Society of Mechanical Engineers*, Vol. 28, pp. 31–45.

Wright, P. K. 1985. "An Ode to a Speaker at Autofact II" in *Poetry for Engineers from Engineers.* Edited by J. T. Black and P. K. Wright. The Society of Manufacturing Engineers, Dearborn, Mich. p. 36.

Wright, P. K., and Holzer, A. J. 1981. A programmable die for the powder metal process. *Proceedings of the 9th North American Manufacturing Research Conference*, Vol. 9, pp. 65–70.

Bibliography

GENERAL MANUFACTURING AND ECONOMICS

DeGarmo, P. E., Black, J. T. and Kohser, R. A. 1984. *Materials and Processes in Manufacturing.* Macmillan, New York. DeGarmo's book has been the long-standing leader of the undergraduate textbooks that introduce students to manufacturing processes. In the recent revision with Black and Kohser, the material has been updated to include reviews of manufacturing systems and robotics.

Engelberger, J. F. 1980. *Robotics in Practice.* Amacom Press (American Management Associations), New York. Because Engelberger is regarded as one of the founding fathers of robotics, his book is an important reference source for robotics in industry. Many case studies are presented, and the style is engaging.

Kalpakjian, S. 1984. *Manufacturing Processes for Engineering Materials.* Addison-Wesley, Reading, Mass. This book is more advanced than DeGarmo, but still an introductory text. The descriptions of individual manufacturing processes are very comprehensive, and some simple analyses are added to stimulate the more advanced reader toward the research literature.

Koenig, D. T., 1987. *Manufacturing Engineering.* Hemisphere Publishing Corporation (Harper and Row), Washington, D. C. This recent book focuses on industrial engineering and factory management issues found in computer aided manufacturing. Topics such as process control and numerical control are covered well. The book is a good contrast to DeGarmo et al and Kalpakjian because Koenig's view is more influenced by his work experience in a large corporation.

Nof, S. Y. 1985. *Handbook of Industrial Robotics.* John Wiley and Sons, New York. This is a handbook for research professionals and industrial engineers building automated systems. The range of topics is very broad, covering all aspects of basic robotics, design, control, sensors, and applications.

ADVANCED MANUFACTURING SOFTWARE

Unfortunately, there are no good reference books on this topic that combine the realities of the manufacturing environment and modern software methods. For this reason, we provide background reading for artificial intelligence systems that sometimes use manufacturing as a case study.

Barr, A., and Feigenbaum, E. A. 1982. *The Handbook of Artificial Intelligence Volumes 1–3.* Heuristech Press, Stanford, Calif. This collection of work is expansive and provides a breakdown of artificial intelligence (e.g., expert systems, knowledge representation, vision, natural language, and robotics) that can be understood by most readers. It provides a good historical perspective on AI, reveals a deep understanding of the literature up until 1982, and conveys many ideas and methods with great brevity.

International Joint Conference on Artificial Intelligence 1967–1987. This bi-annual conference proceedings is perhaps the best source for "new" results in all areas of AI. It is usually where AI research news can be heard first and sometimes best.

Waterman, D. A., and Hayes-Roth, F., *editors.* **1978.** *Pattern Directed Inference Systems.* Academic Press, New York. This book is a collection of enthusiastic papers that describe some of the "guiding lights" for what have become expert systems. It covers many different kinds of pattern-directed systems, one of which is a rule-based system, but most importantly each paper conveys the excitement of its own approach. This book should be read by the future builders of new AI systems and tools.

VISION AND SENSORS

Ballard, D. H., and Brown, C. M. 1982. *Computer Vision.* Prentice Hall, Englewood Cliffs, N. J. This book provides a comprehensive review of approaches to methods in modern computer vision. It is especially a rich resource for references to source materials.

Horn, B. K. P. 1986. *Robot Vision.* MIT Press and McGraw-Hill Book Company. Horn presents a mathematical treatment of modern computer vision, for example, a deep analysis Marr's work. This book is especially valuable for methods that can be used to extract intrinsic images (or shape from X problems). This book requires mathematical maturity on the part of the reader, but it does soften the blow with strong, motivating sections.

Marr, D. 1982. *Vision.* W. H. Freeman Co., San Francisco. This book summarizes a decade of work in computer vision, which has almost single handedly caused a paradigm shift in computer vision research. In particular, he discusses computational results that match some of the phenomena that we observe in humans.

Pugh, A. 1986. *Robot Sensors Volume 2—Tactile and Non-Vision.* IFS (Publications) Ltd., UK. This monograph describes a wide variety of sensor types and their applications. The main sections cover passive and active force sensors, tactile array sensors, tactile transducers, and ultrasonic sensors. The book is primarily for manufacturing engineers and research scientists.

ROBOTIC HANDS AND ARMS

Craig, J. J. 1986. *Introduction to Robotics: Mechanics and Control.* Addison-Wesley, Reading, Mass. This text spans the engineering design and controls aspects of robotics. Although the presentation is mathematically comprehensive, the text's style allows the nonmathematician to grasp the important physical principles in robotics.

Cutkosky, M. R. 1986. *Robotic Grasping and Fine Manipulation.* Kluwer Academic Publishers, Hingham, Mass. This book is based on recent research in the field of manufacturing hands. The introduction analyzes typical tasks that can be performed by sensor-guided manipulators. A deep treatment of grasping stability and task control is provided.

Paul, R. P. 1981. *Robot Manipulators: Mathematics, Programming and Control.* MIT Press, Cambridge, Mass. This a good standard reference text for the mathematics of robotic manipulators. The treatment is very thorough, and it guides a research student or robot designer through the important mathematical details.

KNOWLEDGE ENGINEERING AND EXPERT SYSTEMS

Bobrow, D. G., *editor.* **1985.** *Qualitative Reasoning About Physical Systems.* The MIT Press, Cambridge, Mass. This book is a collection of important papers that constitute a relatively new branch of intelligent systems. In particular, it outlines basic ideas for describing physical systems qualitatively, for example, electrical circuits and hydraulic systems. This general area may provide a missing link to manufacturing engineers who have to cope with systems that are not practical to describe quantitatively. However, the maturity of the work is still very much at the "research" stage.

Gentner, D., Stevens, A. L., *editors.* **1983.** *Mental Models.* Lawrence Erlbaum Associates, Hillsdale, NJ. This book is a collection of papers that describe how to build psychologically motivated expert systems (i.e., *Mental Models*). One important result elaborates the distinction between the way a novice reasons and the way an expert reasons in a range of different applications.

Hayes-Roth, F., Waterman, D. A., and Lenat, D. B., *editors.* **1983.** *Building Expert Systems.* Addison-Wesley, Reading Mass. This is a collection of papers written by leaders in the field. The level of discussion is appropriate for an introductory text, and it gives many case examples.

Waterman, D. A. 1986. *A Guide to Expert Systems.* Addison-Wesley, Reading Mass. This book is a nontechnical introduction to expert systems. It gives a historical perspective to successful, and not so successful, systems. Its strongest feature is that it provides the correct expectation level for current systems and explains the strengths and limitations in nontechnical language.

MACHINING PROCESSES

Shaw, M. C. 1984. *Metal Cutting Principles.* Oxford University Press, Oxford and New York. This book provides "all one needs to know" about the mechanics of metal cutting, short of following the quarterly *Journal of Engineering for Industry,* published by the American Society of Mechanical Engineers. Shaw concentrates on engineering analysis, whereas Trent concentrates on the metallurgical properties found in cutting.

Trent, E. M. 1984. *Metal Cutting.* Butterworths, London and Boston. Trent is well known for research on tool wear and the development of cemented carbide technology. This book provides an easy-to-read introduction to the basics of metal cutting, followed by rich descriptions and photomicrographs of tool wear patterns, chip formation, and chip-tool interactions. There are excellent sections on the machining of aerospace alloys.

Dictionary of Acronyms

ACC Adaptive Control of Constraints

ACO Adaptive Control of Optimi-zation

A/D Analog-to-Digital converter

AGV Automatically Guided Vehi-cle

AI Artificial Intelligence

APT Automatically Programmed Tool

AWC Automatic Work Changer

BNF Backus Naur Form

CAD Computer Aided Design

CAM Computer Aided Manufacturing

CAPP Computer Aided Process Planning

CAPS Computer Aided Planning System

CIM Computer Integrated Manu-facturing

CML Cell Management Language

CNC Computer Numerical Control

CSG Constructive Solid Geometry

DNC Direct Numerical Control

EXSUS Expert's SetUp Sheet

FMC Flexible Manufacturing Cell

FMS Flexible Manufacturing System

IARCC Instrumented Adjustable Remote Center of Compliance

IGES International Graphics Exchange Standard

ISIS Intelligent Scheduling and Information System

LVDT Linear Variable Differential Transformer

MAP Manufacturing Automation Protocol

MDI Manual Data Inputs

MCL Machine Control Language

NC Numerical Control

OPS5 Official Production System version 5

PADL Part and Assembly Description Language

PC Personal Computer or Programmable Controller

PDDI Part Data Definition Interface

PLC Programmable Logic Controller

PLITS Programming Language In The Sky

RAPT Robot Automatically Programmed Tool

RCC Remote Center of Compli-ance

TCM Torque Controlled Machining

VLSI Very Large Scale Integration

Glossary

2 1/2 Dimensions This usually refers to the dimensional information that can be extracted from vision. These dimensions are the x- and y-axes of an image plane as well as the depth information for each point in the image plane. This depth information is only part of the total information needed to describe the third dimension.

Acoustic Emission An acoustic emission sensor (AE) can detect high-frequency emission, for example, from a crack opening in a piece of metal. Current devices aim to detect a breaking drill or cutting tool as the crack is propagating. The intention is to stop the machine tool spindle before serious damage occurs.

Active Lighting Active lighting is the process of projecting a light pattern (e.g., point, stripe, or grid) and then observing how the pattern is deformed by the physical environment. This deformation pattern highlights three-dimensional features during three-dimensional perception.

Active Fingers and Wrists *See* Fingers (active) and Fingers (passive).

Actuator An actuator is a mechanical component that transmits energy from part of the system to another. For example, a gun is actuated by the trigger, which actuates the hammer, which actuates the firing mechanism. Actuation chains are closely related to causality chains.

Agent This term is used in artificial intelligence to describe a program that is waiting for some situation to arise (e.g., information from a sensor in the system). As soon as that situation occurs, the program springs into action. An agent is also called a demon.

AI (Artificial Intelligence) AI is an area of computer science that attempts to re-create intelligence in the body of a machine. Common disciplines within AI include: natural language, computer vision, knowledge representation, expert systems, and methods of automated reasoning. There are three main schools of thought in AI. The first school believes that the only way to re-create intelligence is to duplicate human methods. The second school believes that it is easier to invent intelligent methods for machines and to use those methods to understand how humans work. The third school just builds systems that it hopes are intelligent.

APT (Automatically Programmed Tool) APT is a high-level language designed in the 1950s for programming machine tools. An APT program's structure is in the same family as FORTRAN: fixed columns, goto's, subroutines, and no structured blocks. FORTRAN is designed to make numerical calculations easy, and in that way, APT is designed to make the definition of 2 1/2-dimensional geometrical shapes easy to define (i.e., lines, circles, and curves).

Assembly Assembly is the final phase of manufacturing. By this time, all the components have been fabricated and the remaining problem is to mate the parts into a final configuration.

Autonomous Machines An autonomous machine is a machine that can construct its own goals and effectively satisfy them without human intervention.

AWC (Automatic Work Changer) An AWC is a pallet device on a machine tool that automatically indexes to the next part to be machined and then holds the part for machining.

Backlash Backlash is a description of the slippage in a mechanical system that is usually due to worn or loose gears and other components. Either a human machinist or a computerized control system has to compensate for the resulting position error.

Backlighting Backlighting is a lighting technique to capture the silhouette of an object, which is by nature a binary picture. Many vision systems are designed to work with binary pictures (gray values limited to 0 and 1).

Batch Production In batch production, components are manufactured together within the approximate range of 1 piece per batch to 1,000 pieces per batch. Machinery is then switched over to make another style of component, but often in a similar family to the previous batch.

Batch Shop A batch shop is a factory that produces parts in batches usually between 1 and 1,000 at a time. This is the kind of factory that is being automated in the 1980s.

Billet The term billet arises from steel making practice, but in batch manufacturing it usually refers to a short cylindrical piece of incoming stock that has been sawn off from a longer bar. In the flexible manufacturing cell for forging (Chapter 2), a typical incoming billet size might be 10 inches long by 3 inches in diameter.

Blackboard A blackboard is a common data area that can be used by several expert systems to share and compare their results.

Blobs A blob is a group of adjacent pixels that have the same gray value or are within a range of gray values. Computer vision researchers also call these regions.

Blob tree A picture can be represented as a blob tree: blobs within blobs within blobs. This is a useful way of representing pictures because, once an object is recognized, a blob inside the object can be interpreted as a hole in the object. Also, the original picture data is greatly compressed in a blob tree so that it is easier and quicker to analyze than a raw image.

Blueprint A blueprint contains the engineering drawings for a component to be made. It can be the size of a regular piece of paper 8 1/2 by 11 inches or, for large components such as tools and dies, it may be several feet square. In former times, the machines that reproduced such large drawings gave the paper a blue tint, hence the colloquialism blueprint.

BNF (Backus Naur Form) BNF is the name of a schematic language for defining language grammars. That is, in BNF, it is possible to describe the details of word order without saying anything about the meaning of individual sentences. A sample BNF statement is: "<Sentence>::= <Noun-Phrase> <Verb-Phrase>." Common programming languages are almost always defined in BNF in the back of the language reference manual.

Bread Slice A bread slice, a term used in planning templates, refers to a piece of stock that has been sawn from the end of a long bar. Thus, the outside edges of the stock are still likely to be rolled from the initial supplier. The large faces are rough cut from the end of the stock.

CAD Station A computer aided design (CAD) graphics station enables a designer or draftsman to work with a video screen rather than a blueprint. Two-dimensional graphics, three-dimensional wire frame, or constructive solid geometry (see below) are different levels of representation. In the most advanced systems, stress and thermal analyses are also possible software additions.

Cam A mechanical cam is the projecting part of a wheel in a machine tool that is grooved, toothed, or otherwise adapted to convert circular into reciprocal or variable motion. In older style automatic machines operated without computers, cams were important mechanisms for changing tools, fixtures, or workpieces.

Carbide Carbide is the generic term for all hard metal tool materials made out of tungsten carbide, titanium nitride, or a mixture of both. The carbides may also be coated with hard materials such as titanium nitride. The carbide cutting tools have significantly higher wear resistance and temperature resistance than high speed steel tools. However, they are more brittle. Carbides are used for turning operations and milling operations, but because of their brittle nature, they are less suitable for drilling and tapping.

Casting A casting is a metal product that can be made by melting alloys and pouring them into molds (sand or ceramic). Generally, the mechanical properties of castings are worse than forged or machined products.

Cell A manufacturing cell is a group of machines that work together as a team to carry out a step in the manufacturing process. In human terms, this would duplicate the effort of a group of blue-collar workers. *See also* Flexible Manufacturing Systems.

Chips As a tool cuts metal away from a part, it produces chips. These chips resemble short metallic springs having been formed by the shape of the flutes in the tool. It is important to remove these chips as machining proceeds so they do not clog the tool flutes or get stuck in between the fixture and the part.

CIM (Computer Integrated Manufacturing) CIM is the broadest and most commonly used term to describe the computerization of the factory. A flexible manufacturing system, a computer aided design facility, and a computer aided process planning system all come together to form computer integrated manufacturing.

Closed-loop Control Any mechanical or electrical system that has a feedback sensor in it can be regarded as being closed-loop controlled. The everyday heating system in a house operates under closed-loop control. The input from the user is a set temperature. Gas or electricity fuels the furnace or heating element until the feedback element, the thermostat, indicates a "zero error" between the desired input temperature and the actual temperature of the room. This closing of the loop modulates the system. By contrast, an open-loop system contains no feedback. In this case, there would be no thermostat and the furnace would run constantly. It would be hoped that the heat being produced would be balanced by the heat being lost by the house to the outside air, so that the occupants remained at a comfortable temperature. Obviously this would be wasteful and virtually impossible to regulate.

CML (Cell Management Language) CML is a new programming language that is designed to integrate groups of machines into a system. The language provides primitive mechanisms for understanding and generating programs for different programming languages. This is important because individual machines are generally programmed in different ways. CML also provides a two-dimensional database framework for representing both data and programs.

CNC (Computer Numerical Control) *See* NC.

Cognitive Object A cognitive object is an internal data structure that represents a real-world object. We use this term to describe only those objects that are automatically constructed.

Collet Chuck A collet chuck is a clamp to hold a part. It is made from relatively thin, tubular steel bushings that are split into three longitudinal segments over about two-thirds of their length. Bar stock fits through the hole in the end, and then pressure is applied to the split sections. These sections then clamp down on the workpiece. Thus, clamping relies on the elastic deformations of the split segments of the collet.

Compliant (hardware) A compliant mechanical device deflects without causing damage, whenever it is pushed against a fixed object. This compliance makes it possible to program a robot without having extremely precise information about the position of either the fixed object or the robot.

Compliant (software) Compliant software, a term coined in this book, highlights that there is little difference between hardware and software. The analogy is that a compliant program does not have to be written to exactly match a situation and that the program could adapt to a range of unexpected events.

Conflict Set A conflict set describes the set of rules in rule-based systems that match the current working data. It is called a conflict set because only one of the matching rules can fire at a time; thus, the rules conflict. A general strategy for picking one rule from the conflict set must be then applied.

Conical Punch A conical punch is a conically shaped drive on the end of a gripper actuator.

Constraint A constraint is a rule that limits the possible solutions to a problem. To find a solution, we often must sort through many bad solutions before discovering a good one. A good constraint rules out, in one step, most of the very bad attempts at a solution. For example, a good constraint for fixturing a part would be to put the biggest and flattest side up against the jaw of a vise.

Constructive Solid Geometry (CSG) This is a CAD method that represents the volume, mass, and three dimensionality of an object. The object is constructed by unions and differences of primitive shapes such as blocks and spheres. Without CSG, a CAD system may draw an object that "looks" correct but is just a series of lines that may represent an impossible object (e.g., like M. C. Escher artwork).

Controller A controller is the part of a machine tool that contains the programming facilities, DNC links, and interfacing links to the relays and motors of the machine tool. Most controllers have been built around proprietary computer architectures and have not been based on off-the-shelf CPU chips.

Craft Skills Craft skills represent a different kind of knowledge than that which is usually represented in expert systems. These skills are used to interpret information that originates from a broad range of common sense and sensor-based experience.

Crater Wear Crater wear is the worn area that develops by diffusion in the top face of tool where the forming chip rubs over the tool surface. Its formation weakens the structure of the tool face. *See also* Diffusion Wear.

Cutting Fluid Cutting fluid provides the lubrication that lowers chip/tool friction; it also reduces the temperature of both the tool and the part.

Database A database is a collection of items and their attributes. The items are usually limited to a particular domain. Because databases can grow very large, they are usually located on secondary storage (e.g., disks).

Deburring Many machining processes cause undesirable rough edges on a part. It is almost always necessary to remove (debur) these rough edges.

Deduction This is a particular inference rule that derives a conclusion from a set of premises (facts) resulting in a conclusion that is always valid. For example, from the facts, "Socrates is a man" and "All men are mortal," it can be deduced that "Socrates is mortal."

Demon *See* Agents.

Design for Manufacturability This is a popular area of research that is directed at the redesign of product assemblies, parts, fixtures, and tools to allow easier manufacturing. The goal is to eliminate parts and assembly methods that are difficult to automate.

Dexterity This word describes the grace and competence of a system. It encompasses every element of control from the initial goal of the controller, to the system's final function, and the level of skill with which the mid-course corrections are executed. Chapters 3 and 6 both elaborate on this concept.

Dies (closed) A closed die is the negative of the final part. Raw material is pressed into it, making the shape of the final part. Extra material comes out at the edges as "flash."

Dies (open) An open die is a flat-faced hammer that forms a stock part into the rough shape of the final part configuration. This step is often used to prepare a part for a closed-die forge.

Diffusion Wear Diffusion is a solid-state, atomic transport phenomenon that occurs in solids. If two clean metals are placed in intimate contact and heated in a vacuum, diffusion occurs across the boundary. The metals exchange atoms. In machining, the chip and tool are in intimate contact in a high temperature environment. And, since the chip is moving, it can carry off atoms from the tool material. This results in a crater, or hollow, in the top rake face of the tool. Since diffusion is a slow, solid-state process, the crater surface is smooth.

DNC (Direct Numerical Control) *See* NC.

Drilled Hole When drilling a hole, it is common for a machinist to initially take a spot drill and make a small dimple in the surface in the same location as the final hole. Then the machinist uses the main drill to form the complete hole. Using the spot drill first ensures that the larger drill does not walk on the surface. This technique improves the accuracy of the placement of the center of the drill. After spot drilling, the main hole is formed by a twist-drill. It is then common to chamfer the top of the hole using another tool slightly larger than the drill. This means that the top of the hole will be angled so that a conical headed bolt can sit flush with the surface.

Dynamometer A dynamometer, or load cell, measures force or torque in mechanical engineering research applications. Some are designed to measure the torque or power of an automobile engine on a test bed. Dynamometers are also designed for cutting operations. They measure the forces in turning and milling and the torques in drilling. The active sensor in a dynamometer can be a strain gauge, an LVDT transducer, or a piezo-electric sensor.

Edges Edge is a computer vision term to describe the places in a picture where gray values change significantly in adjacent pixels. There are many approaches for computing edges in a picture, and the resulting edges can be significantly different from one approach to another.

Encoder An encoder is a sensor that gives information on rotary or linear position. Encoders are used in robots and machine tools to sense and report on joint and table positions.

End Effector An end effector is attached to the end of a robot in the place of a hand. It could be any number of working mechanisms including: a gripper, a grinder, a welding torch, or a paint nozzle.

Expert System An expert system is a program that is designed with the express purpose of making decisions that match the decisions made by a human expert(s) in the field. It is not necessary, but an expert system is usually written in a nonprocedural language, such as, a rule-based system.

EXSUS (Expert's SetUp Sheet) EXSUS is a program that replaces the planning, presentation, and explanation of the manufacturing setup sheet. It contains information on tool selection, fixtures, cutting conditions, and planning steps.

Feature A feature in a machined component is a pocket, a hole, a chamfer, or anything that disturbs the flat face of a metal block. There is no exact definition of feature. Generally speaking, if a machinist were to simply remove a 1/4-inch layer from the top of a block, this would not be regarded as a feature.

Feed Rate The feed rate of a milling machine tool is the rate at which a part is driven past the rotating tool. The feed rate is in inches (or meters) per minute for milling operations.

Fingers (active) Fingers for a robotic hand are considered active if they can be controlled in a closed-loop sense. Thus, individual joints should be moveable with tendons, pneumatics, or rare-earth motors. Position and stiffness should be controllable. Finally, sensors in the fingertips should give feedback on the geometry and forces involved in the task being done.

Fingers (passive) Passive fingers open and close by means of actuators, but there is no feedback on intermediate position control. Today, industrial grippers are all passive.

Fixed Jaw A fixed jaw is the side of a parallel vise that is permanently bolted or welded to the base of the vise, hence making a perfect right angle at the joint. The moving jaw, on the other side, may gradually become less than orthogonal to the base of the vise as wear takes place. Thus, the machinist will always try to reference off the fixed jaw.

Fixture A fixture is a special clamp that is designed to hold a part while it is being machined or worked on by any manufacturing process. A fixture can be anything from a vise, to a specially designed piece of hardware, to molten lead that can later be melted away from the final part. *See also* Jig.

Flank Wear Flank wear may be defined as the worn area on the side of the tool where it rubs against the machined surface of the part. Its formation rounds and blunts the cutting edge as shown in Chapter 10.

Flexible Manufacturing Cell *See* Cell.

Flexible Manufacturing System A flexible manufacturing system is a set of workstations and/or cells that can operate and be scheduled independently from each other. The main difference between the control structure of a system and the control of a cell is the amount of predetermined scheduling. In a system, the manufacturing schedule is usually set a priori. In a cell, the sequence of events is adjusted to suit real-time constraints.

Foreman The foreman in a job shop is usually the most experienced machinist and one who has gradually emerged as the leader and spokesman for the other machinists. Generally speaking, the foreman has the most wisdom when it comes to trouble-shooting, setting up machines, and organizing day-to-day shop operations.

Fracture Cutting tools are very brittle, and they can fail by fracture or chipping. Cemented carbide tools are more brittle than high-speed steels because the former consist of fine particles of hard carbide, solid-state bonded with cobalt. Under a microscope they resemble an uneven brick wall, the carbides being the bricks and the cobalt being the mortar. Cracks can form and grow along the internal interfaces. Under a stereomicroscope, a fractured carbide tool looks like a fractured rock sample.

Frame Buffer A frame buffer is computer memory dedicated to storing values that are "read from" an analog visual sensor.

Frame Representation A frame is a collection of slots and values that is used to represent a piece of knowledge. For example, a "person frame" could have slots named: name, sex, age, hair color, eye color, height, and weight. A particular person would then have all the slots filled in with the correct values. Also, most frame-based systems extend this idea to inheritance. For example, a person is an example of a mammal, which has its own slots. A system can then make inferences based on these inherited values (e.g., mammals are warm blooded, therefore "Joe" is warm blooded). Because of the emphasis on CML in Chapter 4, this book sometimes refers to frames as being two dimensional, with rows being individual frames and columns being slots.

Game Tree A game tree is a representation of the possibilities of two players in a game, like chess. At the top of the tree is the current game position, and each arc out represents a legal move for the first player. In turn, these lead to new positions and new outgoing arcs that are the legal moves for the second player. These arcs and positions can continue until each sequence leads to the end of the game.

In practice, most games are too complex to fill out the whole game tree.

Generic Modules Generic modules or programs are the parts of an application program that are not dependent on the task. There is generally a standard way to provide the task information that will drive the generic module in its application.

Gray Scale A vision system outputs a two-dimensional array of numbers. These numbers indicate how much light is present at a given point in the camera frame, and this is represented by one of these numbers (e.g., 1 through 16). That means that the number is a gray value: white, to gray, to black.

Grinding In a grinding operation, the part is clamped on a x-y table like a milling machine. However, instead of using a cutting tool, a rotating grinding wheel is positioned above the part. The spindle of the grinding machine is horizontal. Thus, the grinding wheel makes contact with the part in the same geometrical form as a automobile tire makes contact with the road. The periphery of the grinding wheel contains abrasive particles that gradually cut away the surface of the part. Machine grinding on an x-y table is much more accurate than machining.

Gripper A gripper is the hand of a robot, and it is usually designed for a particular application. This design takes into account the working environment that the gripper will be in (i.e., space constraints, temperature constraints) and the part family that it will have to handle.

Hand Square In some machining situations, it may be necessary to use a 90° set square to ensure that the vertical faces of a block are orthogonal to the bottom of the vise.

Heterarchy A heterarchy is a description of an organization's structure. This particular structure allows for any member of the organization to communicate directly with other members as the need arises. For example, workers (or individual computer controllers) in a factory could work directly with whomever was necessary to carry out the job, without reporting to management (human or CIM system).

Heuristic A heuristic is an inference rule that deduces a result from a recognizable pattern in the data. The heuristic rule is usually based on common sense or expert advice, and it does not always generate a valid result. For example, a heuristic for picking a fixture for parts is: "If the part is prismatic, a vise can be used to hold the part." In some unusual circumstances, the result of applying the heuristic may be invalid; a very thin plate could buckle in a vise.

Hierarchy A hierarchy is a description of an organization's structure. This particular structure allows members to communicate with other members in the structure, but they must always do this through their immediate superior. In turn, each superior may have to report directly to his superior, until the target member is under the authority of a superior with the message. *See also* Heterarchy.

High-Speed Steel This is a tool material that was developed in the early part of this century. It consists of tungsten, chromium, and vanadium in an iron matrix. It is heat treated to have a higher hardness at elevated temperatures than ordinary construction steels, and thus it can be used as a cutting tool. Its predominant use is for drills and taps. Cemented carbide cutting tools have overtaken high-speed steels for turning and milling operations.

Hypothesis and Test This is a problem-solving strategy that is based on first making a good initial hypothesis and then testing to see whether or not it is correct. It is especially useful when a hypothesis can be evaluated for merit and direction. That is, if it can be determined how close to the solution a particular guess came and in which direction the guess must be varied to get even closer to the solution, it is easy to come up with a new and better guess.

Icon An icon is a picture (e.g., of a robot, machine tool, or vision system) that can effectively bring an idea to mind. Icons can be used to define a pictographic language for manufacturing problems, especially when they relate to the physical world.

IGES (International Graphics Exchange Standard) IGES is the current standard for exchanging part descriptions between dissimilar CAD workstations. Unfortunately this standard does not include enough information about the manufacturing process, and this makes it difficult to build a bridge between CAD and CAM.

Induction Induction is a particular inference rule that derives a new conclusion from a series of examples. For example, it can be induced that "All swans are white" from the fact that "I have seen 10 swans and they are all white."

Inference An inference is a logical rule that starts with a set of facts and produces a conclusion. An inference can produce either valid or invalid results.

Interpreter An interpreter is a program that takes source program statements in a language and processes them in such a way that the intended actions occur. This term is usually contrasted with the term compiler, which is a program that strictly translates the statements of one language into another.

Intrinsic Image An intrinsic image is an image that represents only one component of a camera's image. For example, an array of surface orientations, divorced from any other information, is an intrinsic image.

ISIS The ISIS system is an AI constraint-directed reasoning system that addresses the problem of how to construct accurate, timely, realizable schedules and how to manage their use in job shop environments.

Jig A jig is a special mechanical construction, which, by way of reference surfaces or other built-in features, determines location dimensions that are going to be machined into a component. The main feature of a jig is that it determines a priori the precise location of a feature. Although the terms are used interchangeably, a jig is not the same as a fixture. A fixture is a vise or a clamp that holds the work during the machining operation. However, a fixture by itself does not have any location features. Thus, a jig can be incorporated into a fixture so that both accurate location and clamping are done (DeGarmo et al. 1984).

Job Shop A job shop is a colloquialism for a machine shop that will respond to unsolicited customer requests. The economics of operating a job shop are such that batch sizes are usually in the range 1 to 1000 parts.

Knowledge Engineer A knowledge engineer is a programmer who helps a human expert make his or her knowledge explicit enough to represent as a program (or expert system).

Knowledge Representation A knowledge representation is the language and format in which objects can be described. Some knowledge representations make it much easier to solve some problems concerning an object and so it is important to find the most appropriate knowledge representation. For example, a three-dimensional object can be described by its vertices, by its surfaces, or by combinations of simpler solids. If the wrong representation is used, it can be very hard to answer such questions as, "Is this point <*x*, *y*, *z*> inside the object."

Lathe A lathe consists of a long bed with a motor driven chuck (headstock) at one end and a supporting tailstock at the other end. Cylindrical components are clamped between the chuck and the tailstock and made to rotate at high speed. Hard cutting tools are positioned in a tool post and then fed into the rotating bar radially. This sets the depth of cut. Then, movement of the tool along the length of the bar is achieved by a lead screw, and this maintains a constant feed rate along the bar in inches per revolution (see Chapter 10).

Linguistic Knowledge There are many kinds of knowledge. Linguistic knowledge is knowledge that lends itself to verbal descriptions. Therefore, "book knowledge" is mostly linguistic. Most expert systems deal exclusively with linguistic knowledge. *See also* Nonlinguistic Knowledge.

Logic Nets A logic net is a graph structure that represents the connections between ideas. For example, words in the dictionary are defined by other words. In turn, these words have new definitions, some of which point back to the original word, ad infinitum. As a result, this process would generate a huge graph with a wide range of interrelated ideas. In addition to the nodes (in this case, words in the dictionary) and arcs (in this case, word definitions) that define the graph, some researchers add special features to the arcs in an effort to make this representation powerful enough to represent structures in first order logic. Logic nets are also known as "semantic networks."

Lot Size This is synonymous with batch size.

Loop Mechanical and electrical engineers usually use the word loop to describe feedback processes. A block diagram can be drawn showing the feedback element looping back from the output of the open-loop system to the input side.

Machinist A machinist is trained in an apprenticeship program to be intimately familiar with all machining processes, workmaterial characteristics, tooling characteristics, fixturing techniques, and, nowadays, NC programming methods. The machinist has a great deal of craft experience, as opposed to a machine tool operator who may not be highly trained and will just be responsible for overseeing one machine. Over time, an operator can be retrained on other machines in order to become a fully experienced machinist.

Magazine (tools) A machine's tool magazine holds a number of tools that can be automatically loaded into the machine. Tool magazines typically carry between 20 and 100 tools. To increase this capacity, some systems make it possible to interchange the whole magazine, thus giving a machine essentially unlimited tool capacity.

Manufacturing Engineer This job function entails planning, overseeing, and carrying out factory operations. Generally speaking, a manufacturing engineer will not have detailed knowledge of individual machine tool functions.

Manufacturing Intelligence This is the science of creating intelligent systems for manufacturing applications.

MAP (The Manufacturing Automation Protocol) MAP is the name for General Motor's effort to standardize communications in the machine tool and computer industry. MAP is currently organized as seven layers of standards that specify everything from the shape of a plug and a socket, to the shape and duration of electrical pulses, to logic for performing complex routing in a computer network, to application-oriented commands.

Mass Production In mass production, the same component is made by the machinery for an indefinite amount of time. Thus more setup time is economically possible because the setup costs are amortized over an extremely large number of components. In batch production, this is not the case and there is more of an emphasis on reducing setup time.

MCL (Machine Control Language) MCL is a standardized control language dating from the late 1950s and is used by both NC and CNC machine tools. It is a simple assembly-like language with virtually no control structure.

Micrometer A micrometer is a hand-held, mechanical gauging instrument that typically measures the diameter of components across two anvils.

Milling Machine There are many kinds of milling machine, but they can be summarized in the following way. The metal workpiece is clamped to a movable x-y table. The rotating milling cutter is held in a spindle above the workpiece, and this spindle can be raised or lowered in the z direction. Interactions between the rotating tool and the part on the x-y table allow pockets and features to be cut into the workpiece. This description best fits a three-axis vertical milling machine. By tilting the head or the table, two axes of inclination may also be incorporated into the machine. In other designs, the spindle head may be horizontal, and the tool attacks the part from the side. Note that a drilling machine works in the same way as a vertical milling machine, except that the x-y table is not movable.

Natural Parameters This is a term that was first introduced by Marr (1982). However, we have altered the definition slightly to refer to the physical parameters that define objects and are independent of a given sensor.

NC (Numerical Control) In the early days of NC, it was common to complete the programming offline and carry a paper tape to the machine tool. By the late 1960s, it was feasible to install a microcomputer in the machine tool so that computer numerically controlled machine tools (CNC) could be programmed on the factory floor. In recent years, it has been easier to program machine tools from a distance and program them "directly" from a remote terminal. In this way, direct numerical control (DNC) becomes a factory-wide system where computer programmers create programs in a central area and download them to machine tools through DNC links.

Network A network describes the communication system between different computers.

Nonlinguistic Knowledge Unlike linguistic information, nonlinguistic information is hard to capture in the words of common language. This information is usually characterized by sensor phenomena. We have defined this type of knowledge as being related to craft skills needing manual dexterity as well as hand-eye coordination.

Object An object is a set of data and procedures that inherit properties from a superior object and are interfaced by message passing. *See also* Object-Oriented Programming.

Object-Oriented Programming Object-oriented programming is a style of programming where objects can be instantiated once they are defined and pieced together into an object hierarchy, which finally becomes the program. Objects are called with messages instead of the more typical subroutine call.

Offline Programming Many of today's machine tools and robots must be programmed at the machine itself. This programming may involve simply keying in the program, teaching by doing, or a combination of both methods. Offline programming is done away from the machine on another computer and is often done in the comfort of an office rather than on the manufacturing floor. Offline programming usually requires a way to verify that the program is correct, like simulation.

Open-loop Control *See* Closed-loop Control.

OPS5 (Official Production System version 5) OPS5 is a production system that matches the left-hand side of all the rules, puts all of the matched rules into a conflict set, uses a strategy for picking one of these rules, and finally executes the right-hand side. This process continues in a loop.

Optical Flow Field An optical flow field is a two-dimensional array of vectors where each vector corresponds to the direction and velocity of an element in a changing image.

PADL (Part Assembly Description Language) PADL is one of the first constructive solid geometry systems that was designed and built for describing complex parts. *See also* Constructive Solid Geometry.

Part Family A group of parts that are similar enough to manufacture with the same basic steps is a part family. In addition, parts within a family should not require that the setup of any machine in the part's manufacture be substantially changed.

Part Features A part feature is any geometry that has been cut into the original stock. *See also* Feature.

Part Overhang Material stock projecting out of the side of a parallel-sided vise is the overhang.

Part Seating When a part is machined, it is held by a fixture. The part may not be properly seated or squarely held for many reasons, for example, rough edges on the part, chips stuck to the side of the fixture, fixture components not firm, or excess coolant between the fixture and the part. To make very accurate parts, all these problems (and others) must be rectified by a machinist or an automated system.

Pass A pass on a machine tool is more or less synonymous with an individual cut. The tool enters the workpiece, carries out some chip formation work, and then exits from the workpiece. In the broadest sense of the word, this describes a pass. The same cutting tool can be used for several passes, or just used for one pass and then returned to the tool changer.

PDDI (Part Data Definition Interface) The PDDI is an internal representation for product data, an external file format and translation software to translate between the external format and the internal representation. In essence, the internal representation is an uninterpreted, doubly-linked data structure together with a series of management routines, which are currently accessible from Pascal and FORTRAN. The purpose of PDDI is to represent more information about parts than is available in traditional CAD descriptions, so that manufacturing information and other administrative data can be represented as well. To test these ideas, an initial conceptual schema, an interpretation for the internal data structure (e.g., definitions for points, lines, and surfaces), was developed. Unfortunately, this "test" schema also became known as PDDI even though it will be replaced by successively more complete conceptual schemas, while the internal representation remains intact. The plan is to develop GMAP, a more sophisticated conceptual schema, that will then contribute to a final standard proposal, which is called PDES (Part Description Exchange Standard).

Photoelasticity This is a method of stress analysis. For example, suppose a crane hook, gear, or cutting tool is going to be analyzed using photoelasticity. The first step is to make a model of the object from a photoelastic material. Epoxy resins and single crystal alumina can be used. These materials are transparent and exhibit two values of refractive index. When this model of the object is placed under stress, colorful isochromatic fringes arise in the transparent material. These fringes represent stress intensities that are the difference between the two principal stresses in the object. The details cannot be developed here, but analogically the isochromatic lines are like isobars on a weather map and they allow the stress field in the object to be calculated. The stress analysis allows object shape and stress field to be correlated. For example, a crane hook or gear might be redesigned after a photoelasticity analysis to reduce or eliminate stress concentrations at particular design features such as holes, shoulders, or corners.

Pixel A pixel is short for "picture element." A picture element is one of the values in a two-dimensional array that represents a picture.

Planner A planner is a program that automatically generates statements of action from a database of facts and a stated goal. When these statements are executed the system will be brought from its initial state to its goal state.

PLC (Programmable Logic Controller) A PLC is a machine controller that is based on the model of relays. From the late 1940s to the present, some machines have been controlled by large banks of relays, which physically switch on and off to control a machine like a series of light switches. The PLC simulates this function on a computer so that the logic, represented as ladder diagrams, is more familiar to a factory engineer.

Plastic Deformation This metallurgical term describes the permanent deformation of an object beyond its yield point. For example, very small deflections of a paperclip will take place in the elastic range, and the clip will return to its original shape when the load is removed. However, larger deflections cause the metal to plastically flow, resulting in permanent deformation. If the metal is subjected to high temperatures, the plastic deformation occurs more easily, as all blacksmiths know. Thus, if a cutting tool's edge has a high stress and a high temperature acting on it, plastic deformation occurs. If the tool is sectioned after use and viewed under a metallurgical microscope, the grains in the metal can be seen to be affected by plastic flow. Analogically, nonmetals like plastics, clay, and butter also flow under load, but the microstructural changes are very different to those in metals.

Plate A plate is a planning template term that refers to stock that is rolled on the two large flat sides and saw cut around its four edges. *See also* Bread Slice.

Pocket A pocket is a hollow milled into a block. The perimeter of the pocket can have many shapes.

Post-processor A post-processor is a program that formats data after it has been produced by a system such as a CAD station and readies it for either a display or another program.

Pragmatics The pragmatics of a program are the effects of a program. It is not the source program, not the object program, not the intermediate program states, but the results of the program's actions.

Prehension Prehension is the experience of grasping a part. It includes everything from the cognitive understanding of the situation, to the mechanical interaction with the part. In our definition, prehension includes all the processes between the initial desire and the final finished function. For example, the desire begins with a wish to drink coffee. Then, based on experience a grip is chosen, the hand moves to the cup, manipulation occurs, drinking is performed, and the cup is set down again. Thus prehension is more then just the gripping action itself; it involves the complete thought processes from the beginning to the end of the task.

Pre-processor A pre-processor is a program that formats data before it sends it to another system such as a CNC machine tool.

Primal Sketch Marr (1982) introduced this term as the name for the first results that are extracted from a visual image. It describes the intensity changes in an image, labels special locations in an image (e.g., termination points), and makes explicit local two-dimensional geometrical relations (e.g., concavities).

Probe A probe (or touch-trigger probe) is a measurement tool that can be mounted in a machine tool. This probe can then be moved until it touches the edge of a part, and by detecting the deflection of the probe, it is possible to determine the location of the part's edge. This position can then be used to offset the values of a part program so that it executes relative to the actual location of the part.

Procedural Semantics A major problem in AI is how to represent meanings, or semantics. One way to do this is to represent a meaning as a program, and when it is executed the meaning has a pragmatic result. *See also* Pragmatics.

Production Systems A production system is a rule-based system. Computer scientists also call these systems general rewrite systems. Each rule is matched to the input, and the rules that apply rewrite the input according the action part of these rules. *See also* Rules.

Programmable (hardware systems) This term was coined in this book and attempts to show the strong similarity between hardware and software systems. A programmable piece of hardware would be a machine that could take on different configurations based on electrical and mechanical adjustments that tailor it to suit an application.

Programmable (software systems) A programmable software system is a system that can change its behavior based on series of commands that effectively tailor the system to an application.

Program Family A group of programs that are constructed starting with the same generic program as their base is a program family.

Protocol Analysis Protocol analysis is the science of observing how people and/or machines interact. Careful observation can reveal how and why some actions are carried out, and in many cases, this can be used to either reeducate a person or redesign a machine.

RCC (Remote Center of Compliance) An RCC is a point in space for which the force/deflection behavior of a structure (e.g., a robot wrist) becomes decoupled so that a force or torque in any given direction produces a deflection only in the same direction without any side effects. In a passive RCC wrist, springs are angled into the faceplate as shown in Chapter 6. A force on the peg tip causes no tilting but smooth insertion with the peg axis parallel to the hole axis. In an active wrist, sensors gauge the interactions between the peg or tool and the world: Then the robot wrist can be reoriented to better perform the task on the basis of the sensor information.

Rack and Pinion This is a mechanism for translational movement, used in automobile steering and for gripper closing. In the simplest case, a circular gear (pinion), with teeth on its periphery, engages with teeth on a linear rack. Circular rotation of the pinion then results in a sideways translation of the rack.

Rate of Metal Removal This is the rate at which material is cut from the stock, measured in cubic inches per minute. In Chapter 10, it is the product of speed, feed rate, and depth of cut.

Real Time Real time describes the response time of a system. It is always a relative term with respect to a given application. For example, a program is a real-time program if it can keep up with incoming events, and give an apparently immediate response.

Reconstruction A common problem in computer vision is to make (reconstruct) a three-dimensional model of an object from partial information. For example, the information could be visual views of an object with a varying parameter (e.g., position, light source, light pattern, focus, or zoom) or from the nonvisual spectrum (e.g., x rays or infrared).

Reference Model A reference model is a concept used in control theory. It is a parameterized description of a process or a mechanical system. The controls should operate in such a way as to satisfy conditions of the reference model, based on execution "experience."

Reference Surface A reference surface is a plane of known location and smoothness. It could be the surface of a vise, the surface of a part, or the surface of the machine tool bed. In any of these cases, machinists use it to measure from other points in the working environment.

Region Analysis Region analysis is a computer vision term to describe the segmentation process of a picture. Regions are groups of pixels in a picture that have some property in common. This property might be that adjacent pixels have similar gray values or that adjacent pixels represent similar surface normals.

Robot A robot can come in many forms, but in manufacturing it generally refers to a machine that can grasp, move, and manipulate a part. This machine is typically used for loading and unloading a machine tool, and it takes the place of a human arm.

Rolled Side When bar stock is supplied to the factory, it comes from the steelmaker or aluminum company in the form of bars. These bars are finished to size by cold rolling operations in the mill. Such surfaces are relatively smooth and the opposite sides of the bar stock will be reasonably parallel. When working with tight tolerances, the machininst may have to machine these rolled sides. In other cases, if the tolerances are above 0.002 inch, the machinist may be able to use the rolled sides as part of the finished component.

Roughing/Finishing Cut In a typical machining operation, the rough stock is first machined with a series of roughing cuts that take the stock down to the approximate component dimensions. During these roughing cuts, the rates of metal removal are very high, and the machinist is not especially concerned with the surface finish on the component. To obtain the final part, the machinist takes a light finishing cut with a low feed rate and small depth of cut. In the finishing cut, the exact dimensions of the part are established and the machinist strives for an excellent surface finish.

Rules A rule is a programming statement that is roughly equivalent to an "if-then" statement: **If** a condition in a rule matches the current state of working memory, **then** perform an action that changes the state of working memory. A rule-based system is constructed from a series of these rules that are evaluated within a loop.

Run Length Encoding Run length encoding is a method for compacting large amounts of data. For example, a picture is often made up of long runs of the same color. So, instead of notating each point separately, the whole run is notated at once. For example, a run length encoding would read: There are 13 white points in a row, followed by 12 yellow points, followed by 10 blue points, and so on.

Saw-Cut Side This refers to any side of the stock that has been saw cut from an original bar. Thus, saw-cut sides are not good enough for the final component, and at some point during manufacturing they may have to be machined.

Scheduling Scheduling is the process of making time assignments to a list of actions that make up a plan. In the context of the factory, this schedule assigns times to each step of a process plan for every part being made. Of course, the process plan used in scheduling leaves out many details. For example, explicit time assignments are only made for a logical facility (e.g., a machine, a work cell, or the tool room). Small improvements in a schedule can drastically improve machine utilization. *See also* Planner.

Scene Analysis Scene analysis is a branch of computer vision that interprets the meaningful contents of a scene, which could be composed of a series of images.

Scoring, or Slipping, Marks These are very heavy scratches that would be seen on a part that had moved in a chuck, fixture, or clamp as a result of tool forces.

Script A script is a time-sequenced plan that is usually used to guide a discussion between two or more participants. These participants can either be external to the system or can be disjoint subsystems. *See also* Translation Network.

Secondary Hardening Temperature This is a metallurgical term specific to a few highly alloyed materials such as high-speed steel. The term relates to the heat treatment procedures used to harden the steel. An analogy can be drawn with the way a blacksmith heats, water quenches, and then tempers any kind of steel to give it both hardness and toughness. High-speed steels are initially heat treated in this way, but, in addition, they are raised to around 650° C to promote a secondary hardening. At this temperature, fine precipitates are formed in the steel to strengthen it. But, if the steel is heated in later use to a temperature above 650° C, the precipitates grow in size, the steel is weakened (even when it returns to room temperature), and it is no longer an effective cutting tool.

Semantics The semantics of a language captures the meanings of statements in the language. In computer languages, the semantics of a program can be thought of as the assembly language that accurately represents the meaning of the source statements. *See also* Pragmatics.

Semantic Networks *See* Logic Nets.

Setup A machine tool must be prepared (set up) before it can make a new part. This preparation might involve tool selection and mounting, fixture design and mounting, and machine tool programming.

Single-Spindle Automatic In this automatic lathe, cylindrical bar stock is held by a collet chuck and cutting tools form the component. When the component is finished it is cut off from the bar, the bar is advanced through the chuck, and the process is repeated. This form of automatic machine has been used for many years and can be controlled either by cams or by computer.

Source Image A source image is the immediate output of an imaging device. The term is used in this book to represent the first image from which intrinsic images can be extracted.

Speed The cutting speed in feet per minute is the interaction speed between tool and part. In a milling machine, the tool rotates at N revolutions per minute, which translates into a cutting speed at the periphery of the cutter. The speed $S = \frac{\pi D N}{12}$, where D is the cutter diameter in inches.

Step Jaws In a parallel-sided vise on a milling machine, it is common to insert two parallel blocks at the bottom and on either side of the vise opening. In this way, the component sits on the step jaws and protrudes out of the top of the vise so that it can be machined.

Stereo Disparity Stereo disparity is the distance between the computed positions of a matched pattern in two stereo images (i.e., a stereo pair).

Stereo Pair A stereo pair is two images of the same scene simultaneously taken from different perspectives. The perspectives are usually close enough together that most of the same objects are readily visible in both images.

Stock A primary supplier produces stock from raw materials such as iron ore. It is usually in the shape of bars, rods, or plates. The stock is then used by a machine shop or factory to make components.

Strain The engineering strain in a piece of elastically deforming metal is calculated by dividing the extension of the metal by the original length. In a tensile test on a bar of steel, the elastic strain is so small as to be imperceptible to the human eye. As a note to nonengineers, although stress and strain are the same concept to a medical practitioner, in engineering, strain equals extension divided by original length, and stress equals load divided by area. *See also* Stress.

Stress This is calculated by dividing the force, or load, on an object by the cross-sectional area that supports the load. For example, assume that the human foot is 12 inches long and an average of 2 inches wide and that, for simplicity, a standing person makes contact with the ground over an area of 48 square inches. A person weighing 144 pounds applies a stress of 3 pounds per square inch to a bathroom scale. Note that, on a dance floor, if a 144-pound woman pivots on just one stiletto heel of dimensions 1/2 inch by 1/2 inch, the stress increases to 576 pounds per square inch.

Surface Finish Surface finish is the texture of a surface. A glass window has a smooth surface finish compared with, in turn, a gramophone record or a concrete block wall. The quality of the surface on a part being machined can vary considerably. For example, a surface could be saw cut, rolled, or machine cut. Of these surface finishes, only the rolled and machine cut finishes would be smooth enough to use as a reference surface in a fixture.

Syntax The syntax of a language is a description of all the legal statements in that language. However, this does not mean that every statement is meaningful. For example, the syntax of Pascal describes many Pascal programs that do not do anything very useful and in fact, might fatally fail during execution. With some very simple languages, it is possible to make the syntax so narrow that every statement has a meaning. *See also* BNF.

Tail stock The tailstock is the support for the free end of a bar rotating in a lathe.

Tapes Many machine tools are driven by paper tape with the program encoded as holes in the tape. Many engineers discuss computer control and part complexity in terms of the amount of tape in feet.

TCM (Torque Controlled Machining) Strain gauges may be installed on a CNC machine tool spindle to measure the cutting torque during a milling or drilling operation. The output from the gauges may be used to modulate the speed or feed rate of the machine, or to provide information that a tool has just broken.

Teaching by Doing Teaching by doing is a programming methodology that involves recording the "teacher's" actions and then automatically replaying them as needed. This methodology has proven to be useful for programming robots because it simplifies the specification of the robot's position. *See also* Teach Pendant.

Teach Pendant A teach pendant is a hand-held device that lets a "teacher" move each joint separately and, when the robot is in the correct position, to push a record button. Some teach pendants resemble computer terminals and others are restricted to the control of joint position and a few editing functions (i.e., record and delete point).

Template The word template is used in two ways in the book. (1) On a lathe, a template is a sheet of metal that is shaped into the mirror image of the desired part. This piece of metal may be placed behind the tool post of a lathe and traced with a follower pin. In this way, the shape of the template is followed by the tool and its shape is cut into component. (2) The word template also describes the generic types of planning process used by the machinist. These are summarized in Chapter 8.

Three-Fingered Grip This is the most common human grip used for picking up domestic objects, tennis balls, and so on.

Thresholding Thresholding is a method in picture processing for converting multiple gray values into binary values. Usually, it is as simple as picking a number (say, 10), that defines any number less than 10 as 0 and any number greater than or equal to 10 as 1. Outside of carefully illuminated environments, thresholding almost never produces completely satisfactory results, because natural images are almost never evenly lit.

Tolerance The tolerance on a component conveys to the manufacturing engineer or machinist how much variation is allowed in the component dimension. For example, the designer may prescribe a component that is 3 inches long. For machining work, this may be shown on the drawing as 3.0 ± 0.001 inches. This means that the machinist can allow the measured dimension to be between 2.999 inches and 3.001 inches. Obviously, the tighter the tolerance, the more difficult it is to work with the fixtures and tooling to obtain an accurate component. For reference, very tight tolerances in the order of 0.0002 inch will probably need a grinding operation at the end of manufacturing. If the tolerance is between 0.001 and 0.005 inch, an experienced machinist with common milling machines and lathes will be able to successfully obtain the tolerance. Because flexible manufacturing systems have unattended pallet stations and conveyor belts, the designer must relax the tolerances slightly and can only expect 0.002 to 0.005 inch. Closed-die forging operations can obtain tolerances around 0.005 inch; open-die forging operations can only obtain tolerances in the range from 0.015 to 0.02 inch. Sand casting operations exhibit tolerances between 0.02 inch and 0.04 inch. As another point of reference, woodworking and cabinet-making tolerances are around 0.06 inch or 1/16 of an inch.

Tool-Changer Automatic milling machines and lathes need to be able to call up a variety of tools to carry out the machining process. The various tools are stored in a magazine, which indexes the tools around to a loading position ready for use. This system is called a tool-changer.

Tools (hardware) A tool is used in machining work to cut the part. Tools are usually interchangeable so that different kinds of parts can be made with the same set of cutting tools.

Tools (software) Software tools facilitate the job of programming. These tools make up the programming environment in which a programmer works to reduce the overall development time. For example, editors, compilers, and expert system shells are all software tools.

Tool post On a lathe, the cutting tool is held in the tool post for cutting. On a manual machine, the operator winds the tool post and tool into the workpiece to set the depth of cut.

Touch Probe *See* Probe.

TRACE TRACE is the name of software that, if needed, explains why decisions were made and what alternatives are available.

Transformation Matrix A coordinate system can be represented by a matrix of homogeneous equations. A transformation matrix can be multiplied with a coordinate system matrix to yield a "transformed" coordinate system (e.g., rotated, translated, or scaled). This is very convenient for changing viewpoints in robotics, computer vision, and graphics.

Translation Network A translation network is a network of control systems that are designed to solve problems in a task domain. The output of one control system is translated into jargon that is sensible to another system. For example, the manufacturing hand and the manufacturing eye are examples of two control systems where exchanged messages must be at least partially translated.

Unattended Unattended manufacturing is usually a relative term; manufacturing is unattended compared with the way it used to be done. This often means that manufacturing proceeds without human intervention, until there is a problem.

Use Intensity A group of industrial applications is usually dominated by a particular requirement. The use intensity is the level at which this requirement is stressed in order to satisfy all the applications in the use spectrum. *See also* Use Spectrum.

Use Spectrum The use spectrum of a design is the range of intended applications that an implementation is expected to accomplish.

Vibrations Machining vibrations arise from several sources. The chip formation process itself causes low-amplitude vibrations in the frequency range 2,000–10,000 cycles per second. Misalignment in the machine tool, or machine vibrations in the drives, are a second source of vibration. Inadequate clamping, especially if the part is extending out of the vise, leads to another source of vibration. External vibrations from other machines in the factory can also cause transient vibrations. These various factors will be coupled together.

VLSI (Very Large Scale Integration) VLSI is a technology for miniaturizing logic onto computer chips. In fact, it is possible to build computer algorithms on a chip, thus achieving a huge performance increase. For example, language-based machines and chess machines have been built in VLSI.

Welding Welding describes the process by which two surfaces are joined by local melting. Upon solidification, a permanent joint is created. A filler metal is generally needed for welding two plates side to side.

Workmaterial This general term is used to refer to the metal that is being worked on in any metal working process. The term is not only used in cutting (e.g., for the bar in Chapter 10), but also in forging, rolling, and sheet-metal forming.

Workstation A workstation is a self-contained station that can carry out a manufacturing step. In human terms, this replaces the work that would be carried out by one person. *See also* Flexible Manufacturing System.

Wrap-around Grip This is the grip used to hold a hammer handle or large wrench where all four fingers wrap around the object. Depending on the task, the thumb also wraps around or lays along the axis of the handle.

Yield Strength This is a metallurgical term expressing the innate strength of an alloy. The yield strength increases if we compare lead with, in turn, aluminum, medium-carbon steel, high-speed steel, cemented carbide, and diamond. If stressed to a level below the yield strength, deformations are elastic and the material can return to its original unloaded shape. Above the yield strength, plastic deformation additionally occurs and the object is permanently deformed.

Index

Page numbers in italics refer to figures; page numbers followed by t indicate tabular material.